STUDENT'S STUDY GUIDE

LIAL • MILLER • HORNSBY

COLLEGE ALGEBRA

SIXTH EDITION

Prepared with the assistance of

MARJORIE SEACHRIST

HarperCollins*CollegePublishers*

Student's Study Guide to accompany Lial/Miller/Hornsby ***COLLEGE ALGEBRA,*** *Sixth Edition*

ISBN: 0-673-46816-X

92 93 94 95 9 8 7 6 5 4 3 2 1

PREFACE

This book is designed to be used along with College Algebra, sixth edition, by Margaret L. Lial, Charles D. Miller, and E. John Hornsby, Jr.

In this book each of the topics in the textbook is illustrated by a solved example or explained using a modified form of programmed instruction. Practice tests are also provided so that you assess the level of your knowledge of the topics in each chapter.

This book should be used in addition to the instruction provided by your instructor. If you are having difficulty with a topic, find the topic in the Student's Study Guide and carefully read and complete the appropriate section.

Also, in this book you will find a short list of suggestions that may help you to become more successful in your study of mathematics. A careful reading should prove to be a valuable experience.

The following people have made valuable contributions to the production of this Student's Study Guide: Marjorie Seachrist, editor; Judy Martinez, typist; Therese Brown and Charles Sullivan, artists; and Carmen Eldersveld, proofreader.

We also want to thank Tommy Thompson of Seminole Community College for his suggestions for the essay, "To the Student: Success in Mathematics" that follows this preface.

TO THE STUDENT: SUCCESS IN MATHEMATICS

The main reason students have difficulty with mathematics is that they don't know how to study it. Studying mathematics *is* different from studying subjects like English or history. The key to success is regular practice.

This should not be surprising. After all, can you learn to play the piano or to ski well without a lot of regular practice? The same thing is true for learning mathematics. Working problems nearly every day is the key to becoming successful. Here is a list of things you can do to help you succeed in studying mathematics.

1. *Attend class regularly.* Pay attention in class to what your instructor says and does, and make careful notes. In particular, note the problems the instructor works on the board and copy the complete solutions. Keep these notes separate from your homework to avoid confusion when you read them over later.

2. Don't hesitate to ask questions in class. It is not a sign of weakness, but of strength. There are always other students with the same question who are too shy to ask.

3. *Read your text carefully.* Many students read only enough to get by, usually only the examples. Reading the complete section will help you to be successful with the homework problems. Most exercises are keyed to specific examples or objectives that will explain the procedures for working them.

4. Before you start on your homework assignment, rework the problems the instructor worked in class. This will reinforce what you have learned. Many students say, "I understand it perfectly when you do it, but I get stuck when I try to work the problem myself."

5. Do your homework assignment only *after* reading the text and reviewing your notes from class. Check your work with the answers in the back of the book. If you get a problem wrong and are unable to see why, mark that problem and ask your instructor about it. Then practice working additional problems of the same type to reinforce what you have learned.

6. Work as neatly as you can. Write your symbols clearly, and make sure the problems are clearly separated from each other. Working neatly will help you to think clearly and also make it easier to review the homework before a test.

7. After you have completed a homework assignment, look over the text again. Try to decide what the main ideas are in the lesson. Often they are clearly highlighted or boxed in the text.

8. Use the chapter test at the end of each chapter as a practice test. Work through the problems under test conditions, without referring to the text or the answers until you are finished. You may want to time yourself to see how long it takes you. When you have finished, check your answers against those in the back of the book and study those problems that you missed. Answers are referenced to the appropriate sections of the text.

9. Keep any quizzes and tests that are returned to you and use them when you study for future tests and the final exam. These quizzes and tests indicate what your instructor considers most important. Be sure to correct any problems on these tests that you missed, so you will have the corrected work to study.

10. Don't worry if you do not understand a new topic right away. As you read more about it and work through the problems, you will gain understanding. Each time you look back at a topic you will understand it a little better. No one understands each topic completely right from the start.

CONTENTS

CHAPTER 1 PRETEST

Pretest answers are at the back of this study guide.

Let M = $\{-\sqrt{5}, -1/4, 0, 1, \sqrt{3}, 25/53\}$. List the elements of M that are members of the following sets.

1. Whole numbers — 1. ____________

2. Rational numbers — 2. ____________

3. Irrational numbers — 3. ____________

4. Real numbers — 4. ____________

Use the order of operations to evaluate each of the following.

5. $(11 + 5) \div (3 + 1) \cdot 5$ — 5. ____________

6. $\dfrac{5(-2) + (-9)}{-3 \cdot 5 - 4}$ — 6. ____________

Let x = 5, y = −4, and z = −2. Evaluate each of the following.

7. $4x - 7y + z$ — 7. ____________

8. $\dfrac{3(z + 2) - 5y}{x + 1}$ — 8. ____________

Identify the property illustrated by each of the following.

9. $-5 + 5 = 0$ — 9. ____________

10. $6 + (-6) = (-6) + 6$ — 10. ____________

Identify the property illustrated by each of the following.

11. $2 \cdot (3 \cdot 4) = (2 \cdot 3) \cdot 4$ 11. ____________

12. $3(m - 5) = 3m - 15$ 12. ____________

13. $[5(-11)]2 = 2[5(-11)]$ 13. ____________

14. $\frac{1}{\sqrt{6}} \cdot \sqrt{6} = 1$ 14. ____________

Simplify using the distributive property.

15. $-9(2m - 5x)$ 15. ____________

16. $-(3y - 8)$ 16. ____________

Write each of the following in numerical order from smallest to largest.

17. $-5/2, -1/2, -\sqrt{3}, -1$ 17. ____________

18. $-3, -|-5 + 1|, -|-2|, |4|$ 18. ____________

Write without absolute value bars.

19. $|-6| - |-4|$ 19. ____________

20. $|5 - \sqrt{42}|$ 20. ____________

21. $|2y - 8|$, if $y > 4$ 21. ____________

Identify the property used in each of the following.

22. If $p < 4$ and $4 < q$, then $p < q$. 22. ____________

23. If $p < 9$. then $-4p > -36$. 23. ____________

Perform each operation. Assume all variables appearing as exponents represent positive integers.

24. $(-4r^3 - 7r^2 + 11r - 2) - (-5r^3 + 4r^2 - 9)$ 24. ____________

25. $(3x^3 + 2x^2 + 3x) - (x^3 + x^2 + x - 4)$ 25. ____________

26. $-4(-2m^2 - 4m + 6) - 3(m^2 - 5m + 1)$ 26. ____________

27. $(9y - 7x)(3y + 2x)$ 27. ____________

28. $(2r - 3)(4r^3 - 11r^2 + 2r - 5)$ 28. ____________

29. $(r^y - 4)(3r^y + 2)$ 29. ____________

Write out the binomial expansion for each of the following.

30. $(2w + 3z)^6$ 30. ____________

31. $(d - 2f)^5$ 31. ____________

For each of the following, write the indicated term for the binomial expansion.

32. 3rd term of $(4x + y)^9$ 32. ____________

33. 6th term of $(3c - 2d)^7$ 33. ____________

Factor completely.

34. $x^2 + 3x - 18$ 34. ______

35. $2m^2 + mr - 3r^2$ 35. ______

36. $9x^2 - 16m^4$ 36. ______

37. $27 + 8x^3$ 37. ______

38. $125y^6 - 64z^{12}$ 38. ______

39. $16yr - 56r + 2ys - 7s$ 39. ______

40. $x^2 + 2xy + y^2 - 9$ 40. ______

Perform each operation.

41. $\frac{x^2 - x - 6}{x^2 - 2x - 3} \cdot \frac{x^2 + 2x + 1}{x^2 - x - 2}$ 41. ______

42. $\frac{27m^3 - 8n^3}{9m^2 - 12mn + 4n^2} \div \frac{9m^2 + 6mn + 4n^2}{4n^2 - 6mn}$ 42. ______

43. $\left(1 + \frac{2}{x}\right)\left(1 - \frac{2}{x}\right)$ 43. ______

44. $\frac{m - 2}{2m - 3} + \frac{2m + 3}{m + 2}$ 44. ______

45. $\frac{r}{r^2 + 6r + 9} - \frac{2r}{r^2 + r - 6}$ 45. ______

46. $\frac{1 - \frac{1}{m}}{m + \frac{1}{2m}}$ 46. ______

47. $\frac{xy^{-1}}{x^{-1} + y^{-1}}$ 47. ______

Simplify each of the following. Write results without negative exponents. Assume that all variables represent positive numbers.

48. $\left(\frac{-9}{5}\right)^{-2}$ 48. ____________

49. $(6m^5n^{-2})(4m^{-3}n)$ 49. ____________

50. $\frac{8y^3r^4}{(6r^2)(3y^{-2})^2}$ 50. ____________

51. $\frac{(5p^3q^{-2})^{-1}(p^{-2})^{-2}}{(2p^{-1}q^{-2})^2(q^{-1})^3}$ 51. ____________

Write with positive exponents. Assume that all variables represent positive numbers.

52. $(x^2y^{1/2}z^{-1/3})^{-2}$ 52. ____________

53. $\frac{(r^3s^2t^{-2})^{-1/2}}{(r^3st)^{1/2}}$ 53. ____________

54. $\frac{m^{r-1}m^{-1/2}}{m^{r+2}}$ 54. ____________

55. Find the product $y^{-1/2}(6y^{3/2} - 4y^{1/2})$ 55. ____________

Simplify. Assume that all variables represent positive numbers.

56. $\sqrt{1000}$ 56. ____________

57. $\sqrt[5]{64}$ 57. ____________

58. $\sqrt{28r^3s^2t}$ 58. ____________

59. $\frac{x}{\sqrt{x} + 3}$ 59. ____________

60. $(\sqrt[3]{6} + 2)(\sqrt[3]{6^2} - 2\sqrt[3]{6} + 4)$ 60. ____________

CHAPTER 1 ALGEBRAIC EXPRESSIONS

1.1 The Real Numbers

1. The set of natural numbers and 0 make the set of ________ numbers.	whole
2. Counting whole numbers backwards and continuing at 0 creates the set of ________.	integers
3. A(n) ________ number is a quotient of two integers with a nonzero divisor.	rational
4. Any real number which is not a rational number is called a(n) ________ number.	irrational
5. Every rational number can be written as a decimal that ________ or ________.	repeats; terminates

**Let set B = $\{-100, -\sqrt{7}, 0, 1/3, \sqrt{63}, 8\}$.
List the elements of set B that belong to each of the following sets.**

6. Natural numbers ________	8
7. Whole numbers ________	0; 8
8. Integers ________	-100; 0; 8
9. Rational numbers ________	-100; 0; 1/3; 8
10. Irrational numbers ________	$-\sqrt{7}$, $\sqrt{63}$
11. Real Numbers ________	-100, $-\sqrt{7}$, 0, 1/3, $\sqrt{63}$, 8

12. If n is a natural number and a is any real number, then	
a^n = ______________,	$a \cdot a \cdot a \cdots a$
where a appears _____ times.	n
Evaluate each expression in Frames 13–16.	
13. $3^4 = 3 \cdot 3 \cdot$ _____ $\cdot$ _____ = _____	3; 3; 81
The number 4 is the _________ and 3 is the ______.	exponent; base
14. $(-5)^4$ = ____	625
15. $-5^4 = -($_____$)$ = _____	625; -625
16. $\left(\frac{3}{4}\right)^3$ = _____	$\frac{27}{64}$
17. When a problem with real numbers involves more than one operation symbol, use the following	
_______ of operations.	order
If ________ symbols such as parentheses, square brackets, or fraction bars are present:	grouping
(a) Work separately above and below each __________ bar.	fraction
(b) Use the rules below within each set of parentheses or square brackets. Start with the _______________ and work	innermost
_______________.	outward
If no grouping symbols are present:	
(a) Simplify all ________ and roots, working	powers
from _____ to _____.	left; right
(b) Do any _______________ or divisions in the	multiplications
order in which they occur, working from _____ to right.	left
(c) Do any __________ or subtractions from left to right.	additions

Use the order of operations in Frames 18–22.

18. $6 + 5 \cdot 3 - 9 = 6 + \_\_\_ - 9$ (________ first)
$= \_\_\_ - 9$
$= \_\_\_$

15; Multiply
21
12

19. $36 \div 9 - 4 \cdot 5 = \_\_\_ - \_\_\_$
$= \_\_\_$

4; 20
−16

20. $\dfrac{4(-7) - 8(-2)}{-7 - (-1)} = \dfrac{\_\_\_ - \_\_\_}{-7 + \_\_\_}$
$= \dfrac{\_\_\_}{-6}$
$= \_\_\_$

−28, −16
1
−12
2

21. $-5(7 - 4) - (-3)(4)^2 = -5(\_\_\_) - (-3)(\_\_\_)$
$= \_\_\_ - \_\_\_$
$= \_\_\_$

3; 16
−15; −48
33

22. $\dfrac{[4 + (-2)(3)]}{15 \div 3 \cdot 2 + 3} = \dfrac{[4 + (\_\_\_)]}{\_\_\_ \cdot 2 + 3}$
$= \_\_\_$

−6
5
$-\dfrac{2}{13}$

Evaluate the expressions in Frames 23–27 if $p = -2$, $q = 5$, and $r = -3$.

23. $7p + 9q - r = 7(\_\_\_) + 9(\_\_\_) - (\_\_\_)$
$= -14 + \_\_\_ + \_\_\_$
$= \_\_\_$

−2; 5; −3
45; 3
34

24. $\dfrac{p + 6}{q - 1} = \_\_\_$

1

25. $\dfrac{6q + 5r}{7 + p} = \_\_\_$

3

26. $\dfrac{3p - q + r}{q + r + 5} = \_\_\_$

−2

27. $\dfrac{p^2 + r(2q - p)}{q + \dfrac{2r}{p}} = \dfrac{\_\_\_ + \_\_\_(\_\_\_)}{\_\_\_ + \_\_\_}$

$= \_\_\_$

4; -3; 12

5; 3

-4

Name the properties in Frames 28–33. Assume a, b, and c repreent real numbers.

28. $a + b = b + a$, $ab = ba$ ____________ property

commutative

29. $a + (b + c) = (a + b) + c$

$a(bc) = (ab)c$ ____________ property

associative

30. There exists a real number 0 such tht $0 + a = a$ and $a + 0 = a$.
There exists a real number 1 such that $a \cdot 1 = a$ and $1 \cdot a = a$

____________ property

identity

31. The number 0 is the __________ element for addition, and 1 is the __________ element for ________________.

identity
identity
multiplication

32. For every real number a, there exists a number $-a$ such that $a + (-a) = 0$ and $-a + a = 0$.

For every nonzero real number a, there exists a number $1/a$ such that

$$a \cdot \frac{1}{a} = 1 \quad \text{and} \quad \frac{1}{a} \cdot a = 1.$$

____________ property

inverse

33. $a(b + c) = ab + ac$ ____________ property

distributive

Each of the statements of Frames 34–45 illustrate one of the basic properties. Name the property.

34. $-9 + 4 = 4 + (-9)$ ________________

commutative

35. $8 \cdot (4 \cdot 3) = (8 \cdot 4) \cdot 3$ ________________

associative

36. $-6(8 - p) = -48 + 6p$ ____________ distributive

37. $\frac{8}{11} \cdot \frac{11}{8} = 1$ ____________ inverse

38. $6 + (-6) = 0$ ____________ inverse

39. $\pi + 0 = \pi$ ____________ identity

40. $(r + 5) \cdot \frac{1}{r + 5} = 1$ (if $r \neq -5$) ____________ inverse

41. $3(4m - 9) = 3(4m) - 3 \cdot 9$ ____________ distributive

42. $42 \cdot 1 = 42$ ____________ identity

43. $-\frac{5}{3} + \frac{5}{3} = 0$ ____________ inverse

44. $-8 + 8 = 8 + (-8)$ ____________ commutative

45. $0 + 0 = 0$ ____________ identity or inverse

Identify all properties used in the statements of Frames 46–48.

46. $[3 + (-3)] \cdot 5 =$ ____ $\cdot 5$ by ____________ property.
$= 5 \cdot$ ____ by ____________ property

0; inverse
0; commutative

47. $[4 + (-4)] \cdot 8 = (-4) \cdot 8 + 4 \cdot 8$ by ____________ and ____________ properties.

distributive
commutative

48. $[2(-3)]4 = [4(-3)]2$
Use the ____________ property to get $[2(-3)]4 =$ $2[$________$]$. Then use the ____________ property to get $2[(-3)4] = [(-3)4]$ ___. Use the ____________ property to get $[(-3)4] \cdot 2 = [$________$] \cdot 2$.

associative
(-3)4; commutative
2; commutative
4(-3)

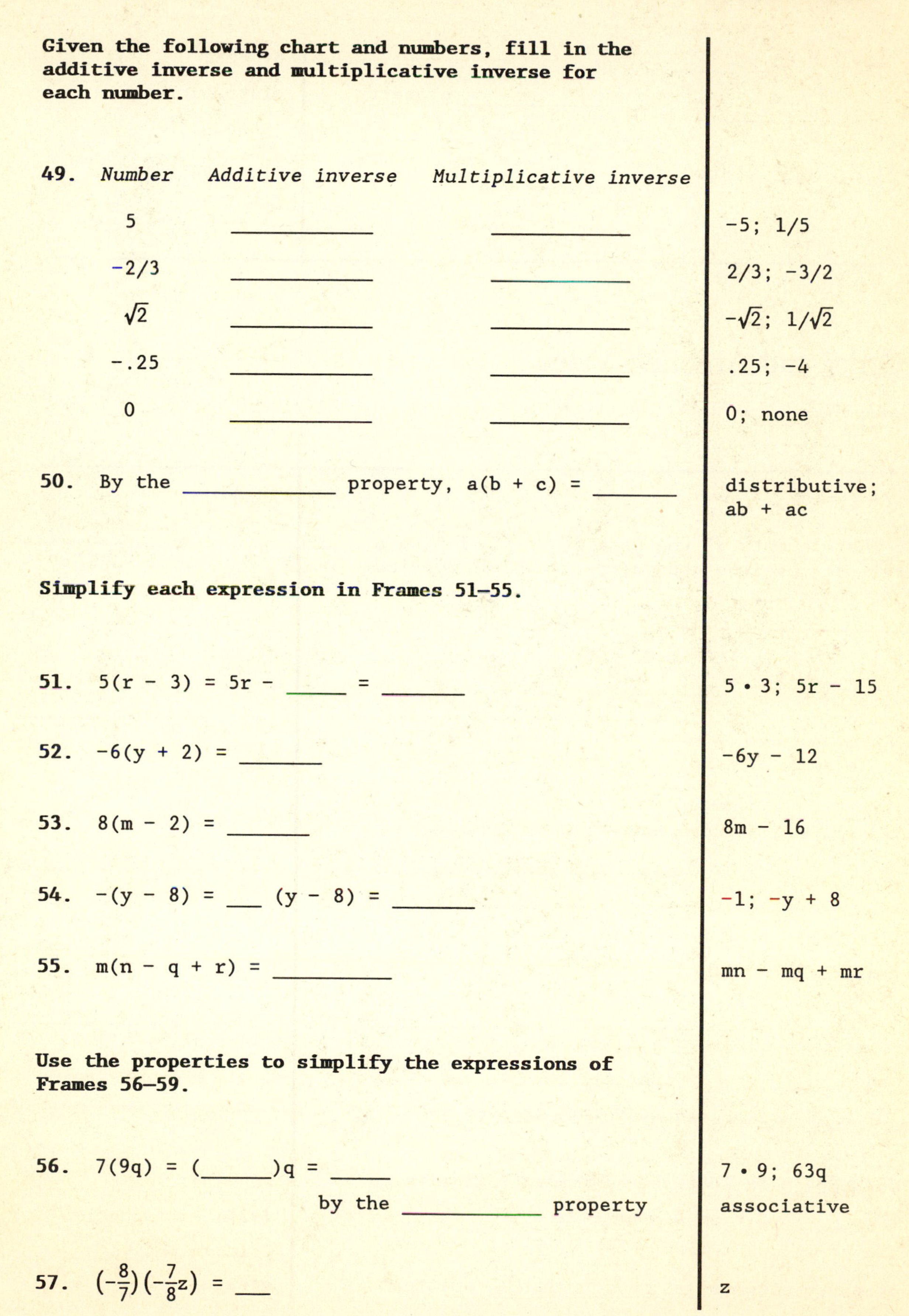

Given the following chart and numbers, fill in the additive inverse and multiplicative inverse for each number.

49.

Number	*Additive inverse*	*Multiplicative inverse*	
5	______	______	-5; $1/5$
$-2/3$	______	______	$2/3$; $-3/2$
$\sqrt{2}$	______	______	$-\sqrt{2}$; $1/\sqrt{2}$
$-.25$	______	______	$.25$; -4
0	______	______	0; none

50. By the ______ property, $a(b + c) =$ ______ — distributive; $ab + ac$

Simplify each expression in Frames 51–55.

51. $5(r - 3) = 5r -$ ____ $=$ ______ — $5 \cdot 3$; $5r - 15$

52. $-6(y + 2) =$ ______ — $-6y - 12$

53. $8(m - 2) =$ ______ — $8m - 16$

54. $-(y - 8) =$ ___ $(y - 8) =$ ______ — -1; $-y + 8$

55. $m(n - q + r) =$ ______ — $mn - mq + mr$

Use the properties to simplify the expressions of Frames 56–59.

56. $7(9q) = ($______$)q =$ ____ by the ______ property — $7 \cdot 9$; $63q$ associative

57. $(-\frac{8}{7})(-\frac{7}{8}z) =$ ___ — z

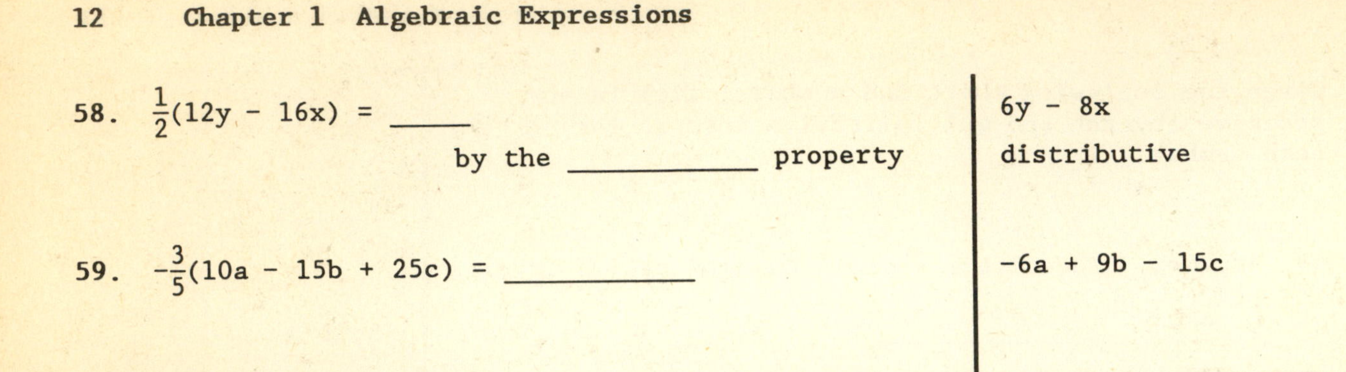

58. $\frac{1}{2}(12y - 16x) =$ _____ by the __________ property

$6y - 8x$
distributive

59. $-\frac{3}{5}(10a - 15b + 25c) =$ __________

$-6a + 9b - 15c$

1.2 Order and Absolute Value

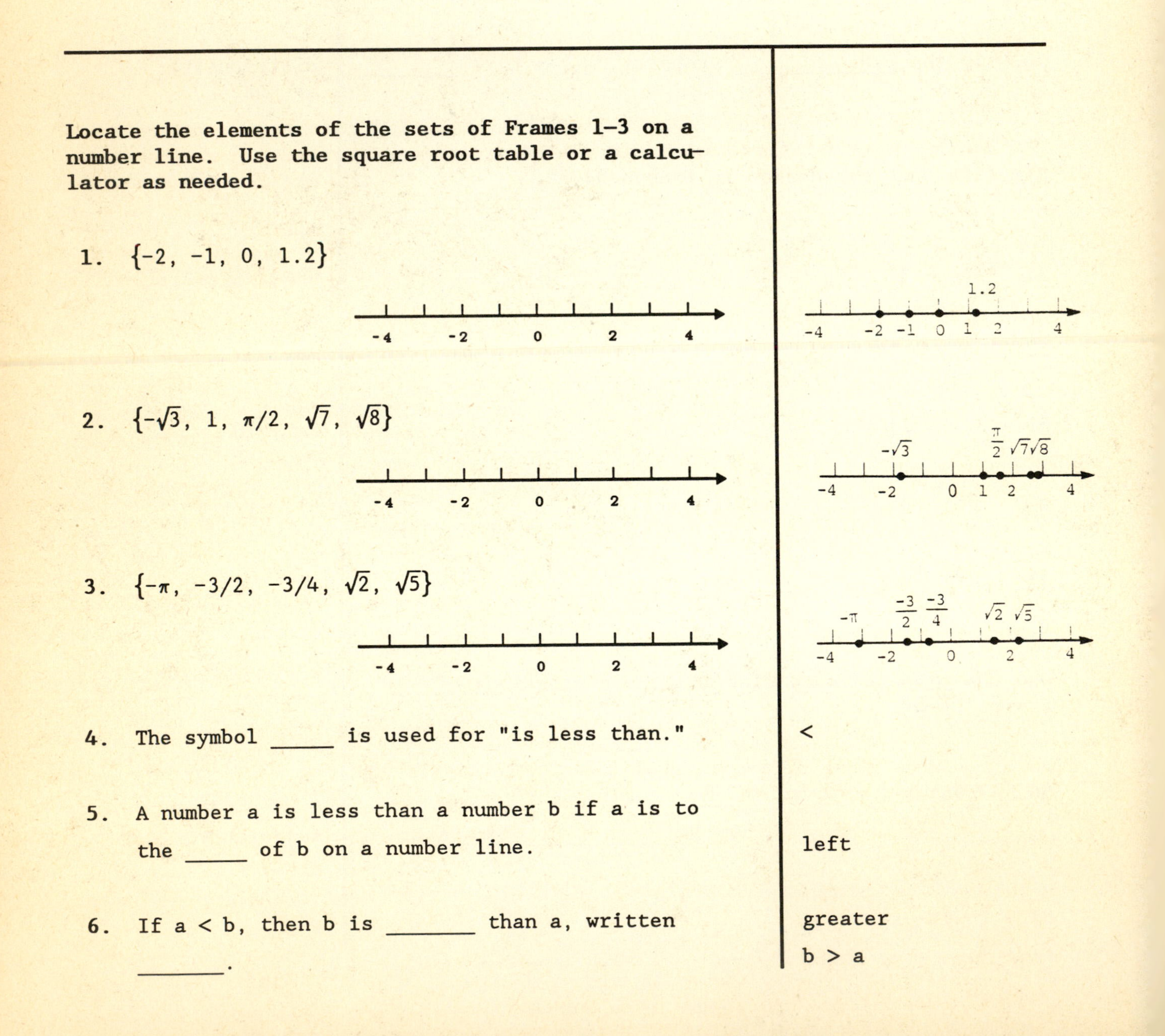

Locate the elements of the sets of Frames 1–3 on a number line. Use the square root table or a calculator as needed.

1. $\{-2, -1, 0, 1.2\}$

2. $\{-\sqrt{3}, 1, \pi/2, \sqrt{7}, \sqrt{8}\}$

3. $\{-\pi, -3/2, -3/4, \sqrt{2}, \sqrt{5}\}$

4. The symbol _____ is used for "is less than."

$<$

5. A number a is less than a number b if a is to the _____ of b on a number line.

left

6. If $a < b$, then b is _______ than a, written _______.

greater
$b > a$

Write < or > in the blanks of Frames 7–12.

7. -4 ____ -1 | <

8. 3 ____ -9 | >

9. 0 ____ -1 | >

10. -7 ____ $\frac{-13}{2}$ | <

11. $-(-2)$ ____ -1 | >

12. $-(-11)$ ____ $-(-13)$ | <

13. To say that a is less than or equal to b, write ______, while $a \geq b$ means that a is ________ than or ________ to b. | $a \leq b$; greater equal

Identify the property illustrated by the statements of Frames 14–17.

14. If $6x \leq 30$, then $x \leq 5$. ______________ | Multiplication property

15. If $r < 2$ and $2 < p$, then $r < p$. ______________ | Transitive property

16. If $a + 2 < 9$, then $a < 7$. ______________ | Addition property

17. If $-\frac{2}{3}y \geq 6$, then $y \leq -9$. ______________ | Multiplication property

Multiply both sides of the inequalities of Frames 18–20 by the indicated number.

18. $4x < 12$, $\frac{1}{4}$

$$4x \cdot \frac{1}{4} < 12\left(\frac{1}{4}\right)$$
$$x < \_\_\_.$$

| 3

19. $-3x < 15,\ -\frac{1}{3}$ ________ | $x > -5$

20. $-2x > -6,\ -\frac{1}{2}$ ________ | $x < 3$

21. The absolute ______ of a number is the ________ on the number line from the number to ___. | value; distance 0

Find the absolute values in Frames 22–25.

22. $|-9|$ = ___, since the distance from ___ to 0 is ___. | 9; −9 9

23. $|-2| - |-7|$ = ___ − ___ = ___ | 2; 7; −5

24. $-|3|$ = ___ | −3

25. $-|-9| - |2|$ = ___ − ___ = ___ | −9; 2; −11

26. By definition, $|x|$ = ________ or = ________ | x if $x \geq 0$; $-x$ if $x < 0$

Write the expressions of Frames 27–31 without absolute value bars.

27. $-|-8 + 2|$ = ___ | −6

28. $|1 - \sqrt{5}|$

We know $\sqrt{5}$ ___ 1. For this reason, $1 - \sqrt{5}$ ___ 0. By the definition of absolute value, this means that $|1 - \sqrt{5}| = -(1 - \sqrt{5})$ = ______. | >; < $\sqrt{5} - 1$

29. $|7 - \pi|$ = ______ (since π ___ 7.) | $7 - \pi$; <

30. $|r - q|$, where $q < r$

If $q < r$, then $r > q$, and $r - q$ ___ 0. Because $r - q$ is positive,

$|r - q|$ = ______.

Answers: $>$; $r - q$

31. $|3a - 9|$, $a < 3$

If $a < 3$, then $3a - 9$ ___ 0, and

$|3a - 9|$ = ______.

Answers: $<$; $9 - 3a$

Write the numbers of Frames 32 and 33 in order, from smallest to largest.

32. $|-6|$, $-|2|$, $-|-7|$, $|4|$, 0

Removing the absolute value signs we have 6, −2, −7, 4, 0. The order for this is −7, −2, 0, 4, 6. The answer is $-|-7|$, _____, 0, _____, _____.

Answers: $-|2|$; $|4|$; $|-6|$

33. $-|0|$, $|-3|$, $-|-9|$, $|-16|$, $-|1|$ ______________

Answer: $-|-9|$, $-|1|$, $-|0|$, $|-3|$, $|-16|$

Justify each statement of Frames 34–39 with the correct property from the box of properties of absolute value in the textbook.

34. $|-21| = |21|$ ______________ Answer: $|-a| = |a|$

35. $|\pi| \geq 0$ ______________ Answer: $|a| \geq 0$

36. $\left|\frac{2r}{5a}\right| = \frac{|2r|}{|5a|}$ ______________ Answer: $\left|\frac{a}{b}\right| = \frac{|a|}{|b|}$

37. $|7r + 8y| \leq |7r| + |8y|$ ______________ Answer: triangle inequality

38. $|11 + m| \geq 0$ ______________ Answer: $|a| \geq 0$

39. $|2k| \cdot |9y| = |18ky|$ ______________ Answer: $|a| \cdot |b| = |ab|$

Let $x = -4$ and $y = 6$. Evaluate the expressions in Frames 40 and 41.

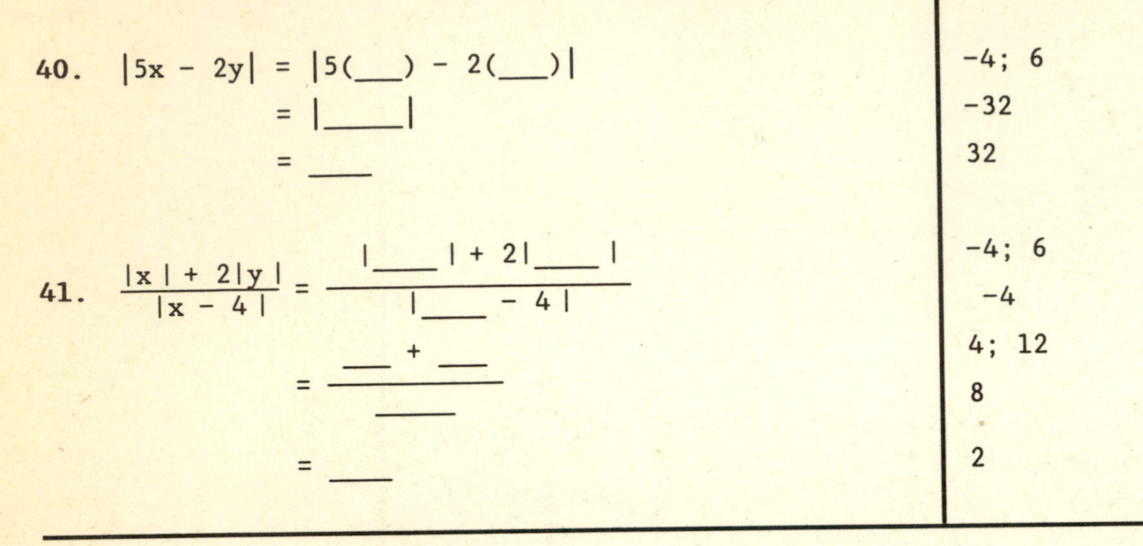

40. $|5x - 2y| = |5(\_\_\_) - 2(\_\_\_)|$ | $-4;\ 6$

$= |\_\_\_\_|$ | -32

$= \_\_\_\_$ | 32

41. $\dfrac{|x| + 2|y|}{|x - 4|} = \dfrac{|\_\_\_\_| + 2|\_\_\_\_|}{|\_\_\_\_ - 4|}$ | $-4;\ 6$ / -4

$= \dfrac{\_\_\_ + \_\_\_}{\_\_\_\_}$ | $4;\ 12$ / 8

$= \_\_\_\_$ | 2

1.3 Polynomials

1. To find the product of two expressions with exponents, use the __________ rule:

$$a^m \cdot a^n = \_\_\_\_\_\_\_\_.$$

product; a^{m+n}

Find each product in Frames 2–5.

2. $5^3 \cdot 5^9 = 5^{\_\_\_+\_\_\_} = \_\_\_\_\_$ | $3;\ 9;\ 5^{12}$

3. $7^6 \cdot 7^3 \cdot 7^5 \cdot 7 = 7^{6+3+5+\_\_\_} = \_\_\_\_\_\_\_$ | $1;\ 7^{15}$

4. $(4y^8)(-3y^7) = (\_\_\_\_)y^{8+7} = \_\_\_\_\_\_\_$ | $-12;\ -12y^{15}$

5. $(-7a^p)(-9a^{1+p}) = \_\_\_\_\_\_\_$ | $63a^{1+2p}$

6. Zero can be used as an exponent: by definition, if $a \neq 0$,

$$a^0 = \_\_\_.$$

1

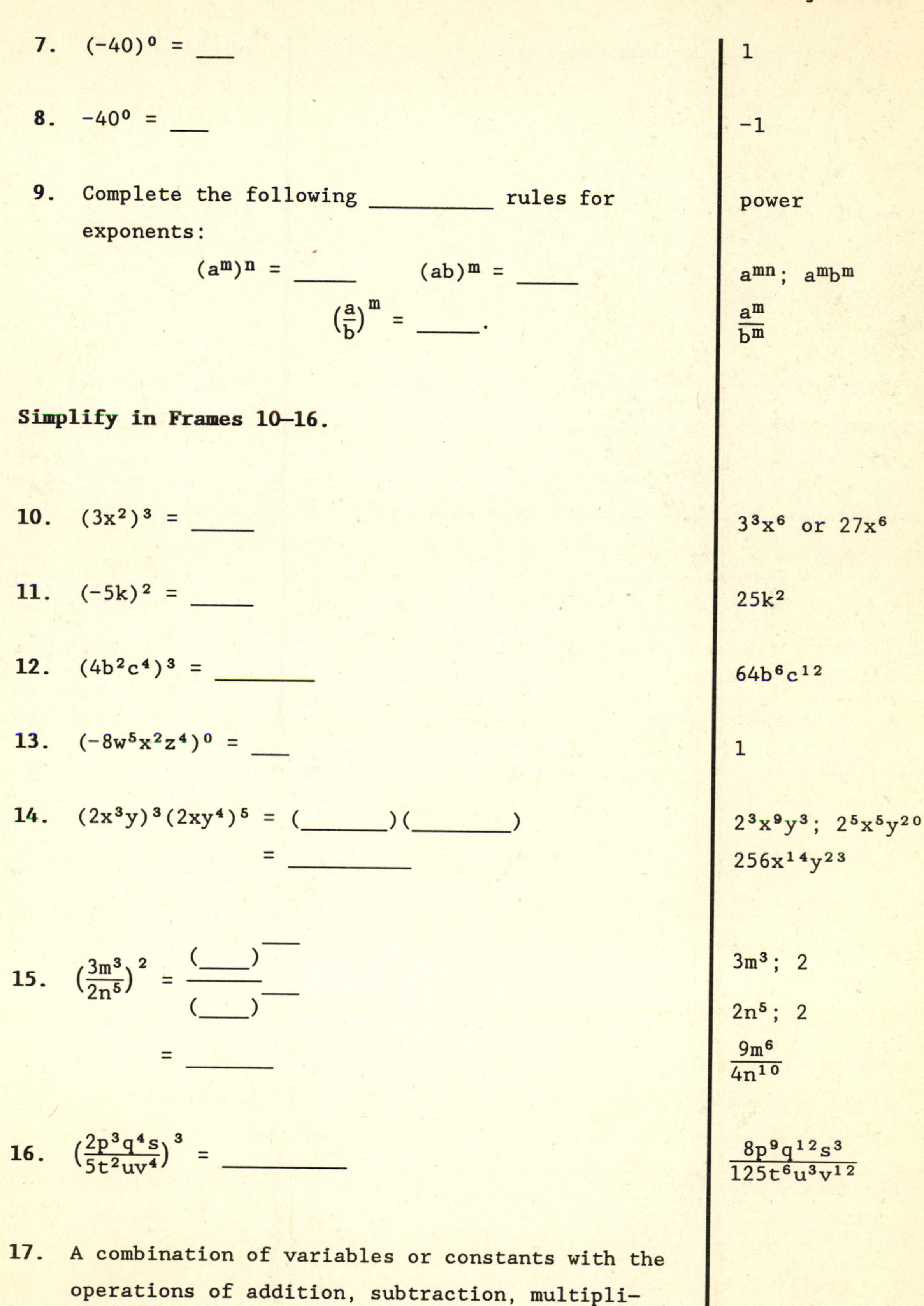

Frame	Answer
7. $(-40)^0$ = ___	1
8. -40^0 = ___	-1
9. Complete the following ________ rules for exponents:	power
$(a^m)^n$ = _____ $(ab)^m$ = _____	a^{mn}; $a^m b^m$
$\left(\frac{a}{b}\right)^m$ = _____.	$\frac{a^m}{b^m}$

Simplify in Frames 10–16.

Frame	Answer
10. $(3x^2)^3$ = _____	3^3x^6 or $27x^6$
11. $(-5k)^2$ = _____	$25k^2$
12. $(4b^2c^4)^3$ = ________	$64b^6c^{12}$
13. $(-8w^5x^2z^4)^0$ = ___	1
14. $(2x^3y)^3(2xy^4)^5$ = (_______)(________)	$2^3x^9y^3$; $2^5x^5y^{20}$
= __________	$256x^{14}y^{23}$
15. $\left(\frac{3m^3}{2n^5}\right)^2 = \frac{(\_\_\_\_)^{\_\_\_}}{(\_\_\_\_)^{\_\_\_}}$	$3m^3$; 2 $2n^5$; 2
= _______	$\frac{9m^6}{4n^{10}}$
16. $\left(\frac{2p^3q^4s}{5t^2uv^4}\right)^3$ = __________	$\frac{8p^9q^{12}s^3}{125t^6u^3v^{12}}$
17. A combination of variables or constants with the operations of addition, subtraction, multiplication, division (except by zero), or extraction of roots is called a(n) _________ ____________.	algebraic; expression

18. A product of a real number and __________ raised to powers is called a _______. The real number is called the ____________.	variables term coefficient
19. Find the coefficient of $-11m^4$. ____	-11
20. Find the coefficient of $-r^9$. ____	-1
21. A finite sum of _______ with ________ number exponents is a ____________.	terms; whole polynomial
22. For example, $-11z^4 + 2z^3 - 5z^2$ is a polynomial of ___ terms.	3
23. The largest ____________ on a polynomial of one variable is called the _______ of the polynomial.	exponent degree
24. The degree of $-11z^4 + 2z^3 - 5z^2$ is ___.	4
25. The degree of $21y^8 - 14y^{10}$ is ____.	10
26. The degree of 12 (or $12x^0$) is ___.	0
27. A polynomial of one term is a ____________. Two terms are contained in a __________, while three terms are contained in a __________.	monomial binomial trinomial
28. Polynomials are added with the ______________ property. For example, $11y^4 + 2y^4 = ($_________$)y^4 =$ ______.	distributive $11 + 2$; $13y^4$
29. Also, $3z^2 - 12z^2 = ($_________$)z^2 =$ ______.	$3 - 12$; $-9z^2$

Add or subtract in Frames 30–35.

30. $(-9r^2 - 11r + 4) + (12r^2 + 5r - 9)$
Rewrite with like terms (terms with the same variable and ___________) together.

$(-9r^2 + \_\_\_\_\_) + (-11r + \_\_\_) + (\_\_\_\_\_\_)$
$= \_\_\_\_\_\_\_\_\_\_\_\_$

exponent

$12r^2$; $5r$; $4 - 9$

$3r^2 - 6r - 5$

31. $(3x^3 - 6x^2 + 4x + 8) + (x^3 + 3x^2 - 3x + 2)$
$= \_\_\_\_\_\_\_\_\_\_\_\_\_\_\_\_\_\_\_\_$

$4x^3 - 3x^2 + x + 10$

32. $(2y^2 - 7y + 4) - (-3y^2 + 5y - 2)$
To subtract, change all the _______ on the second polynomial, and ____.

$(2y^2 - 7y + 4) - (-3y^2 + 5y - 2)$
$= 2y^2 - 7y + 4 + \_\_\_ - \_\_\_ + \_\_\_$
$= \_\_\_\_\_\_\_\_\_\_\_\_$

signs

add

$3y^2$; $5y$; 2

$5y^2 - 12y + 6$

33. $(3x^4 - 2x^2 - 11x) - (-5x^4 - 9x^2 + 12x)$
$= \_\_\_\_\_\_\_\_\_\_\_\_$

$8x^4 + 7x^2 - 23x$

34. $-(4m^2 - 5m + 2) + (m^2 + 7m - 9)$
$= \_\_\_\_\_\_\_\_\_\_\_\_$

$-3m^2 + 12m - 11$

35. $2(5y^2 - 4y + 6) - 3(2y^2 + 7y + 1)$
$= \_\_\_\_\_\_\_\_\_\_\_\_$

$4y^2 - 29y + 9$

36. We can also use the _______________ property, together with the associative property and the properties of exponents, to multiply polynomials. For example,

$3x^2(5x - 9) = (3x^2)(\_\_\_\_) - (3x^2)(\_\_\_\_)$
$= \_\_\_\_\_\_\_\_\_\_\_\_$

distributive

$5x$; 9

$15x^3 - 27x^2$

37. Find $-2m^4(-6m^3 + 2m^2 - 5)$. _______________

$12m^7 - 4m^6 + 10m^4$

38. To find $(5y - 2)(3y + 4)$, start with

$(5y - 2)(3y + 4) = (5y - 2)(\_\_\_) + (5y - 2)(\_\_\_)$

$= \_\_\_\_\_\_\_\_\_\_\_\_$

$= \_\_\_\_\_\_\_\_\_\_\_\_$.

Answers: $3y$; 4 — $15y^2 - 6y + 20y - 8$ — $15y^2 + 14y - 8$

39. It is often easier to write such products vertically. For example, to find the product of $4r - 3$ and $2r^2 - 5r + 7$, start as shown.

$$\begin{array}{r} 2r^2 - 5r + 7 \\ 4r - 3 \\ \hline \end{array}$$

To begin, multiply $2r^2 - 5r + 7$ by ____.

Answer: -3

$$\begin{array}{r} 2r^2 - 5r + 7 \\ 4r - 3 \\ \hline \\ \hline \end{array}$$

Answer: $-6r^2 + 15r - 21$

Now multiply $2r^2 - 5r + 7$ by ____. Place _____ terms in columns.

Answers: $4r$; like

$$\begin{array}{rrrr} & 2r^2 & - \ 5r & + \ 7 \\ & & 4r & - \ 3 \\ \hline & -6r^2 & + \ 15r & - \ 21 \\ \hline \end{array}$$

Answer: $8r^3 - 20r^2 + 28r$

Finally, add like terms.

$$\begin{array}{rrrr} & 2r^2 & - \ 5r & + \ 7 \\ & & 4r & - \ 3 \\ \hline & -6r^2 & + \ 15r & - \ 21 \\ 8r^3 & - \ 20r^2 & + \ 28r & \\ \hline \\ \hline \end{array}$$

Answer: $8r^3 - 26r^2 + 43r - 21$

40. The FOIL method (for F_______, O________, I________, L_______) can be used to find the product of two binomials.

Answers: First; Outside; Inside; Last

41. Use the FOIL method to find the product.

$$(9y - 7)(2y + 5) = \overset{F}{(\_\_\_)(\_\_\_)} + \overset{O}{(\_\_\_)(\_\_\_)}$$

$$+ \overset{I}{(\_\_\_)(\_\_\_)} + \overset{L}{(\_\_\_)(\_\_\_)}$$

$$= \_\_\_\_\_\_\_\_\_\_\_\_$$

Answers: $9y$; $2y$; $9y$; 5 — -7; $2y$; -7; 5 — $18y^2 + 31y - 35$

Find each product in Frames 42–46.

42. $(8k - 5m)(2k + 7m) =$ ______ | $16k^2 + 46km - 35m^2$

43. $(3m^p - 2)(4m^p + 1) =$ ______ | $12m^{2p} - 5m^p - 2$

44. $(2k - 3)(k^2 - 5) =$ ______ | $2k^3 - 10k - 3k^2 + 15$

45. $(3r - 4)(2r^2 - 5r + 9) =$ ______ | $6r^3 - 23r^2 + 47r - 36$

46. $(r + s - t)(r - 2s + t) =$ ______ | $r^2 - rs - 2s^2 + 3st - t^2$

47. Some products occur so often that they are given special names. For example,

$(x + y)(x - y) =$ ______ | $x^2 - y^2$

is called the ______ of two ______. | difference; squares

Also, $(x + y)^2 =$ ______ is the square of | $x^2 + 2xy + y^2$

a ______. | binomial

Find each product in Frames 48–53.

48. $(6m - 5r)(6m + 5r) = (\_\_\_\_)^2 - (\_\_\_\_)^2$ | 6m; 5r

$=$ ______ | $36m^2 - 25r^2$

49. $(8a + 11b)(8a - 11b) =$ ______ | $64a^2 - 121b^2$

50. $(7p - 2q^2)(7p + 2q^2) =$ ______ | $49p^2 - 4q^4$

51. $(2z + 3y)^2 = (\_\_\_\_)^2 + 2(\_\_\_\_)(\_\_\_\_) + (3y)^2$ | 2z; 2z; 3y

$=$ ______ | $4z^2 + 12zy + 9y^2$

52 $(5m - 7n)^2 =$ ______ | $25m^2 - 70mn + 49n^2$

53. $(a - 3b^2)^2 =$ ______ | $a^2 - 6ab^2 + 9b^4$

1.4 The Binomial Theorem

1. To expand $(x + y)^n$, use the ________ theorem.

binomial

2. The variables in the terms of the expansion have the following pattern.

x^n, $x^{n-1}y$, ______, $x^{n-3}y^3$, ..., xy^{n-1}, ___

$x^{n-2}y^2$; y^n

Remember that the sum of the exponents in each term equals ____.

n

3. The coefficients can be found with ________ Triangle.

Pascal's

Complete the first eight rows.

```
              1
           1     1
         1    ___    1
       1   ___   ___   1
     1   4    ___    4   1
   1   5   ___   ___   5   1
 1   6   ___   ___   15   ___   1
1   7  ___   ___   ___   21   7   1
```

2
3; 3
6
10; 10
15; 20; 6
21; 35; 35

Expand each binomial in Frames 4 and 5.

4. $(x + y)^7$

Use the binomial theorem with n = ____. The ______________ come from the eighth row of Pascal's Triangle.

7
coefficients

$(x + y)^7 = x^7 + 7$ ______ $+ 21$ ______ $+$ ______ x^4y^3
$+ 35$ ______ $+$ ______ x^2y^5
$+ 7$ ______ $+$ ______

x^6y; x^5y^2; 35
x^3y^4; 21
$xy^6 + y^7$

5. $(p - q)^5$

Think of this expression as $[p + (\_\_\_\_)]^5$. The second term of the expansion is negative because $-q$ is raised to an ____ power. After that the signs of the terms will __________.

$(p - q)^5$ = ____________________________

Answers: $-q$; odd; alternate; $p^5 - 5p^4q + 10p^3q^2 - 10p^2q^3 + 5pq^4 - q^5$

6. Pascal's triangle becomes unwieldy to find coefficients in binomial expansions when n is large. A more efficient way uses factorial notation. For any positive integer n, n! is read ______________ and is defined as follows.

$n! = n(n - 1)(\_\_\_\_\_) \cdots (\_\_\_)(\_\_\_)\, 1$, and

$0! = \_\_\_$.

Answers: n-factorial; $n - 2$; 3; 2; 1

7. For example, $6! = 6 \cdot \_\_ \cdot \_\_ \cdot \_\_ \cdot \_\_ \cdot 1 = \_\_\_\_\_$.

Answers: 5; 4; 3; 2; 720

8. The __________ coefficient, often written as $\binom{n}{r}$, for the term of an expansion of $(x + y)^n$, in which the variable part is __________ where $r \leq n$, will be

$$\frac{n!}{r!(n - r)!}.$$

Answers: binomial; $x^r y^{n-r}$

Expand each binomial in Frames 9–11.

9. $(2x + y)^6 = (\_\_\_)^6 + \binom{6}{5}(2x)^5y + \binom{6}{\_\_}(\_\_)^4y^2$

$+ \binom{6}{3}(2x)^3y^3 + \binom{6}{\_\_}(2x)^2y^4$

$+ \binom{6}{\_\_}(\_\_\_\_\_) + \_\_\_$

= ______________________________

Answers: 2x; 4; 2x; 2; 1; $2xy^5$; y^6; $64x^6 + 192x^5y + 240x^4y^2 + 160x^3y^3 + 60x^2y^4 + 12xy^5 + y^6$

10. $(3r - s)^4 =$ ______________________

$81r^4 - 108r^3s + 54r^2s^2 - 12rs^3 + s^4$

11. $(\frac{p}{2} - \frac{q}{3})^3 =$ ______________________

$\frac{p^3}{8} - \frac{p^2q}{4} + \frac{pq^2}{6} - \frac{q^3}{27}$

12. The rth term of the binomial expansion of $(x + y)^n$, where $n \geq$ ______, is given by

$$\binom{n}{\_\_} = x^{\_\_}y^{\_\_}.$$

r - 1

n - (r - 1); r - 1

n - (r - 1)

Write the indicated term of each binomial expansion in Frames 13–16.

13. 4th term of $(x + y)^9$

Here n = ___, r = ___ so r - 1 = ___. Write the 4th term of the binomial expansion as

$\frac{9!}{3!(9-3)!}$ __________, which is __________.

9; 4; 3

$x^{9-3}y^3$; $84x^6y^3$

14. 8th term of $(x + y)^{13}$ ______________

$1716x^6y^7$

15. 7th term of $(2x - 3y)^{12}$

First, find r - 1 to avoid careless sign mistakes.

r - 1, for r = 7, is 6. Then

$\frac{12!}{6!(12-6)!}(2x)^6(-3y)^6 =$ ______________

$43,110,144x^6y^6$

16. 11th term of $(4x - 3y)^{12}$ ______________________

$62,355,744x^2y^{10}$

1.5 Factoring

1. To factor a polynomial, write it as the ______ of other __________.

product

polynomials

2. For example, the polynomial $9x + 45$ can be written as $9(\_\_\_\_)$. Both 9 and $x + 5$ are ________ of $9x + 45$, and $9(x + 5)$ is the ________ ____ of $9x + 45$.

$x + 5$
factors
factored; form

3. A polynomial that cannot be factored, such as $5x + 9$, or $m^2 + 4$, is a ______, or irreducible, polynomial.

prime

4. To begin to factor a polynomial, always look for the greatest ________ factor. For example, the greatest common factor for $8y^2 - 12y^3$ is _____, the largest term that will divide without remainder into both $8y^2$ and ______.

common
$4y^2$
$12y^3$

5. $8y^2 - 12y^3$ can now be factored using the ____________ property.

distributive

$$8y^2 - 12y^3 = (4y^2)(\_\_\_) - (4y^2)(\_\_\_)$$
$$= \_\_\_\_\_\_\_\_\_\_$$

2; 3y
$4y^2(2 - 3y)$

Factor out the greatest common factor in Frames 6–8.

6. $24z^5 - 16z^4 + 4z^3 = \_\_\_\_\_\_\_\_\_\_\_\_$
(When looking for a common factor, always use the smallest ________ on any of the variables.)

$4z^3(6z^2 - 4z + 1)$
exponent

7. $32m^4n^3 - 24m^2n^5 + 36m^5n^2 = \_\_\_\_\_\_\_\_\_\_\_\_$

$4m^2n^2(8m^2n-6n^3+9m^3)$

8. $12r^3(r - 1) - 8r^2(r - 1) + 16r^4(r - 1)$
$= \_\_\_\_\_\_\_\_ (3r - 2 + 4r^2)$

$4r^2(r - 1)$

9. We can also factor by grouping. For example, to factor $3r + 3t + br + bt$, collect the terms into groups.

$$3r + 3t + br + bt = (3r + 3t) + (\_\_\_\_\_\_)$$

$br + bt$

Then factor out the greatest common factor in each group.	
$= 3(\_\_\_\_\_) + b(\_\_\_\_\_)$	r + t; r + t
$= \_\_\_\_\_\_\_\_$.	(r + t)(3 + b)

Factor by grouping in Frames 10–13.

10. $mr - 4r + 7m - 28 = \_\_\_\_\_\_\_\_$	$(m - 4)(r + 7)$
11. $a^2 + ab^2 - ab - b^3 = \_\_\_\_\_\_\_\_$	$(a + b^2)(a - b)$
12. $y^2 - 5y - 2y + 10 = \_\_\_\_\_\_\_\_$	$(y - 5)(y - 2)$
13. $x^2 + 6x + 9 - z^2$ Group $x^2 + 6x + 9$ to get	
$x^2 + 6x + 9 - z^2 = (\_\_\_\_\_)^2 - z^2$	$x + 3$
$= (x + 3 + z)(\_\_\_\_\_\_)$.	$x + 3 - z$
14. To factor $6x^2 + 5x - 4$, use FOIL backwards. We must find integers a, b, c, and d such that $6x^2 + 5x - 4 = (ax + b)(cx + d)$. By FOIL, $ac = 6$ and $bd = -4$. The positive factors of 6 are 3 and 2, and 6 and 1. The factors of -4 are 2 and -2, -4 and 1 or	
________.	4 and -1
Use trial and error to find the factors.	
$(3x + 2)(2x - 2) = \_\_\_\_\_\_$ *Incorrect*	$6x^2 - 2x - 4$
Notice that $2x - 2$ has a factor ___. Since	2
$6x^2 + 5x - 4$ does not have this factor, $2x - 2$ cannot be a ______.	factor
$(3x - 4)(2x + 1) = \_\_\_\_\_\_$ *Incorrect*	$6x^2 - 5x - 4$
$(3x + 4)(2x - 1) = \_\_\_\_\_\_$ *Correct*	$6x^2 + 5x - 4$

Factor in Frames 15–26.

15. $x^2 + 9x + 18 =$ ______	$(x + 3)(x + 6)$
16. $p^2 - 3p - 10 =$ ______	$(p - 5)(p + 2)$
17. $x^2 - 5x - 24 =$ ______ Hint: Work with the factors of -24; pick the two whose sum is -5. This will become easier with practice.	$(x - 8)(x + 3)$
18. $q^2 - 11qp + 30p^2 =$ ______	$(q - 6p)(q - 5p)$
19. $z^2 - zx - 72x^2 =$ ______	$(z - 9x)(z + 8x)$
20. $2y^4 - 6y^3 - 56y^2$ First factor out the ______ common factor:	greatest
$2y^4 - 6y^3 - 56y^2 =$ ____ (______)	$2y^2$; $y^2 - 3y - 28$
$= 2y^2($______$)($______$)$.	$y - 7$; $y + 4$
Don't forget to write _____ as part of the answer.	$2y^2$
21. $6a^4 + 72a^3 + 210a^2 =$ ______	$6a^2(a + 5)(a + 7)$
22. $6a^2 - a - 12 =$ ______	$(2a - 3)(3a + 4)$
23. $6m^2 + 5mz - 25z^2 =$ ______	$(2m + 5z)(3m - 5z)$
24. $25a^2 - 30ab + 9b^2 =$ ______	$(5a - 3b)^2$
25. $28p^2 - 90pr - 28r^2 =$ ______	$2(2p - 7r)(7p + 2r)$
26. $40a^3 + 155a^2b - 20ab^2 =$ ______	$5a(8a - b)(a + 4b)$

27. Each special product of Section 1.3 leads to a pattern for factoring. For example, the ________ square ________:

$$x^2 + 2xy + y^2 = \_\_\_\_\_\_$$
$$x^2 - 2xy + y^2 = \_\_\_\_\_\_.$$

perfect; trinomials

$(x + y)^2$

$(x - y)^2$

Factor in Frames 28–32.

28. $y^2 + 14y + 49 = y^2 + 2(\_\_\_)(\_\_\_) + (\_\_\_)^2$
$= \_\_\_\_\_\_$

y; 7; 7

$(y + 7)^2$

29. $16z^2 - 24z + 9 = (\_\_\_)^2 - 2(\_\_\_)(3) + 3^2$
$= \_\_\_\_\_\_$

$4z$; $4z$

$(4z - 3)^2$

30. $36r^2 - 60rs + 25s^2 = \_\_\_\_\_\_$

$(6r - 5s)^2$

31. $25y^4 - 30y^2z + 9z^2 = \_\_\_\_\_\_$

$(5y^2 - 3z)^2$

32. $x^{2p} - 4x^py^q + 4y^{2q} = \_\_\_\_\_\_$

$(x^p - 2y^q)^2$

33. Another special pattern of factoring is the ________ of two squares:

$$x^2 - y^2 = \_\_\_\_\_\_\_\_.$$

difference

$(x + y)(x - y)$

Factor in Frames 34–38.

34. $121y^2 - 9z^2 = (\_\_\_)^2 - (\_\_\_)^2$
$= \_\_\_\_\_\_\_\_$

$11y$; $3z$

$(11y + 3z)(11y - 3z)$

35. $100p^2 - 81q^4 = \_\_\_\_\_\_\_\_$

$(10p + 9q^2)(10p - 9q^2)$

36. $256y^4 - 81x^8 = (16y^2 + 9x^4)(\_\_\_\_\_\_\_\_)$
$= (16y^2 + 9x^4)(\_\_\_\_\_\_)(\_\_\_\_\_\_)$

$16y^2 - 9x^4$

$4y + 3x^2$; $4y - 3x^2$

37. $(2k + p)^2 - 9q^2 = (2k + p + 3q)(\_\_\_\_\_\_\_\_)$

$2k + p - 3q$

38. $a^{10p} - b^{4r} = \_\_\_\_\_\_\_\_\_\_$

$(a^{5p} + b^{2r})(a^{5p} - b^{2r})$

39. Another special pattern of factoring is the sum or difference of two ______.

$x^3 - y^3 = (x - y)(\_\_\_\_\_\_\_\_)$

$x^3 + y^3 = (x + y)(\_\_\_\_\_\_\_\_)$

cubes

$x^2 + xy + y^2$

$x^2 - xy + y^2$

Factor in Frames 40–45.

40. $m^3 - 8 = m^3 - \_\_\_$

$= (m - 2)(\_\_\_\_\_\_\_\_)$

2^3

$m^2 + 2m + 4$

41. $y^3 - 8x^3 = y^3 - (\_\_\_)^3$

$= \_\_\_\_\_\_\_\_\_\_$

$2x$

$(y-2x)(y^2+2yx+4x^2)$

42. $27m^3 + n^3 = (\_\_\_)^3 + n^3$

$= \_\_\_\_\_\_\_\_\_\_$

$3m$

$(3m+n)(9m^2-3mn+n^2)$

43. $(z + 1)^3 - 125$

$= (z + 1)^3 - \_\_\_$

$= (z + 1 - \_\_\_) \cdot [(z + 1)^2 + 5(\_\_\_\_) + \_\_\_\_]$

$= (z - 4)(\_\_\_\_\_\_\_\_)$

5^3

$5;\ z + 1;\ 5^2$

$z^2 + 7z + 31$

44. $(m - 1)^3 + (m + 1)^3$

$= (m - 1 + m + 1)[\_\_\_\_ - (m - 1)(m + 1) + \_\_\_\_]$

$= (2m)(m^2 - 2m + 1 - m^2 + \_\_\_ + \_\_\_\_\_)$

$= (2m)(\_\_\_\_)$

$(m - 1)^2;\ (m + 1)^2$

$1;\ m^2 + 2m + 1$

$m^2 + 3$

45. $x^{3p} - y^{6q} = (x^p - y^{2q})(\_\_\_\_\_\_\_\_\_\_)$

$x^{2p} + x^p y^{2q} + y^{4q}$

46. Factor $m^6 - p^{18}$.

This can be factored using two different methods. We can consider this to be the difference of two squares with $(m^3)^2 - (p^9)^2 = (\_\_\_\_\_\_)(\_\_\_\_\_\_)$.

$m^3 + p^9;\ m^3 - p^9$

Then we have the product of the sum of two cubes and the difference of two cubes. Factoring each of these gives

$(m + p^3)(\_\_\_\_\_\_)(m - p^3)(\_\_\_\_\_\_)$.

$m^2 - mp^3 + p^6$; $m^2 + mp^3 + p^6$

Suppose we consider $m^6 - p^{18}$ to be the difference of two cubes. Then we have

$(\_\_)^3 - (\_\_)^3$

$= (m^2 - p^6)(\_\_\_\_\_\_)$

$= (m + p^3)(m - p^3)(m^4 + m^2p^6 + p^{12})$

m^2; p^6

$m^4 + m^2p^6 + p^{12}$

But $m^4 + m^2p^6 + p^{12} = (m^2 - mp^3 + p^6)(m^2 + mp^3 + p^6)$. (Check it for yourself.) The answers from the two methods are the same but the use of the difference of two squares gives us a fully factored answer.

Factor in Frames 47–49.

47. $2p^4 + 9p^2 - 35$

We can use ________ to factor this trinomial. Let x = ___, so that x^2 = ___. Then $2p^4 + 9p^2 - 35$ becomes

___ + ___ $- 35$

Factor as

(____)(____).

Replace x with ___ to get

$2p^4 + 9p^2 - 35 = (\_\_\_\_)(\_\_\_\_)$.

substitution

p^2; p^4

$2x^2$; $9x$

$2x - 5$; $x + 7$

p^2

$2p^2 - 5$; $p^2 + 7$

48. $3(m - 1)^2 + 11(m - 1) - 4$

= ________

$(3m - 4)(m + 3)$

49. $x^6 + 10x^3 + 25$ = ______

$(x^3 + 5)^2$

1.6 Rational Expressions

1. The quotient of two polynomials is called a ____________ expression.

rational

2. In the rational expression

$$\frac{7x - 1}{4x + 12},$$

x can take on any value at all, except a value that makes the ______________ equal 0. Here $4x + 12$ is 0 if $x =$ ____. The restriction on the variable is ________.

denominator

-3

$x \neq -3$

Find the restrictions on the variable in Frames 3 and 4.

3. $\dfrac{9x}{(x - 4)(x + 3)}$ ____________

$x \neq 4$; $x \neq -3$

4. $\dfrac{6x + 18}{x^2 + 1}$ ____________

no restrictions

5. In order to write rational expressions in lowest terms, use the _______________ principle:

$$\frac{ac}{bc} = \_\_\_ \quad (b \neq 0,\ c \neq 0).$$

fundamental

$\frac{a}{b}$

Write in lowest terms in Frames 6 and 7.

6. $\dfrac{9k + 15}{6k + 12} = \dfrac{3(\_\_\_ + \_\_\_)}{3(\_\_\_ + \_\_\_)}$

$= \_\_\_\_\_\_\_\_\_\_$

$3k$; 5

$2k$; 4

$\dfrac{3k + 5}{2k + 4}$

7. $\dfrac{t^2 - 4}{t + 2} = \dfrac{(\_\_\_\_\_)(\_\_\_\_\_)}{t + 2} = \_\_\_\_\_$

$t + 2$; $t - 2$

$t - 2$

Multiply or divide in Frames 8–15.

8. $\frac{3(r + 2)}{5} \cdot \frac{10}{r + 2} = \frac{3(r + 2) \cdot 10}{5(\_\_\_)}$

$= \_\_\_$

r + 2

6

9. $\frac{x(x + y)}{3} \cdot \frac{9}{x + y} = \_\_\_\_$

3x

10. $\frac{m^2 + m - 6}{m^2 + 4m + 4} \cdot \frac{m + 2}{m - 2}$

Start by factoring where possible.

$$\frac{m^2 + m - 6}{m^2 + 4m + 4} \cdot \frac{m + 2}{m - 2} = \frac{(\_\_\_\_)(\_\_\_\_)(m + 2)}{(\_\_\_\_)(\_\_\_\_)(m - 2)}$$

$= \_\_\_\_\_\_$

m + 3; m − 2

m + 2; m + 2

$\frac{m + 3}{m + 2}$

11. $\frac{6p^2 + p - 2}{8p^2 - 6p + 1} \cdot \frac{32p - 8}{9p - 6} = \_\_\_\_\_\_\_\_$

$\frac{8(3p + 2)}{3(3p - 2)}$

12. $\frac{x^3 - 8z^3}{x^2 - 2zx} \cdot \frac{9xz + 10x}{x^2 + 2xz + 4z^2} = \_\_\_\_\_\_\_\_$

9z + 10

13. $\frac{x(y + 3)}{y} \div \frac{2y(y + 3)}{y^2}$

Use the definition of division to get

$$\frac{x(y + 3)}{y} \div \frac{2y(y + 3)}{y^2} = \frac{x(y + 3)}{y} \cdot \_\_\_\_\_\_\_\_$$

$= \_\_\_\_.$

$\frac{y^2}{2x(y + 3)}$

$\frac{y}{2}$

14. $\frac{r^2(2m + 6)}{m} \div \frac{r^3(m + 3)}{m} = \_\_\_\_$

$\frac{2}{r}$

15. $\frac{r^2 + 4r - 5}{r^2 - 1} \div \frac{r^2 + 5r}{r^2 + 4r + 3}$

$= \frac{(r + 5)(r - 1)}{(r + 1)(r - 1)} \div \frac{\_\_\_\_\_\_}{(\_\_\_\_)(\_\_\_\_)}$

$= \_\_\_\_\_\_\_\_$

r(r + 5)

r + 3; r + 1

$\frac{r + 3}{r}$

16. To add $\frac{8}{3p}$ and $\frac{7}{4p}$, we need a common denominator, which is _____. Write each fraction with this denominator.

12p

$$\frac{8}{3p} + \frac{7}{4p} = \frac{8 \cdot 4}{3p \cdot 4} + \frac{7 \cdot \_\_}{4p \cdot 3} = \_\_\_\_ + \_\_\_\_$$

3; $\frac{32}{12p}$; $\frac{21}{12p}$

Now add numerators.

$$= \_\_\_\_$$

$\frac{53}{12p}$

Add or subtract in Frames 17–21.

17. $\frac{9}{7k} - \frac{5}{3k} = \_\_\_\_\_$

$\frac{-8}{21k}$

18. $\frac{x + 1}{x + 2} + \frac{x}{x + 1}$

Here the common denominator is the product of individual denominators. We get

$$\frac{(x + 1)(\_\_\_\_)}{(x + 2)(x + 1)} + \frac{(\_\_\_\_)x}{(x + 2)(x + 1)}$$

x + 1; x + 2

(Note that we have essentially multiplied each of the rational expressions by 1.)

$$= \frac{(x + 1)(\_\_\_\_) + (x + 2)\_\_\_}{(x + 2)(x + 1)}$$

x + 1; x

$$= \frac{x^2 + 2x + 1 + \_\_\_\_\_}{(x + 2)(x + 1)}$$

$x^2 + 2x$

$$= \frac{\_\_\_\_\_\_\_\_}{(x + 2)(x + 1)}.$$

$2x^2 + 4x + 1$

19. $\frac{x}{x - 1} + \frac{x + 1}{x - 2} = \_\_\_\_\_\_\_\_$

$\frac{2x^2 - 2x - 1}{(x - 1)(x - 2)}$

20. $\dfrac{3}{x^2 - 1} - \dfrac{1}{x^2 + 2x + 1}$

(Hint: First factor the denominators, then find the common denominator.)

= ____________

$\dfrac{2x + 4}{(x+1)(x-1)(x+1)}$

21. $\dfrac{1}{x^2 - x - 6} - \dfrac{1}{x^2 + x - 2} =$ ____________

$\dfrac{2}{(x-3)(x+2)(x-1)}$

22. The quotient of two __________ expressions is called a ________ fraction.

rational

complex

Simplify the complex fractions of Frames 23 and 24.

23. $\dfrac{\dfrac{3}{y} - \dfrac{2}{y^2}}{\dfrac{5}{y^2} - \dfrac{1}{y}}$

Multiply both numerator and denominator by the least common denominator of all the fractions.

$= \dfrac{y^2(\_\_) - y^2(\_\_)}{y^2(\_\_) - y^2(\_\_)}$

$\dfrac{3}{y}$; $\dfrac{2}{y^2}$

$\dfrac{5}{y^2}$; $\dfrac{1}{y}$

= ________

$\dfrac{3y - 2}{5 - y}$

24. $\dfrac{\dfrac{1}{r + 1} + \dfrac{1}{r}}{\dfrac{r}{r + 1} + \dfrac{2}{r}}$

An alternative method is to perform the indicated additions in the numerator and denominator, and then divide.

$= \dfrac{\dfrac{\_\_ + \_\_}{r(r + 1)}}{\dfrac{\_\_ + \_\_}{r(r + 1)}}$

r; r + 1

r^2; 2r + 2

= ____________

$\dfrac{2r + 1}{r^2 + 2r + 2}$

25. $\dfrac{\frac{1}{p} - \frac{1}{p+1}}{\frac{1}{p} + \frac{1}{p+1}} =$ ________

$\dfrac{1}{2p+1}$

1.7 Rational Exponents

1. In order to work division problems, we need to define a negative exponent: if $a \neq 0$ and n is an integer,

$$a^{-n} = \_\_\_\_.$$

$\dfrac{1}{a^n}$

Evaluate each expression in Frames 2-5.

2. $4^{-3} = \dfrac{1}{\_\_\_} = \_\_\_\_\_$

4^3; $\dfrac{1}{64}$

3. $-5^{-2} = -(\_\_\_) = \_\_\_\_\_$

$\dfrac{1}{5^2}$; $-\dfrac{1}{25}$

4. $y^{-5} = \_\_\_\_$ (if _____)

$\dfrac{1}{y^5}$; $y \neq 0$

5. $3^{-1} - 4^{-1} = \_\_\_ - \_\_\_ = \_\_\_$

$\dfrac{1}{3}$; $\dfrac{1}{4}$; $\dfrac{1}{12}$

6. To divide, use the ____________ rule: For all integers m and n,

$$\frac{a^m}{a^n} = \_\_\_\_\_. \quad (a \neq 0)$$

quotient

a^{m-n}

Simplify in Frames 7–10.

7. $\frac{5^9}{5^3} = 5^{9-\_\_\_} = \_\_\_\_$

3; 5^6

8. $\frac{r^{15}}{r^{19}} = r^{\_\_\_-\_\_\_} = \_\_\_\_ = \_\_\_\_$

15; 19; r^{-4}; $\frac{1}{r^4}$

9. $\frac{25z^5}{10z^7} = (\_\_\_)(\_\_\_) = \frac{5}{2}z^{\_\_\_} = \_\_\_\_$

$\frac{25}{10}$; $\frac{z^5}{z^7}$; -2; $\frac{5}{2z^2}$

10. $\frac{m^{3r}}{m^{5r}} = m^{3r-\_\_\_} = \_\_\_\_ = \_\_\_\_$

$5r$; m^{-2r}; $\frac{1}{m^{2r}}$

11. $\frac{5^{-2} \cdot 5^3 \cdot 5^8}{5^{-1} \cdot 5^{-4}} = \_\_\_\_$

5^{14}

12. $6m^{-1}(2m^{-3})^3 = \_\_\_\_$

$\frac{48}{m^{10}}$

13. $4r^{-1}(5^2r^{-3})^{-2} = \_\_\_\_$

$\frac{4r^5}{5^4}$ or $\frac{4r^5}{625}$

14. $\frac{(4y^3)^{-2}(4y^{-2})^3}{(4^{-1}y^{-3})^3} = \_\_\_\_$

$\frac{4^4}{y^3}$ or $\frac{256}{y^3}$

15. $\frac{(3r^{-1})^{-1}(r^{-2})^3}{(r^2)^{-3}(3^{-2}r^4)^{-1}} = \_\_\_\_$

$\frac{r^5}{3^3}$ or $\frac{r^5}{27}$

16. $\frac{(a^2bc^3)^{-1}(ab^{-4}c^2)^2}{(abc)^{-3}(abc^{-2})^2} = \_\_\_\_$

$\frac{ac^8}{b^8}$

17. By definition, if n is a positive integer greater than 1 and a is a real number, $a^{1/n}$ is the ____ root of a. That is,

nth

$$(a^{1/n})^n = \_\_\_$$

a

However, if n is even and $a < 0$, then ____ is not defined.

$a^{1/n}$

Simplify each expression in Frames 18-24.

18. $100^{1/2}$ = ____ since (____)^2 = 100	10; 10
19. $-81^{1/2}$ = -(____) = ____	9; -9
20. $1000^{1/3}$ = ________	10
21. $(-1000)^{1/3}$ = ________	-10
22. $(-16)^{1/4}$ = ________	Not defined
23. $-64^{1/6}$ = ________	-2
24. $-216^{1/3}$ = ________	-6
25. Rational exponents are defined as follows. If m is an integer, n is a natural number, and $a^{1/n}$ exists, then $a^{m/n}$ = (_____)^m or $a^{m/n}$ = (____)$^{1/n}$.	$a^{1/n}$, a^m

Simplify in Frames 26-32.

26. $16^{3/4}$ = (______)^3 = (___)^3 = ___	$16^{1/4}$; 2; 8
27. $64^{3/2}$ = (_____)^3 = ____	$64^{1/2}$; 8^3 or 512
28. $125^{2/3}$ = ____	25
29. $32^{8/5}$ = _____	256
30. $81^{3/4}$ = ____	27
31. $(-1000)^{2/3}$ = _____	100

32. $\left(\frac{8}{125}\right)^{2/3} =$ ____ | $\frac{4}{25}$

33. The rules of exponents can also be used with rational exponents. For example,

$5^{1/4} \cdot 5^{-5/4} = 5^{\_\_+\_\_} = 5^{\_\_} =$ ____ | 1/4; −5/4; −1; $\frac{1}{5}$

Rewrite the expressions of Frames 34–44 using only positive exponents. Assume that all variables represent positive real numbers and that variables used as exponents represent rational numbers.

34. $9^{2/3} \cdot 3^{-1/3} = (\_\_\_)^{2/3} \cdot 3^{-1/3}$ | 3^2

$= 3^{\_\_} \cdot 3^{-1/3}$ | 4/3

$=$ ___ | 3

35. $(x + 1)^{3/4}(x + 1)^{-5/4} =$ ____ | $\frac{1}{(x + 1)^{1/2}}$

36. $(r^{4/5})(r^{3/5}) =$ ____ | $r^{7/5}$

37. $(x^{2/3}y^{1/2})(x^{-1/3}y^{3/2}) =$ ____ | $x^{1/3}y^2$

38. $\frac{x^{1/4} \cdot y^{1/3} \cdot z^{2/5}}{x^{5/4} \cdot y^{-2/3} \cdot z^{12/5}} =$ ____ | $\frac{y}{xz^2}$

39. $\frac{r^{7/3} \cdot s^{-5/2} \cdot t^{3/2}}{r^{1/3} \cdot s^{1/2} \cdot t^{1/2}} =$ ____ | $\frac{r^2t}{s^3}$

40. $\left(\frac{x^3y^2z^6}{8x^{-3}yz^{-6}}\right)^{-1/3} =$ ____ | $\frac{2}{x^2y^{1/3}z^4}$

41. $y^{2/3}(y^{7/3} - 2y^{1/3}) =$ ____ | $y^3 - 2y$

42. $(m - m^{1/2})(m + m^{1/2}) =$ ____ | $m^2 - m$

43. $(2y^{1-r})(-6y^2) =$ ____ | $-12y^{3-r}$

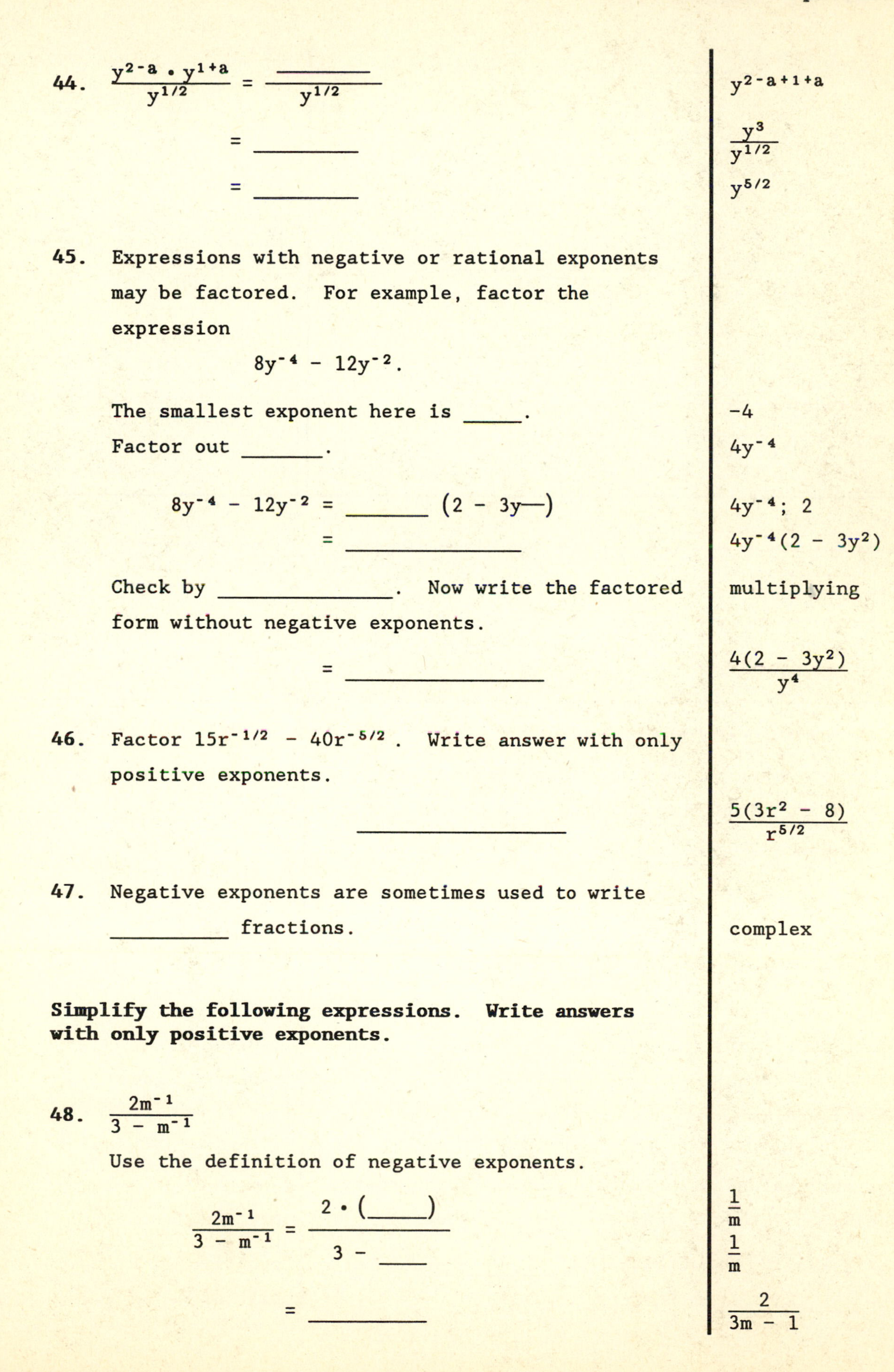

44. $\dfrac{y^{2-a} \cdot y^{1+a}}{y^{1/2}} = \dfrac{\_\_\_\_\_\_}{y^{1/2}}$

$= \_\_\_\_\_\_$

$= \_\_\_\_\_\_$

$y^{2-a+1+a}$

$\dfrac{y^3}{y^{1/2}}$

$y^{5/2}$

45. Expressions with negative or rational exponents may be factored. For example, factor the expression

$$8y^{-4} - 12y^{-2}.$$

The smallest exponent here is _____.

Factor out _______.

$$8y^{-4} - 12y^{-2} = \_\_\_\_\_\_\_ (2 - 3y^{\_\_})$$

$$= \_\_\_\_\_\_\_\_\_\_\_\_$$

Check by ____________. Now write the factored form without negative exponents.

$$= \_\_\_\_\_\_\_\_\_\_\_\_$$

-4

$4y^{-4}$

$4y^{-4}$; 2

$4y^{-4}(2 - 3y^2)$

multiplying

$\dfrac{4(2 - 3y^2)}{y^4}$

46. Factor $15r^{-1/2} - 40r^{-5/2}$. Write answer with only positive exponents.

____

$\dfrac{5(3r^2 - 8)}{r^{5/2}}$

47. Negative exponents are sometimes used to write _________ fractions.

complex

Simplify the following expressions. Write answers with only positive exponents.

48. $\dfrac{2m^{-1}}{3 - m^{-1}}$

Use the definition of negative exponents.

$$\frac{2m^{-1}}{3 - m^{-1}} = \frac{2 \cdot (\_\_\_\_)}{3 - \_\_\_\_}$$

$$= \_\_\_\_\_\_\_\_$$

$\dfrac{1}{m}$

$\dfrac{1}{m}$

$\dfrac{2}{3m - 1}$

49. $(a^{-1} - b^{-1})^{-1} =$ ________ | $\frac{ab}{b - a}$

50. $\frac{x - 4y^{-1}}{(x + 2y^{-1})(x - 2y^{-1})}$

Rewrite this as

$$\frac{x - \frac{4}{y}}{(x + \frac{2}{y})(x - \frac{2}{y})} = \frac{\_\_\_\_\_\_}{(\frac{xy + 2}{y})(\frac{xy - 2}{y})}$$

| $\frac{xy - 4}{y}$

$$= \_\_\_\_\_\_\_\_\_\_\_\_.$$

| $\frac{y(xy - 4)}{(xy + 2)(xy - 2)}$

51. $\frac{x^2 - 9y^{-2}}{(x + 3y^{-1})(x - 3y^{-1})} =$ ___ | 1

1.8 Radicals

1. In the previous section we defined $a^{1/n}$ as the _____ root of a. The nth root of a can also be written with radical notation as

$$a^{1/n} = \_\_\_\_\_,$$

where a is a ______ number and n is a positive integer greater than 1.

| nth | $\sqrt[n]{a}$ | real

2. The symbol $\sqrt[n]{\ }$ is called a __________ sign. The number a is the __________, and n is the ______ of the radical.

| radical | radicand; index

3. For even values of n (square roots, fourth roots, and so on), there (*is/are*) _____ nth root(s). the symbol $\sqrt[n]{\ }$ represents the __________ root, or the ____________ nth root.

| are; 2 | positive | principle

Evaluate each root in Frames 4–9.

4. $\sqrt[3]{8}$ = ____ since ____ = 8. | $2; 2^3$

5. $\sqrt[4]{81}$ = ____ | 3

6. $\sqrt{100}$ = ____ | 10

7. $\sqrt[3]{-1000}$ = ____ | -10

8. $\sqrt[4]{\frac{16}{625}}$ = ____ | $\frac{2}{5}$

9. $\sqrt[5]{-\frac{32}{243}}$ | $-\frac{2}{3}$

10. Using a radical symbol, $a^{m/n}$ can be written as ____. | $\sqrt[n]{a^m}$

Write in radical form in Frames 11–13.

11. $m^{2/3}$ | $\sqrt[3]{m^2}$

12. $-3r^{3/4}$ | $-3\sqrt[4]{r^3}$

13. $(r + 2)^{1/6}$ | $\sqrt[6]{r + 2}$

Write in exponential form in Frames 14 and 15.

14. $\sqrt[4]{x^3}$ | $x^{3/4}$

15. $6\sqrt{r^5}$ | $6r^{5/2}$

16. There are three key rules for radicals:

$$\sqrt[n]{a} \cdot \sqrt[n]{b} = \_\_\_\_$$

$$\sqrt[n]{\frac{a}{b}} = \_\_\_\_ \quad (b \neq 0)$$

$$\sqrt[m]{\sqrt[n]{a}} = \_\_\_\_.$$

For example:

$$\sqrt{3} \cdot \sqrt{4} = \_\_\_\_$$

$$\sqrt[3]{\frac{5}{8}} = \_\_\_\_$$

$$\sqrt[3]{\sqrt[4]{6}} = \_\_\_\_.$$

$\sqrt[n]{ab}$

$\dfrac{\sqrt[n]{a}}{\sqrt[n]{b}}$

$\sqrt[mn]{a}$

$\sqrt{12}$ or $2\sqrt{3}$

$\dfrac{\sqrt[3]{5}}{2}$

$\sqrt[12]{6}$

17. To simplify radicals, the following conditions must be satisfied.

(a) The radicand has no factor raised to a power greater than or equal to the _____.

(b) The ______________ has no fractions.

(c) No __________________ contains a radical.

(d) Exponents in the radicand and the index of the radical have no common _________.

(e) All indicated __________________ have been performed.

index

radicand

denominator

factor

operations

Simplify the radicals of Frames 18–30. Assume all variables represent positive real numbers.

18. $\sqrt{200} = \sqrt{\_\_\_\_ \cdot 2} = \sqrt{\_\_\_\_} \cdot \sqrt{2} = \_\_\_\_\_\_$

100; 100; $10\sqrt{2}$

19. $\sqrt{12m^4} = \sqrt{12} \cdot \sqrt{\_\_\_\_} = \_\_\_\_\_\_$

m^4; $2\sqrt{3}m^2$

20. $\sqrt[3]{16} = \sqrt[3]{\_\_\_\_ \cdot 2} = \_\_\_\_\_\_$

8; $2\sqrt[3]{2}$

21. $\sqrt[4]{243} = \_\_\_\_\_$

$3\sqrt[4]{3}$

22. $\sqrt[3]{16x^7y^5z^9} = \sqrt[3]{2^3 2^1 x^6 x^1 y^3 y^2 z^9}$

Note that we rewrite the expression by separating out powers of three. Then we have

$$2 \cdot x^2 \cdot \_\_\_\_ \cdot \_\_\_\_ \sqrt[3]{2 \cdot x \cdot \_\_\_\_}.$$

y; z^3; y^2

23. $\sqrt[4]{81r^7y^4z^8} = \_\_\_\_\_\_\_\_$

$3ryz^2 \sqrt[4]{r^3}$

24. $\sqrt{147x^5y^3} = \_\_\_\_\_\_\_\_$

$7x^2y\sqrt{3xy}$

25. $\sqrt{50} + \sqrt{32} + \sqrt{18} = \_\_\_\_ + \_\_\_\_ + \_\_\_\_$

$= \_\_\_\_$

$5\sqrt{2}$; $4\sqrt{2}$; $3\sqrt{2}$

$12\sqrt{2}$

26. $-5\sqrt{27} + 7\sqrt{12} - \sqrt{108} = \_\_\_\_\_\_$

$-7\sqrt{3}$

27. $(\sqrt{6} + 1)(\sqrt{6} - 1) = \sqrt{6} \cdot \sqrt{6} - \sqrt{6} + \sqrt{6} - 1 = \_\_\_$

5

28. $(2\sqrt{3} - 3)(3\sqrt{3} + 4)$

$= (2\sqrt{3})(3\sqrt{3}) + \_\_\_\_\_ - 9\sqrt{3} - \_\_\_\_$

$= \_\_\_ - \sqrt{3} - 12$

$= \_\_\_\_\_\_$

$8\sqrt{3}$; 12

18

$6 - \sqrt{3}$

29. $(4\sqrt{2} + 1)^2 = \_\_\_\_\_\_\_$

$33 + 8\sqrt{2}$

30. $(\sqrt[3]{10} - 1)(2\sqrt[3]{10} + 1) = \_\_\_\_\_\_\_\_$

$2\sqrt[3]{10^2} - \sqrt[3]{10} - 1$

31. For a radical to be __________, there must be no __________ in the __________.

The process of removing radicals is called __________ the denominator.

simplified, radicals; denominator

rationalizing

Rationalize the denominators in Frames 32–38.

32. $\dfrac{9}{\sqrt{3}}$

Multiply numerator and denominator by ______.

$$\frac{9}{\sqrt{3}} = \frac{9 \cdot \sqrt{3}}{\sqrt{3} \cdot \sqrt{3}} = \frac{9\sqrt{3}}{\_\_\_} = \_\_\_\_\_$$

Answers: $\sqrt{3}$; 3; $3\sqrt{3}$

33. $\sqrt{\dfrac{4}{3x}} =$ ______

Answer: $\dfrac{2\sqrt{3x}}{3x}$

34. $\sqrt[3]{\dfrac{2}{3}} = \dfrac{\sqrt[3]{2} \cdot (\_\_\_)}{\sqrt[3]{3} \cdot (\_\_\_)}$

$= \dfrac{\sqrt[3]{\_\_\_}}{\sqrt[3]{\_\_\_}}$

$=$ ________

Answers: $\sqrt[3]{3^2}$; $\sqrt[3]{3^2}$; 18; 27; $\dfrac{\sqrt[3]{18}}{3}$

35. $\sqrt[4]{\dfrac{5}{9}} =$ ______

Answer: $\dfrac{\sqrt[4]{45}}{3}$

36. $\sqrt[3]{\dfrac{27x^5y^6}{2z}} = \dfrac{\sqrt[3]{27x^5y^6 2^2 z^2}}{\sqrt[3]{2^3z^3}}$

Simplifying gives __________.

Answer: $\dfrac{3xy^2\sqrt[3]{4x^2z^2}}{2z}$

37. $\dfrac{x}{\sqrt{x} - 1}$

Multiply numerator and denominator by _______,

the _________ of $\sqrt{x} - 1$.

$$\frac{x}{\sqrt{x} - 1} = \frac{x(\sqrt{x} + 1)}{(\sqrt{x} - 1)(\sqrt{x} + 1)} = \_\_\_\_\_\_\_\_\_\_$$

Answers: $\sqrt{x} + 1$; conjugate; $\dfrac{x(\sqrt{x} + 1)}{x - 1}$

38. $\dfrac{\sqrt{m} + \sqrt{m + 1}}{\sqrt{m} - \sqrt{m + 1}} =$ ________________

Answer: $-(\sqrt{m} + \sqrt{m + 1})^2$ or $-2m - 2\sqrt{m(m + 1)} - 1$

Rationalize the numerator of the following. (This means write the numerator without radicals.)

39. $\frac{\sqrt{x}}{1 + 2\sqrt{x}} = \frac{(\sqrt{x})(\sqrt{x})}{(1 + 2\sqrt{x})(\sqrt{x})} = \frac{x}{\_\_\_\_}$

$\sqrt{x} + 2x$

40. $\frac{\sqrt{x} - \sqrt{x^2 - 1}}{\sqrt{x} + \sqrt{x^2 - 1}} = \_\_\_\_\_\_\_\_$

$\frac{-x^2 + x + 1}{x^2+x+2\sqrt{x(x^2 - 1)} - 1}$

CHAPTER 1 TEST

Test answers are at the back of this study guide.

Let set R = $\{-\sqrt{11}, -3/8, 0, 5, \sqrt{28}, 39/3\}$. List all the elements of set R that belong to each of the following sets.

1. Whole numbers 1. ____________

2. Irrational numbers 2. ____________

3. Evaluate the following when $w = 5$, $x = 3$, and $z = -2$.

$$\frac{-4(x + w) + 2z}{w + x + 1}$$

3. ____________

Identify the property illustrated in each of the following.

4. $(-8 + 16) + 4 = 4 + (-8 + 16)$ 4. ____________

5. $\frac{2}{3} \cdot \frac{3}{2} = 1$ 5. ____________

Perform each operation. Assume that all variables appearing as exponents represent integers.

6. $(3r^2 - 9r - 4) - (-2r^2 - 5r + 1)$ 6. ____________

7. $(2k - 7z)(3k + 2z)$ 7. ____________

8. $(3r - 1)(2r^3 - 5r^2 + r - 1)$ 8. ____________

9. Use the binomial theorem to expand $(3a - 2b)^5$.

9. ____________

10. Find the fourth term in the expansion of $(2k - m)^7$. 10. ____________

Factor as completely as possible.

11. $10r^2 + 9rx - 9x^2$

11. ____________________

12. $216a^9 - 125r^3$

12. ____________________

13. $9x^2 - 6x + 1 - y^2$

13. ____________________

14. What is wrong with the following simplification?

$$\frac{3x^2 - 4x - 4}{x^2 - 4} = \frac{3x^2 - 4x}{x^2}$$
$$= \frac{x(3x - 4)}{x^2}$$
$$= \frac{3x - 4}{x}$$

14. ____________________

Perform each operation.

15. $\dfrac{z^3 + 125y^3}{3z + 15y} \div \dfrac{z^2 - 5zy + 25y^2}{9z + 45y}$

15. ____________________

16. $\dfrac{y - 1}{y + 2} + \dfrac{y - 3}{y - 2}$

16. ____________________

17. $\dfrac{m}{m^2 - 4m + 3} - \dfrac{2}{m^2 - 5m + 6}$

17. ____________________

18. $\dfrac{3 - \dfrac{2}{r}}{r + \dfrac{5}{4r}}$

18. ____________________

Simplify each of the following. Write results without negative exponents. Assume that all variables represent positive numbers.

19. $\dfrac{12m^9n^{-1}}{(3m^2)(2n^{-3})^2}$

19. ____________________

20. $\dfrac{(6z^5r^{-1})(r^{-4})^{-1}}{(3z^{-2}r^{-3})^{-1}(z^{-2})^2}$

20. ____________________

Simplify. Assume that all variables represent positive numbers.

21. $\sqrt{500}$ 21. ______________________

22. $\sqrt[4]{48}$ 22. ______________________

23. $\sqrt{12x^4st^3}$ 23. ______________________

24. $\dfrac{\sqrt{x}}{\sqrt{x} - 2}$ 24. ______________________

25. $(\sqrt{r} - 3\sqrt{s})(2\sqrt{r} + \sqrt{s})$ 25. ______________________

CHAPTER 2 PRETEST

Pretest answers are at the back of this study guide.

Solve each of the following equations.

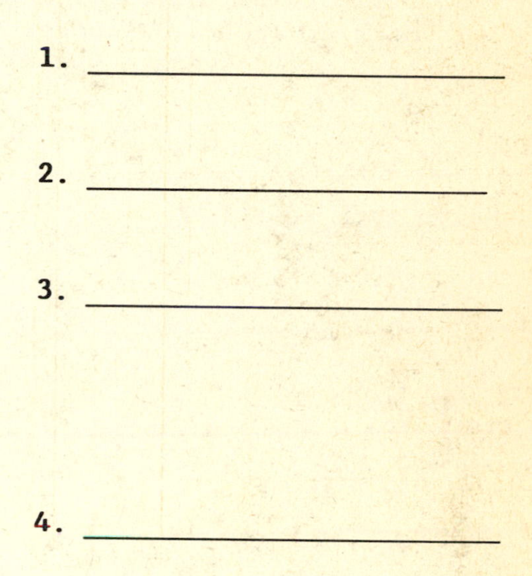

1. $2y - 3(y + 1) = 2(3y + 1)$ 1. ____________

2. $\frac{2r + 1}{2} - \frac{3r}{r - 1} = r$ 2. ____________

3. Solve $M\left(\frac{x}{k} + a\right) = x$ for x. 3. ____________

4. Dolores and John are painting the living room. Working alone, Dolores would need 6 days, while John would require 8. If they work together, how long will it take them to complete the project? 4. ____________

Perform each operation. Write each result in standard form.

5. $(-1 + 5i) + (-6 - 4i)$ 5. ____________

6. $(2 - 5i)(3 + 7i)$ 6. ____________

7. $(10 - 4i)(10 + 4i)$ 7. ____________

8. $\frac{6 - 3i}{1 + 2i}$ 8. ____________

Solve each equation.

9. $2x^2 - x - 6 = 0$ 9. ____________

10. $3m^2 - 5m + 1 = 0$ 10. ____________

11. $\frac{4}{r^2} - \frac{7}{r} = 2$ 11. ____________

12. Evaluate the discriminant of $4x^2 - 3x + 10 = 0$. Use it to predict the type of solution for the equation.

12. ______________________

13. Mark drives 10 mph slower than Cindy. Both start at the same time on a trip to Miami from Stuart, which is a distance of 100 mi. Mark drove 1/3 hr longer than Cindy. What is Cindy's average speed?

13. ______________________

Solve each equation.

14. $r^4 - 6r^2 + 8 = 0$

14. ______________________

15. $(2m - 1)^2 - 3(2m - 1) + 2 = 0$

15. ______________________

16. $\sqrt{x + 3} - \sqrt{x - 2} = 1$

16. ______________________

17. $(3x^2 - 2)^{1/3} = x^{1/3}$

17. ______________________

Solve each inequality. Write each solution in interval notation.

18. $6 + 2y \leq 4 - y$

18. ______________________

19. $-8 \leq 2x - 1 \leq 4$

19. ______________________

20. $x^2 - 2x - 15 \geq 0$

20. ______________________

21. $\dfrac{x + 3}{x + 2} \leq 0$

21. ______________________

Solve each equation.

22. $|x - 3| = 5$

22. ______________________

23. $|5 - 4k| = 2$

23. ______________________

Solve each inequality.

24. $|x| \geq 7$

24. ______________________

25. $|x - 4| < 3$

25. ______________________

CHAPTER 2 EQUATIONS AND INEQUALITIES

2.1 Linear Equations

1. An equation that is true for all meaningful values of the variable is a(n) _______. Is

$$5x + 5 = 5(x + 1)$$

an identity? (*yes/no*)

An equation true for only some values of the variable is a(n) ____________ equation. Is $4x + 1 = 3x + 5$ a conditional equation? (*yes/no*)

An equation which is false for all values of x is a(n) ______________. Is $2(x + 2) = 3 + 2x$ a contraction? (*yes/no*)

identity

yes

conditional

yes

contradiction

yes

Identify the equations in Frames 2–4 as an identity, conditional equation, or a contradiction.

2. $2r(r + 5) = 2r^2 + 10r$ ____________

identity

3. $4(x + 1) - 3(x - 2) = x + 3$ ____________

contradiction

4. $6p - 11 = 5p - 11$ ____________

conditional equation

5. To solve equations, we find a chain of simpler ______________ equations. Two equations are equivalent if they have the same ________ set.

equivalent

solution

6. Are the equations $5x - 1 = 14$ and $7x - 3 = 18$ equivalent? The equations both have solution set ____, so they (*are/are not*) equivalent. | $\{3\}$; are

7. Are $m = 9$ and $m^2 = 81$ equivalent? (*yes/no*) | no

8. To solve linear equations, we use two properties of equality:

The same expression may be ________ to both sides of an equation without changing the solution set. | added

The same nonzero expression may be ___________ on both sides of an equation without changing the solution set. | multiplied

9. A linear __________ in one variable is one that can be simplified to the form ____________. | equation; $ax + b = 0$

Solve the equations in Frames 10–17.

10. $5(2m - 1) = 10 + 5m$

Use the _______________ property on the left. | distributive

__________ $= 10 + 5m$ | $10m - 5$

Now we can add $-5m$ to both sides to get ________. | $5m - 5 = 10$

Add ___ to get | 5

$5m =$ ____. | 15

Multiply both sides by _____. | 1/5

$m =$ ____ | 3

The solution set is _____. | $\{3\}$

11. $8r - 12 = -(r - 1) - 4$ Solution set: ___ | $\{1\}$

12. $-7(k - 1) + 2 = 3(2k - 3) + 9$

Solution set: ___ | $\{9/13\}$

13. $\frac{m}{5} - 3 = 2 - \frac{4m}{5}$

To eliminate __________, multiply both sides by ___. | fractions; 5

$$5\left(\frac{m}{5} - 3\right) = 5(\_\_\_\_\_\_)$$ | $2 - \frac{4m}{5}$

$$5\left(\frac{m}{5}\right) - 15 = \_\_\_\_\_\_$$ | $10 - 4m$

$$m - 15 = \_\_\_\_\_\_$$ | $10 - 4m$

Solve the equation to get the solution set ____. | $\{5\}$

14. $\frac{2p + 3}{2} - \frac{3p}{p - 1} = p$

Multiply by ________, assuming that $p \neq 1$. | $2(p - 1)$

$$2(p - 1)\left(\frac{2p + 3}{2}\right) - 2(p - 1)\left(\frac{3p}{p - 1}\right) = \_\_\_\_\_\_$$ | $2(p - 1)p$

$$(p - 1)(\_\_\_\_\_\_) - 2(\_\_\_\_) = 2p(p - 1)$$ | $2p + 3$; $3p$

$$\_\_\_\_\_\_\_\_ - 6p = \_\_\_\_\_\_$$ | $2p^2 + p - 3$; $2p^2 - 2p$

Add $-2p^2$ to both sides and simplify to get

$$-3 - 5p = \_\_\_\_$$ | $-2p$

$$p = \_\_\_.$$ | -1

Check that -1 is a solution. The solution set is _____. | $\{-1\}$

15. $\frac{3a + 1}{3} - \frac{2a}{a + 1} = a$ Solution set: _______ | $\{1/5\}$

16. $\frac{r}{r - 4} = \frac{4}{r - 4} + 3$	
Multiply by ______.	r - 4
$(r - 4)(\frac{r}{r - 4}) = (r - 4)(\frac{4}{r - 4}) +$ ______	3(r - 4)
$r = 4 +$ ______	3r - 12
$r =$ ______	-8 + 3r
______ $= -8$	-2r
$r =$ ___	4
Check this solution by substituting ___ for r in in the original equation to get	4
$\frac{4}{4 - 4} = \frac{4}{4 - 4} + 3$	
or $\frac{4}{0} = \frac{4}{0} + 3.$	
Since division by ___ is not defined, the solu-	0
tion set for the original equation is ___.	Ø
17. $\frac{2}{x - 1} + \frac{3}{x + 2} = \frac{11}{x^2 + x - 2}$	
Find a common denominator.	
$x^2 + x - 2$ can be factored as (_____)(_____).	x - 1; x + 2
Multiply each term by ______________.	(x - 1)(x + 2)
2(_____) + 3(_____) = 11	x + 2; x - 1
x = ___	2
Check the answer in the original equation to verify that the solution set is _____.	{2}
18. Sometimes an equation which contains more than one __________ must be solved for a specified	variable
variable. This kind of equation is called a __________ equation.	literal
For example, let us solve the following equation for x.	
$5x - 2y = 3xy + 5$	

To solve for x, get all the terms containing ___ on one side of the equals sign. In our example, we should add _____ and ______ to both sides. This gives

x

2y; $-3xy$

__________ = $2y + 5$.

$5x - 3xy$

On the left, the common factor is ___.

x

__________ = $2y + 5$

$x(5 - 3y)$

Solve for x.

x = ________

$\frac{2y + 5}{5 - 3y}$

Solve for x in Frames 19–21.

19. $9m - 3xm + 5x = 2m - 5mx$ x = ________

$\frac{-7m}{2m + 5}$

20. $y^2x - 3x = 8y$ x = ________

$\frac{8y}{y^2 - 3}$

21. $R(\frac{x}{3} - q) = x$

Multiply both sides by ___.

3

$3R(\frac{x}{3} - q)$ = _____

$3x$

Use the ____________ property on the left.

distributive

_________ = $3x$

$Rx - 3Rq$

Add _____ to both sides.

$-Rx$

$-3Rq$ = ________

$3x - Rx$

_________ = x

$\frac{-3Rq}{3 - R}$

2.2 Applications of Linear Equations

1. The methods learned in Section 2.1 for solving a literal equation for a specified ______________ may be extended to solving a formula for a specified variable.

   variable

2. The formula

$$A = \frac{1}{2}bh$$

gives the area of a(n) ___________.

   triangle

3. To solve

$$A = \frac{1}{2}bh$$

for b, we would first multiply both sides by ___, getting

   2

$$2A = \_\_\_\_.$$

   bh

Multiply by _____ to get

   1/h

$$\_\_\_\_\_ = b.$$

   $\frac{2A}{H}$

Solve for the specified variable in Frames 4–6.

4. $A = \frac{1}{2}(B + b)h$ for h

Multiply by ___.

   2

$$2A = \_\_\_\_\_\_\_\_$$

   (B + b)h

Solve for h.

$$h = \_\_\_\_\_\_\_\_$$

   $\frac{2A}{B + b}$

5. Solve $\frac{1}{R} = \frac{1}{r_1} + \frac{1}{r_2}$ for R.

Multiply both sides by ______. | Rr_1r_2

$(Rr_1r_2)\frac{1}{R} = (Rr_1r_2)(\frac{1}{r_1}) + (Rr_1r_2)(\_\_\_\_)$ | $\frac{1}{r_2}$

______ = ______ + ______ | r_1r_2; Rr_2; Rr_1

$r_1r_2 = R(\_\_\_\_\_\_)$ | $r_2 + r_1$

______ = R | $\frac{r_1r_2}{r_2 + r_1}$

6. Solve $\frac{1}{Q} = \frac{1}{x} + \frac{1}{y}$ for y.

y = ______ | $\frac{Qx}{x - Q}$

7. Applied problems can be solved using six steps:

(a) Decide on an an unknown, and name it with a(n) ______. | variable

(b) Draw a(n) ______ or ______, if appropriate, showing the information given in the problem. | sketch; chart

(c) Decide on variable expressions to represent other ______ of the problem. | unknowns

(d) Using the information you have collected, write a(n) ______. You may be able to use related formulas. | equation

(e) Solve the ______. | equation

(f) Check the solution in the words of the original ______. | problem.

8. **The length of a rectangle is 4 cm longer than the width. The perimeter is 76 cm. Find the length and width of the rectangle.**

Step a Let x represent the width of the rectangle.

Step b Draw a sketch.

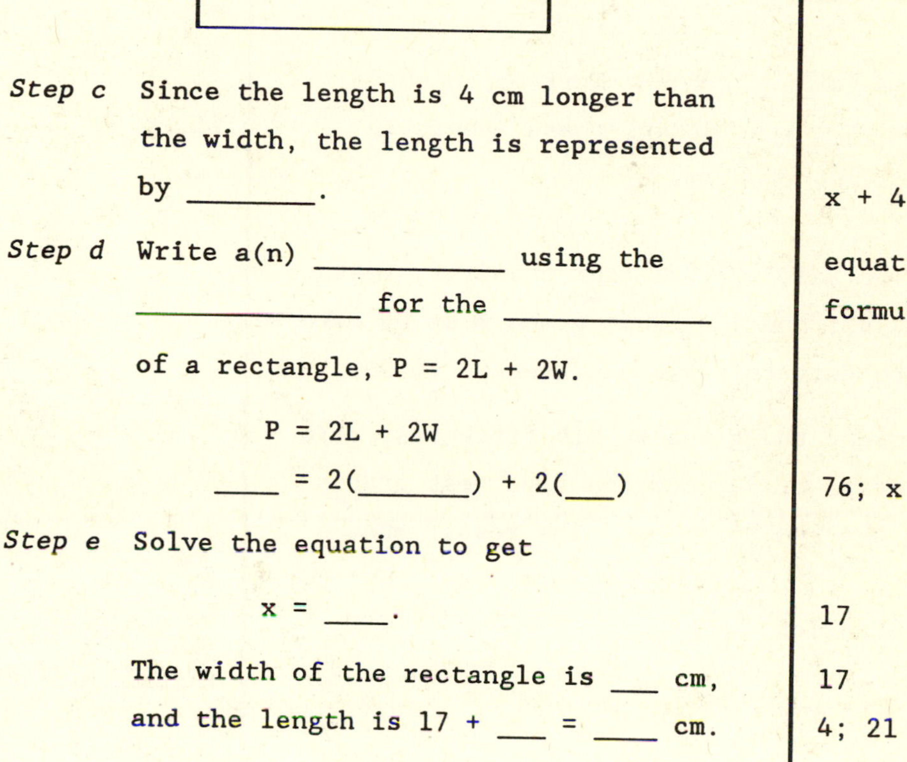

x

Step c Since the length is 4 cm longer than the width, the length is represented by ________.

$x + 4$

Step d Write a(n) ____________ using the ______________ for the ______________ of a rectangle, $P = 2L + 2W$.

equation

formula; perimeter

$$P = 2L + 2W$$

$$\_\_\_\_ = 2(\_\_\_\_\_\_\_) + 2(\_\_\_)$$

76; $x + 4$; x

Step e Solve the equation to get

$$x = \_\_\_\_.$$

17

The width of the rectangle is ___ cm, and the length is 17 + ___ = ____ cm.

17

4; 21

Solve each word problem in Frames 9–13.

9. **Kerry drives 25 kph faster than Fred. Fred goes 300 km in the same time that Kerry goes 400 km. Find the speed of each.**

Use the six steps for solving applied problems. Let x represent Fred's speed. Then Kerry's speed is _________. Use the formula $d =$ _____, noticing that $t =$ _____, and complete a chart showing the information of the problem.

$x + 25$; rt

d/r

	d	r	t
Fred	300	x	$\frac{300}{x}$
Kerry	400	___	_____

$x + 25$; $\frac{400}{x + 25}$

Since the times for the drivers are the same,

_______ = _______.

$\frac{300}{x}$; $\frac{400}{x + 25}$

Solve the equation to get

x = _______.

75

Fred's speed is ____ kph and Kerry's speed is 75 + ____ = ____ kph. Check these speeds in the words of the problem.

75

25; 100

10. **The grades on Dale's first four chemistry tests are 88, 94, 82, and 90. What grade must he make on his fifth test to raise his average to 90?**

Let x represent the grade on his fifth test. To have an average of 90, add the five test grades and divide the sum by _____. The result must equal _____.

5

90

Write the equation.

_______________ = _____

$\frac{88+94+82+90+x}{5}$; 90

Solve the equation to get

x = _____.

96

Check the answer. Dale should get a score of ____ on his fifth test.

96

11. **One person can do a job in 5 hr while another person can do the same job in 4 hr. How long would it take them to do the job working together?**

Let t represent the time it would take them working together. In one hour, the first person does _____ of the job, while the second person does _____ of the job. These are their _____ for doing the job. Multiplying each rate by ___ gives the part of the job each accomplishes working together. Complete the chart.

1/5

1/4; rates

t

	Rate	*Time*	*Part of the job accomplished*
First person	$\frac{1}{5}$	t	______
Second person	$\frac{1}{4}$	t	______

$\frac{1}{5}t$ or $\frac{t}{5}$

$\frac{1}{4}t$, or $\frac{t}{4}$

Write an equation showing that the sum of the parts of the job each does is ___.

1

_____ + _____ = 1

$\frac{t}{5}$; $\frac{t}{4}$

Solve this equation to get

t = _______ hr.

20/9 or 2 2/9

12. **A chemist needs to mix a 70% solution with 12 liters of a 40% solution to get a 50% solution. How many liters of the 70% solution are needed?**

Let x represent the number of liters of ______ solution.

70%

Make a chart showing the information of the problem.

Strength	*Liters of solution*	*Liters of pure chemical*
40%	12	12(.40) = ____
70%	x	______________
50%	____	______________

4.8

.70x

12 + x; .50(12 + x)

The number of liters of pure chemical in the 40% solution plus the number in the 70% solution must equal the number in the 50% solution, or

4.8 + ______ = ____________.

.70x; .50(12 + x)

Solve this equation, to get

$x =$ ____. | 6

____ liters of 70% solution are needed. | 6

13. **Tom Cameron inherits $80,000 from his uncle. He invests part at 12% and part at 10%, earning $9000 a year in interest. How much is invested at each rate?**

Let x represent the amount invested at 12%. Then the amount invested at 10% is ________. | 80,000 − x

The formula for interest is $i =$ _____. Organize this information in a chart. | prt

Amount invested (p)	*Interest rate (r)*	*Interest earned (prt)**
________	________	________
________	________	________

x; 12%; .12x

80,000 − x; 10%; .10(80,000 − x)

*Note: time(t) = 1 year

Since the total interest is $9000,

_____ + __________ = 9000. | .12x; .10(80,000 − x)

Solve for x. $x =$ ________ | 50,000

Cameron has __________ invested at 12%, and __________ at 10%. | $50,000 | $30,000

2.3 Complex Numbers

1. The number i is defined as

$i =$ _____. | $\sqrt{-1}$

Frame	Answer
2. Also, i^2 = ____	-1
3. The number a + bi is called a(n) ________ number.	complex
4. The numbers 5 - 2i, -4 + i, 8i, and 11 are all examples of __________ numbers.	complex
5. A complex number of the form a + bi where $b \neq 0$ is called a(n) ____________ number.	imaginary
6. The numbers 3 + 4i, -11i, and $-2 - i\sqrt{2}$ are all ____________ numbers. The numbers 7, 3 + 0i, and $-\pi$, however, are all ________ numbers.	imaginary real
7. Every real number and every ____________ number are also __________ numbers.	imaginary complex
8. A complex number written as a + bi is in ________ form.	standard
9. If $a > 0$, then $\sqrt{-a}$ = _____	$i\sqrt{a}$

Simplify each expression in Frames 10–12.

Frame	Answer
10. $\sqrt{-25} = i\sqrt{\_\_\_} =$ ___	25; 5i
11. $\sqrt{-12}$ = ______	$2i\sqrt{3}$
12. $\sqrt{-300}$ = ______	$10i\sqrt{3}$
13. Products or quotients with negative radicands are simplified by rewriting $\sqrt{-a}$ as _____ for nonnegative numbers ___.	$i\sqrt{a}$ a

Multiply or divide as indicated.

14. $\sqrt{-3} \cdot \sqrt{-5} =$ _____ • _____

$= i^2($_____$)$

$=$ _______ $=$ _______

Answers: $i\sqrt{3}$; $i\sqrt{5}$ | $\sqrt{15}$ | $-1\sqrt{15}$; $-\sqrt{15}$

15. $\sqrt{-14} \cdot \sqrt{-2} =$ ______ • ______

$=$ ______

Answers: $i\sqrt{14}$; $i\sqrt{2}$ | $-2\sqrt{7}$

16. $\dfrac{\sqrt{-105}}{\sqrt{-21}} = \dfrac{\_\_\_\_}{\_\_\_\_}$

$= \sqrt{\dfrac{\_\_\_}{\_\_\_}}$

$=$ ___

Answers: $i\sqrt{105}$ / $i\sqrt{21}$ | 105 / 21 | $\sqrt{5}$

17. The sum of a + bi and c + di is

$(a + bi) + (c + di) = (a + c) + ($_______$)i$.

Answer: $b + d$

Add or subtract in Frames 18–22.

18. $(-6 + 2i) + (3 - 5i) =$ ________ Answer: $-3 - 3i$

19. $(12 - 5i) + (-3 + 7i) =$ ________ Answer: $9 + 2i$

20. $(2 - 5i) - (5 - 9i) = 2 - 5i + ($________$)$

$=$ ________

Answers: $-5 + 9i$ | $-3 + 4i$

21. $(-1 - 8i) - (-2 - i) =$ ________ Answer: $1 - 7i$

22. $(3 - 11i) - (8 - 14i) =$ ________ Answer: $-5 + 3i$

23. Multiply complex numbers just as we did with binomials.	
For example, the product $(2 - 5i)(3 + 4i)$ is	
$(2 - 5i)(3 + 4i)$	
$= 2(\_\_\_) + 2(\_\_\_) - 5i(\_\_\_) - 5i(\_\_\_)$	3; 4i; 3; 4i
$= 6 + 8i - \_\_\_\_ - 20i^2$.	15i
Use the fact that $i^2 = \_\_\_$.	-1
$= 6 + 8i - 15i - 20(\_\_\_)$	-1
$= \_\_\_\_\_\_\_\_$	26 - 7i

Find each product in Frames 24–26.

24. $(8 - 6i)(4 + i) = \_\_\_\_\_\_\_\_$	38 - 16i
25. $(2 + 7i)(3 + 5i) = \_\_\_\_\_\_\_\_$	-29 + 31i
26. $(11 - 3i)(11 + 3i) = \_\_\_\_$	130
27. To find powers of i, use the fact that $i^4 = \_\_\_$, notice that the values repeat as	1
$i^1 = i$, $i^2 = \_\_\_\_$, $i^3 = \_\_\_\_$, $i^4 = \_\_\_\_$,	-1; -i; 1
$i^5 = \_\_\_\_$, $i^6 = \_\_\_\_$, $i^7 = \_\_\_\_$, $i^8 = \_\_\_\_$,	i; -1; -i; 1
and so on. Therefore, we can find powers of i if we remember only the first _____ powers.	four

Find each power of i.

28. $i^{11} = (i^4)^2 \cdot i^{\_\_} = 1^2 \cdot i^{\_\_}$	3; 3
$= 1(\_\_\_) = \_\_\_$	-i; -i
29. $i^{53} = \_\_\_$	i

30. $i^{-13} = \frac{1}{i^{13}} = \frac{1}{\_\_\_} = \_\_\_$ | i; i^{-1}

31. Quotients of complex numbers are found by using ____________. For example, the conjugate of $2 - 5i$ is ________, and the conjugate of $-1 + 4i$ is _______. The product of a complex number and its conjugate is always a ______ number. For real numbers a and b

$$(a + bi)(a - bi) = \_\_\_\_\_\_\_\_.$$

conjugates
$2 + 5i$
$-1 - 4i$
real
$a^2 + b^2$

Find each quotient in Frames 32–34.

32. $\frac{2 - i}{3 + 2i}$

Multiply numerator and denominator by ________. | $3 - 2i$

$$\frac{2 - i}{3 + 2i} = \frac{(2 - i)(3 - 2i)}{(3 + 2i)(\_\_\_\_\_\_)}$$

$= \_\_\_\_\_\_\_\_$

$= \_\_\_\_\_\_\_\_$ (in standard form)

$3 - 2i$
$\frac{4 - 7i}{13}$
$\frac{4}{13} - \frac{7}{13}i$

33. $\frac{5 - 4i}{1 + 2i} = \_\_\_\_\_\_\_\_\_\_$

$= \_\_\_\_\_\_\_\_\_\_$ (in standard form)

$\frac{-3 - 14i}{5}$
$-\frac{3}{5} - \frac{14}{5}i$

34. $\frac{9}{i} = \_\_\_\_\_$ | $-9i$

2.4 Quadratic Equations

1. A quadratic equation is an equation that can be written in the form ______ = 0, where $a \neq 0$.

   Answer: $ax^2 + bx + c$

2. Some quadratic equations can be solved by the ______ factor property: if $ab = 0$, then $a = 0$ or ______.

   Answer: zero–; $b = 0$

Use the zero–factor property in Frames 3–8.

3. $m^2 + 3m - 18 = 0$

   Factor to get

   (______)(______) = 0.

   Answer: $m + 6$; $m - 3$

   By the zero–factor property,

   ______ = 0 or ______ = 0.

   Answer: $m + 6$; $m - 3$

   Solve each equation separately to get

   m = ______ or m = ______.

   Answer: -6; 3

   The solution set is ______.

   Answer: $\{-6, 3\}$

4. $a^2 + a = 6$

   Rewrite the equation as ______ = 0.

   Answer: $a^2 + a - 6$

   Solution set: ______

   Answer: $\{-3, 2\}$

5. $2y^2 = 7y + 4$ Solution set: ______

   Answer: $\left\{-\frac{1}{2}, 4\right\}$

6. $2r^2 + r - 3 = 0$ Solution set: ______

   Answer: $\left\{-\frac{3}{2}, 1\right\}$

7. $1 + \frac{14}{x} + \frac{49}{x^2} = 0$

First put the equation in quadratic form by multiplying both sides by ____ to get | x^2

________________ = 0. | $x^2 + 14x + 49$

By the zero-factor property,

$(x + 7)(x + \_\_\_\_) = 0$ | 7

$x = \_\_\_.$ | -7

There is only one solution to this equation. The solution set is ______. | $\{-7\}$

8. $2 + \frac{8}{x} + \frac{6}{x^2} = 0$ Solution set: ________ | $\{-1, -3\}$

9. Some quadratic equations can be solved by taking the ________ root of both sides. For example, | square

the solutions of $x^2 = 11$ are

$x = \_\_\_\_\_\_\_\_$ | $\pm\sqrt{11}$

with solution set: __________. | $\{\pm\sqrt{11}\}$

Solve the equations in Frames 10–12.

10. $(r + 7)^2 = 24$

Take the square root of both sides.

$r + 7 = \_\_\_\_\_\_\_\_$ | $\pm\sqrt{24}$

$r + 7 = \_\_\_\_\_\_\_\_$ (Simplify.) | $\pm 2\sqrt{6}$

Add ____ on both sides. | -7

$r = \_\_\_\_\_\_\_\_$ | $-7 \pm 2\sqrt{6}$

Solution set: __________ | $\{-7 \pm 2\sqrt{6}\}$

11. $(3m - 1)^2 = 50$ Solution set: __________ | $\left\{\frac{1 \pm 5\sqrt{2}}{3}\right\}$

12. $(8m + 7)^2 = 6$ Solution set: ________

$\left\{\frac{-7 \pm \sqrt{6}}{8}\right\}$

13. Any quadratic equation can be solved if it is first written in the form $(x + n)^2 = k$. To see how to do this, let us solve the quadratic equation

$$x^2 - 2x - 1 = 0.$$

First, add ___ to each side.

1

$$x^2 - 2x = \_\_\_$$

1

Take half of ___, which equals ___, and square this result.

−2; −1

$$(\_\_\_)^2 = \_\_\_$$

−1; 1

Add 1 to each side.

$$x^2 - 2x + \_\_\_ = 1 + \_\_\_$$

1; 1

Factor on the left.

$$\_\_\_\_\_ = 2$$

$(x - 1)^2$

Take the square root of each side.

$$x - 1 = \_\_\_\_$$

$\pm\sqrt{2}$

$$x = \_\_\_\_$$

$1 \pm \sqrt{2}$

The solution set is ________.

$\{1 \pm \sqrt{2}\}$

14. Solve $2k^2 - 2k - 1 = 0$ by completing the square. Start by dividing both sides by 2.

$$\frac{2k^2 - 2k - 1}{2} = \_\_\_\_$$

$\frac{0}{2}$ or 0

$$k^2 - k - \_\_\_ = 0$$

$\frac{1}{2}$

Add 1/2 to both sides.

$$\_\_\_\_ = \frac{1}{2}$$

$k^2 - k$

Take half of ____, which equals ____, and square this result. | -1; $-\frac{1}{2}$

$(\_\_\_\_)^2 = \_\_\_\_$ | $-\frac{1}{2}$; $\frac{1}{4}$

Add 1/4 to both sides.

$\_\_\_\_\_\_ = \frac{1}{2} + \_\_\_$ | $k^2 - k + \frac{1}{4}$; $\frac{1}{4}$

Factor on the left.

$(\_\_\_\_)^2 = \frac{3}{4}$ | $k - \frac{1}{2}$

Take the square root of each side.

$k - \frac{1}{2} = \_\_\_\_$ | $\pm\frac{\sqrt{3}}{2}$

$k = \_\_\_\_$ | $\frac{1}{2} \pm \frac{\sqrt{3}}{2}$

The solution set is ______. | $\left\{\frac{1 \pm \sqrt{3}}{2}\right\}$

15. As an alternative method for solving quadratic equations, use the quadratic ______: The solutions of $ax^2 + bx + c = 0$ (where $a \neq 0$) are given by | formula

$$x = \frac{-b \pm \sqrt{\_\_\_\_}}{\_\_\_}.$$ | $b^2 - 4ac$; $2a$

Use the quadratic formula in Frames 16–20.

16. $m^2 - 3m - 40 = 0$

Here a = ___, b = ___ and c = ___. | 1; −3; −40

Substitute these values into the quadratic formula.

$$m = \frac{-(\_\_) \pm \sqrt{(\_\_)^2 - 4(\_\_)(\_\_)}}{2(\_\_)}$$ | −3; −3; 1; −40; 1

$$= \frac{3 \pm \sqrt{9 + \_\_\_}}{2} = \frac{3 \pm \sqrt{\_\_\_}}{2}$$ | 160; 169

$$m = \frac{3 \pm \_\_\_}{2}$$ | 13

Use the + sign: $m = \dfrac{3 + \_\_\_}{2} = \_\_\_.$ | 13; 8

Use the - sign: $m = \dfrac{3 - \_\_\_}{2} = \_\_\_.$ | 13; -5

The solution set is ________. | $\{8, -5\}$

17. $m^2 - 3m = 3$

Write the equation as ________ = 0, with | $m^2 - 3m - 3$

a = ___, b = ___, c = ___. Then | 1; -3; -3

$$m = \frac{-(-3) \pm \sqrt{(-3)^2 - 4(1)(-3)}}{2(1)} = \frac{3 \pm \sqrt{\_\_\_}}{2}.$$ | 21

Solution set: ________ | $\left\{\dfrac{3 \pm \sqrt{21}}{2}\right\}$

18. $x^2 + 7x + 5 = 0$ Solution set: ________ | $\left\{\dfrac{-7 \pm \sqrt{29}}{2}\right\}$

19. $2r^2 + 3r = 1$ Solution set: ________ | $\left\{\dfrac{-3 \pm \sqrt{17}}{4}\right\}$

20. $x^2 + x + 5 = 0$

$$x = \frac{-1 \pm \sqrt{1 - 4(1)(5)}}{2(1)}$$

$$= \frac{-1 \pm \sqrt{\_\_\_}}{2}$$ | -19

$$x = \frac{-1 \pm \_\_\_}{2}$$ | $i\sqrt{19}$

Solution set: ________ | $\left\{-\dfrac{1}{2} \pm \dfrac{i\sqrt{19}}{2}\right\}$

21. $2r^2 - 3r + 6 = 0$ Solution set: ________ | $\left\{\dfrac{3}{4} \pm \dfrac{i\sqrt{39}}{4}\right\}$

22. $m^2 - im + 2 = 0$ Solution set: ________ | $\{2i, -i\}$

23. $\sqrt{5}x^2 + 2x + \sqrt{5} = 0$

$a =$ ___, $b =$ ___, $c =$ ___ | $\sqrt{5}$; 2; $\sqrt{5}$

$$m = \frac{-(\_\_\_) \pm \sqrt{(\_\_\_)^2 - 4(\_\_\_)(\_\_\_)}}{2(\_\_\_)}$$

2; 2; $\sqrt{5}$; $\sqrt{5}$

$\sqrt{5}$

$=$ ______________ | $\dfrac{-2 \pm 4i}{2\sqrt{5}}$

Write in lowest terms and rationalize the denominator of each solution.

Solution set: ____________ | $\left\{-\dfrac{\sqrt{5}}{5} \pm \dfrac{2i\sqrt{5}}{5}\right\}$

24. Solve $x^3 + 125 = 0$

The first step in solving this ________ equation is to __________ on the left side. | cubic; factor

$$x^3 + 125 = 0$$

$(x + 5)($______________$) = 0$ | $x^2 - 5x + 25$

________ $= 0$ or $x^2 - 5x + 25 = 0$ | $x + 5$

$x =$ ___ | -5

Use the ________________ formula to solve $x^2 - 5x + 25 = 0$. | quadratic

$x^2 - 5x + 25 = 0$

$$x = \frac{-(\_\_\_) \pm \sqrt{(\_\_\_)^2 - 4(\_\_\_)(\_\_\_)}}{2(\_\_\_)}$$

-5; -5; 1; 25

1

$x =$ ______________ | $\dfrac{5 \pm 5i\sqrt{3}}{2}$

Solution set: ____________ | $\left\{-5, \dfrac{5}{2} \pm \dfrac{5i\sqrt{3}}{2}\right\}$

25. Solve $V = \pi r^2 h$ for r.

Multiply both sides by __________. | $1/(\pi h)$

$$V = \pi r^2 h$$

$$\frac{V}{\pi h} = r^2$$

Use the square root property.

$\_\_\_\_\_ = r$ $\pm\sqrt{\dfrac{V}{\pi h}}$

Rationalize the denominator on the left.

$\_\_\_\_\_ = r$ $\pm\dfrac{\sqrt{V\pi h}}{\pi h}$

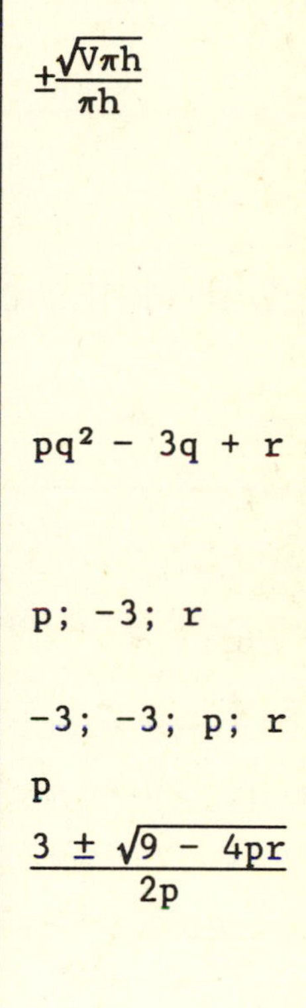

26. Solve $pq^2 = 3q - r$ for q.

Rewrite in the form of a general quadratic equation.

$\_\_\_\_\_\_\_\_\_\_ = 0$ $pq^2 - 3q + r$

Use the quadratic formula with

a = ___, b = ____, and c = ____. p; -3; r

$$q = \frac{-(\_\_\_) \pm \sqrt{(\_\_\_)^2 - 4(\_\_\_)(\_\_\_)}}{2(\_\_\_\_)}$$

-3; -3; p; r

p

$q = \_\_\_\_\_\_\_\_\_\_$ $\dfrac{3 \pm \sqrt{9 - 4pr}}{2p}$

27. The quantity under the square root sign in the quadratic formula, __________, is called the ___________. When the numbers a, b, and c are integers, the discriminant is used to predict whether the solutions will be ______________, _______________, or _______________.

$b^2 - 4ac$

discriminant

rational

irrational; imaginary

Use the discriminant to predict the type of solutions in Frames 28–30.

28. $2k^2 + 3k + 10 = 0$

a = ___, b = ___, c = ___ 2; 3; 10

$b^2 - 4ac = (\_\_\_)^2 - 4(\_\_\_\_)(\_\_\_\_) = \_\_\_\_$ 3; 2; 10; -71

The discriminant is negative, so the solutions are __________. imaginary

Frame	Answer
29. $y^2 + 2y - 11 = 0$ Discriminant = ________ The solutions are ________.	48 irrational
30. $z^2 + 2z + 1 = 0$ Discriminant = ________ The solutions are ________.	0 rational

2.5 Applications of Quadratic Equations

Solve the word problems in Frames 1–5.

Frame	Answer
1. **Mr. Rainer wants to fence a rectangular pasture for his Guernsey cows. He has 250 m of fencing. He wants to enclose an area of 2500 sq m. Find the length and width of this pasture.**	
Let W be the width of the pasture. We know that the distance around the pasture is _____ m; now use the formula for the perimeter of a rectangle, ___________, to find the length in terms of the width.	250 $P = 2L + 2W$
________ $= 2L + 2($____$)$	250; W
Solve for L. $L =$ __________	$125 - W$
The area is to be ______ sq m. To find the area, multiply the length by the width since $A =$ ____.	2500 LW
$2500 =$ __________	$W(125 - W)$
Multiply on the right. $2500 =$ __________	$125W - W^2$
Solve this equation to get $W =$ _____ or $W =$ _____	100; 25

Check each solution in the words of the original problem. If the width were _____ m, the _______ would be shorter than the width, so reject this potential solution. The width is _______. The length is 125 - _____ or _____.

100; length

25 m

25; 100 m

2. **Raul has a rectangular lot 48 ft by 80 ft in which he wishes to plant a garden surrounded by a border of grass of equal width on the four sides. If he wishes to cover only 960 sq ft with grass, what should be the width of the border?**

Let x = the width of the border.
Draw a sketch of the lot.

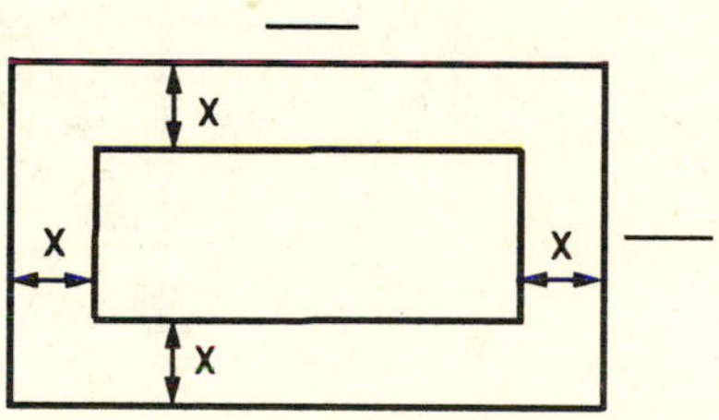

x x x x

80 ft

48 ft

The length of the garden will be __________ and the width will be __________. Write an equation saying that the sum of the areas of the ________ and the grass border will equal the area of the lot. The equation is

$$(\_\_\_\_\_\_\_\_)(\_\_\_\_\_\_\_\_) + 960 = (\_\_\_)(\_\_\_).$$

80 - 2x

48 - 2x

garden

80 - 2x; 48 - 2x; 80; 48

Solve the equation to get

$$x = \_\_\_\_ \text{ or } x = \_\_\_\_.$$

60; 4

The solution ____ is not reasonable so the width of the border should be ___ ft.

60

4

3. **Sam can clean the house in 3 hr less time than his roommate Joe. Working together they can do the job in 2 hr. How long would it take each one to do the job alone?**

Let x = the time for Joe to do the job. Then ______ = the time for Sam. In one hour the part of the job Joe can do is _____ and the part of the job Sam can do is _________.

Make a chart to organize the information.

x − 3

1/x

1/(x − 3)

	Rate	*Time*	*Part of the job accomplished*
Joe	___	2	_________
Sam	___	2	_________

$\frac{1}{x}$; $\frac{2}{x}$

$\frac{1}{x-3}$; $\frac{2}{x-3}$

Write an equation.

_____ + _______ = 1

$\frac{2}{x}$; $\frac{2}{x-3}$

Multiply by the common denominator ____________ and solve. The solution set is {___, ___} but only ____ is reasonable in the problem.

Joe needs ___ hr and Sam needs ___ hr to do the job alone.

x(x − 3)

6; 1

6

6; 3

4. **John and Elizabeth rode the Delta Queen down the Mississippi River for 15 mi. The return trip took 45 min longer since the rate of the current was 5 mph. What was the speed of the vessel each way?**

Let x = the rate of the Delta Queen in still water. Then _______ = the rate of the vessel downstream and _______ = the rate of the vessel upstream. Make a chart.

x + 5

x − 5

	d	r	$t = \frac{d}{r}$
Downstream	______	______	______
Upstream	______	______	______

$15;\ x + 5;\ \frac{15}{x + 5}$

$15;\ x - 5;\ \frac{15}{x - 5}$

Write an equation using the fact that it took longer going upstream.

Time to go downstream + ____________ = Time to go upstream

__________ + ____________ = __________

3/4 hr

$\frac{15}{x + 5};\ \frac{3}{4};\ \frac{15}{x - 5}$

A common denominator is ________________.

$4(x + 5)(x - 5)$

Solve for x: x = ____ or x = ____.

15; −15

Only the solution ____ makes sense. The vessel traveled ____ mph downstream and ____ mph upstream.

15

20; 10

2.6 Other Types of Equations

1. Some equations are not quadratic but can be solved by methods similar to those used to solve ______________ equations.

quadratic

An equation is quadratic in _____ if it can be written as

form

$$au^2 + bu + c = 0,$$

where $a \neq 0$, and u is an __________ expression.

algebraic

Solve the equations of Frames 2–18.

2. $x^4 + 2x^2 - 8 = 0$

Make the substitution $u =$ ____, so that $u^2 =$ ____. | x^2; x^4

The original equation becomes

________________ $= 0$. | $u^2 + 2u - 8$

Factor to get

(______)(______) $= 0$ | $u + 4$; $u - 2$

from which

$u + 4 = 0$ or ____ $= 0$ | $u - 2$

$u =$ ____ or $u =$ ____. | -4; 2

Since $u =$ ____, | x^2

$x^2 =$ ____ or $x^2 =$ ____. | -4; 2

If $x^2 = -4$; then $x = \pm\sqrt{\_\_\_\_}$, or $x =$ _____. | -4; $\pm 2i$

If $x^2 = 2$, then $x =$ _____. | $\pm\sqrt{2}$

The solution set is _____________. | $\{\pm 2i, \pm\sqrt{2}\}$

3. $m^4 - 5m^2 + 6 = 0$ Solution set: ________ | $\{\pm\sqrt{3}, \pm\sqrt{2}\}$

4. $2(r - 2)^2 + 5(r - 2) = 3$ Solution set: ________ | $\left\{\frac{5}{2}, -1\right\}$

5. $(b + 2)^{2/3} - (b + 2)^{1/3} - 2 = 0$

Let $u =$ __________. Then | $(b + 2)^{1/3}$

__________ $= 0$ | $u^2 - u - 2$

(______)(______) $= 0$ | $u - 2$; $u + 1$

$u =$ ____ or $u =$ ____. | 2; -1

Then $(b + 2)^{1/3} = 2$ or $(b + 2)^{1/3} = -1$.

Cube both sides.

______ $= 8$ or ______ $= -1$ | $b + 2$; $b + 2$

$b =$ ______ or $b =$ ______ | 6; -3

Solution set: ________ | $\{6, -3\}$

6. $\sqrt{x} - 2 = 1$

First add ____ to both sides, to get | 2

$\sqrt{x} =$ ____. | 3

Now square both sides.

$x =$ ____ | 9

We must _______ this proposed solution. A check | check

confirms that the solution set is _____. | $\{9\}$

7. $\sqrt{m + 2} = 4$ Solution set: _________ | $\{14\}$

8. $\sqrt{2y - 4} = \sqrt{y + 4}$ Solution set: _________ | $\{8\}$

9. $\sqrt{3p - 1} = \sqrt{2p + 2}$ Solution set: _________ | $\{3\}$

10. $\sqrt{2k - 2} + 4 = k + 3$

Add ____ to both sides, to get | -4

$\sqrt{2k - 2} =$ _______ | $k - 1$

Square both sides.

$2k - 2 =$ __________ | $k^2 - 2k + 1$

$0 =$ __________ | $k^2 - 4k + 3$

$0 =$ (______)(______) | $k - 3$; $k - 1$

$k =$ ____ or $k =$ ____ | 3; 1

Check each answer; the solution set is _______. | $\{3, 1\}$

11. $\sqrt{3r + 1} - 2 = r - 1$ Solution set: _________ | $\{0, 1\}$

12. $\sqrt{4x - 1} + 1 = \sqrt{6x + 1}$

Square both sides.

_______ $+ 2\sqrt{4x - 1} +$ ____ $=$ _______ | $4x - 1$; 1; $6x + 1$

$2\sqrt{4x - 1} =$ _______ | $2x + 1$

Square both sides again.	
$4(4x - 1) =$ ______	$4x^2 + 4x + 1$
______ $= 4x^2 + 4x + 1$	$16x - 4$
$0 =$ ______	$4x^2 - 12x + 5$
Factor.	
$0 = (2x -$ ___$)($______$)$	1; $2x - 5$
$x =$ ____ or $x =$ ____	1/2; 5/2
Solution set: ______	{1/2, 5/2}
13. $\sqrt{2p + 1} + 1 = \sqrt{3p + 4}$ Solution set: ______	{0, 4}
14. $\sqrt[3]{2x - 3} = \sqrt[3]{x + 4}$	
Remember the cube root of a number is the same thing as the number raised to the 1/3 power. Thus we should raise both sides of the equation to the ______ power.	third
$2x - 3 =$ ______	$x + 4$
$x =$ ____	7
Check your solution; the solution set is ____.	{7}
15. $\sqrt[5]{3m - 4} = \sqrt[5]{2m + 3}$ Solution set: ______	{7}
16. $(a + 7)^{1/4} = (3a + 1)^{1/4}$ Solution set: ______	{3}
17. $(3k + 4)^{1/3} = 2$ Solution set: ______	{4/3}
18. $(2x^2 + 3)^{1/4} = -3$ Solution set: ______	$\emptyset$

2.7 Inequalities

1. Inequalities can be solved in a manner very similar to that used for equations; for example, any number may be _______ to both sides of an _____________ without changing the solution set.	added inequality
2. Multiplication is not so simple: we can multiply both sides of an inequality by a ______________ number without changing the direction of the ____________ symbol, but if we ____________ both sides by a ____________ number, the direction of the inequality symbol must be ____________.	positive inequality; multiply negative reversed
3. A linear inequality in one variable is an inequality that can be written in the form __________________, where ____ $\neq 0$. (Any of the symbols ____, <, or ____ may also be used.)	$ax + b > 0$ a; $\geq$ $\leq$
4. Solve $3m + 2 \geq 8$ Add _____ to both sides, getting $3m \geq$ ____ Multiply by the positive number ______. $m \geq$ ____ The solution set is $\{m \| \_\_\_\_\_\_\_\}$. Draw a graph of the solution set; use a square bracket at ____, since 2 (*is/is not*) part of the solution.	-2 6 1/3 2 $m \geq 2$ 2 is
⟶	number line: 0, 2

5. The solution set of an inequality is often written in ________ notation. For example, the set $\{m|-1 < m < 4\}$ is written as the _____ interval ________. From now on, we write solution sets in interval notation.

interval
open
$(-1, 4)$

Write each set of Frames 6–8 in interval notation.

6. $\{r|r > -2\}$ ________

$(-2, \infty)$

7. $\{z|-5 \le z \le 7\}$ ________

$[-5, 7]$

8. $\{a|a \ge 0\}$ ________

$[0, \infty)$

Solve the inequalities of Frames 9 and 10.

9. $-2x + 7 < 19$

Add _____ to both sides to get

__________.

Multiply by the negative number _______. This forces us to ________ the inequality symbol.

____

Using interval notation, we have the solution set __________. Graph the solution set. Use a __________ to show that -6 is not part of the solution.

-7
$-2x < 12$
$-1/2$
reverse
$x > -6$
$(-6, \infty)$
parenthesis

-6 -2 0

10. $2(r + 3) - 4r \ge 9 - r$ Solution set: ________

$(-\infty, -3]$

11. $-3 < 4 + 2y < 5$

Work with all three expressions at the same time.

Add _____ to each part.

_____ < _____ < _____

-4
-7; 2y; 1

Multiply by ______.

____

Draw the graph of the solution set, ____________,

1/2

$-7/2 < y < 1/2$

$(-7/2, 1/2)$

12. A product will at least ______ _____, or produce a profit, when ___________ (R) at least equals ______ (C). The break-even point is where ______.

break even

revenue

cost; R = C

13. If $C = 50x + 200$ and $R = 75x$, find the interval where the product will at least break even. (x is the number of units produced.)

Set $R \geq C$ and solve for x.

_______ $\geq$ _______ + _______

75x; 50x; 200

$x \geq$ ____ so the product will at least break even when the number of units produced is in the interval __________.

8

$[8, \infty)$

14. The inequality $m^2 - 4m - 5 \leq 0$ is an example of a __________ inequality. To solve this inequality, first factor as

quadratic

(_______)(_______) ≤ 0.

m − 5; m + 1

We want the product to be ________. The product will be negative if the two factors have (*the same/opposite*) signs. The factor $m - 5$ is positive if _____, and negative if _____, while $m + 1$ is positive if ______, and negative if _____. Show this information on a sign _____.

nonpositive

opposite

$m > 5$; $m < 5$

$m > -1$

$m < -1$; graph

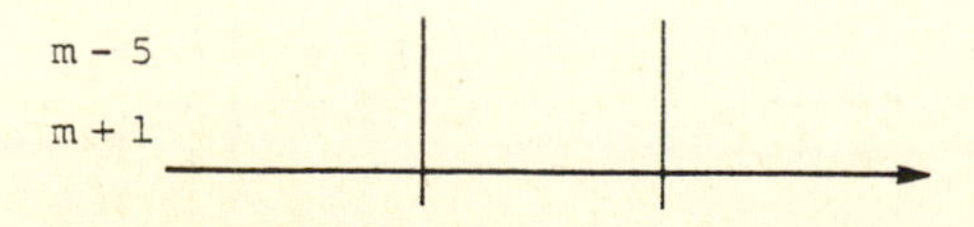

The factors have opposite signs or equal zero if ________, giving the interval ________. Graph the solution.

$-1 \le m \le 5$; $[-1, 5]$

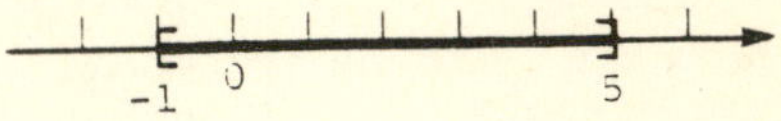

-1 0 5

Solve the quadratic inequalities of Frames 15–18.

15. $2m^2 + 7m > 4$

The values of m which satisfy $2m^2 + 7m - 4 = 0$ are _____ and _____. Use these points to divide the number line into three regions.

$m + 4$

$m - \frac{1}{2}$

-4; 1/2

-4 1/2

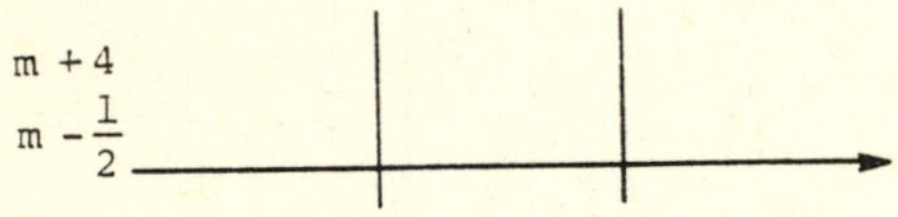

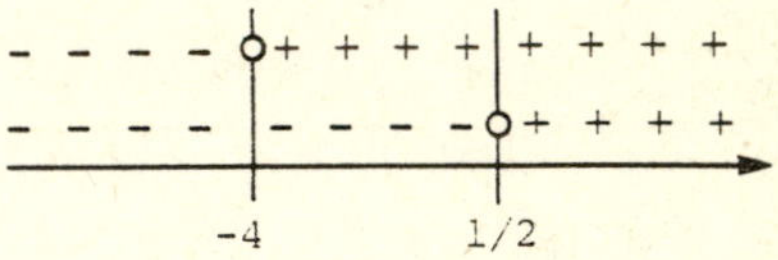

Show where the factors are positive or negative in each region. Since the inequality is ________, choose the intervals where the ________ of the factors is ________.

Solution set: ________

positive

product; positive

$(-\infty, -4) \cup (1/2, \infty)$

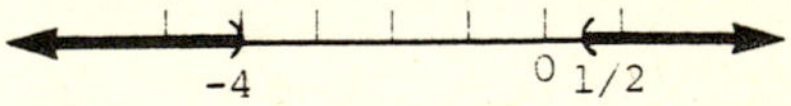

Graph the solution set.

-4 0 1/2

16. $r^2 \ge 2r + 24$

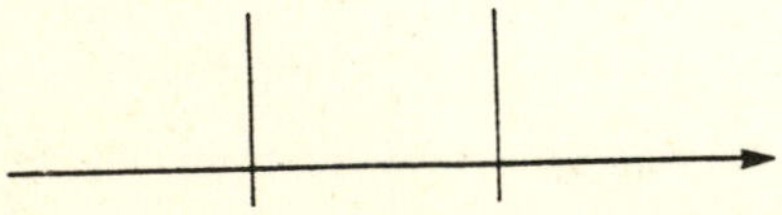

$r - 6$

$r + 4$

-4 6

Solution set: ________

$(-\infty, -4] \cup [6, \infty)$

Graph this solution.

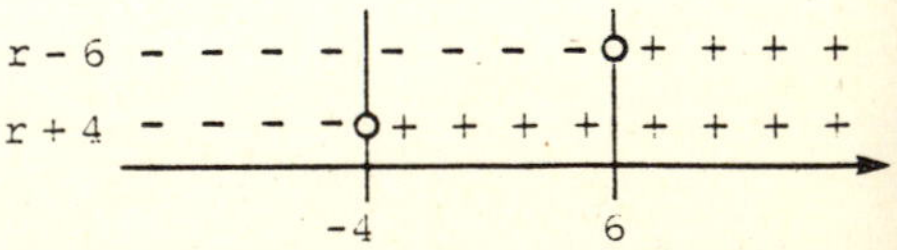

-4 0 6

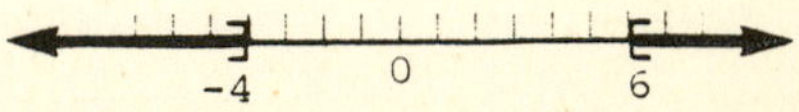

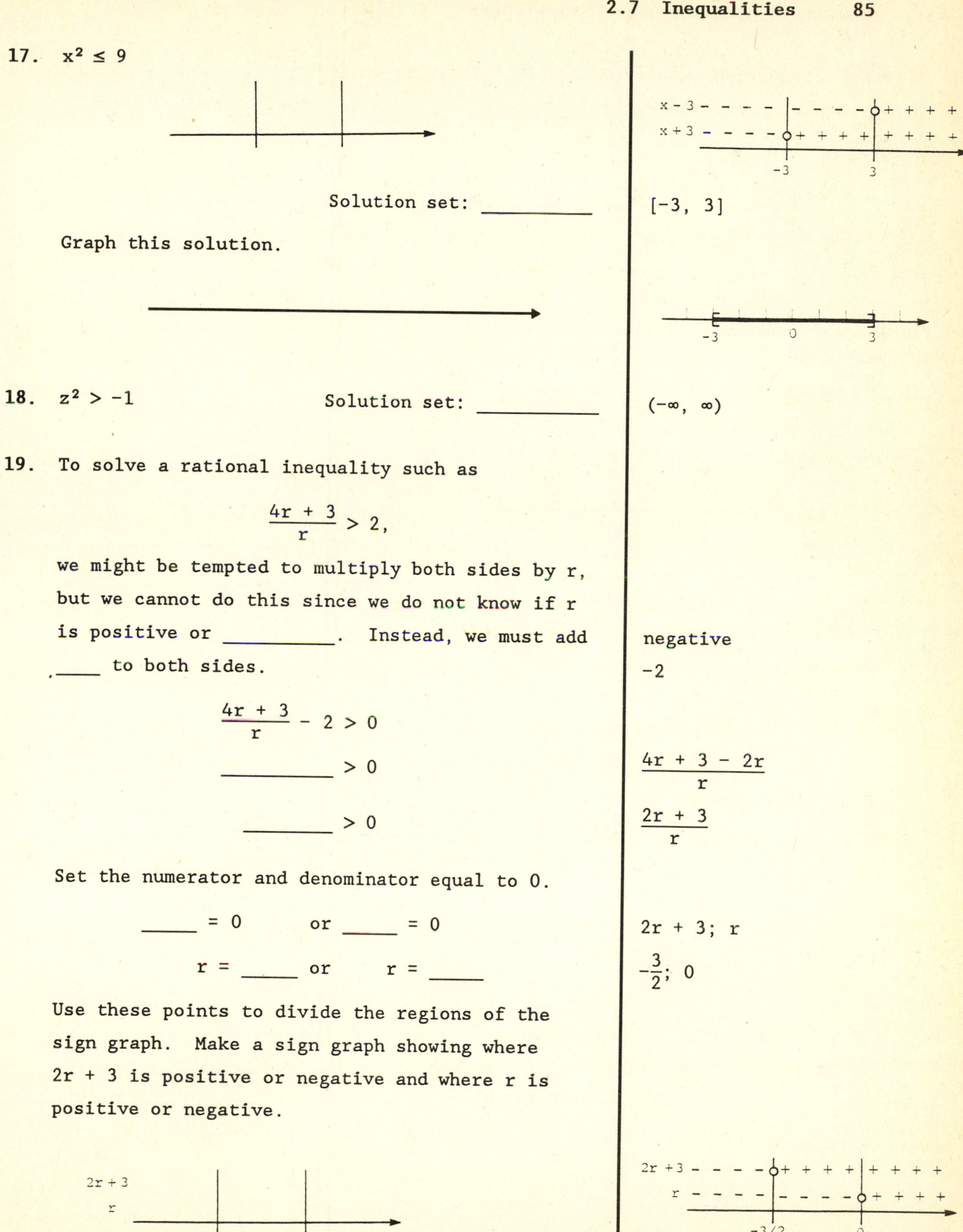

17. $x^2 \le 9$

Solution set: ________

Graph this solution.

18. $z^2 > -1$ Solution set: ________

19. To solve a rational inequality such as

$$\frac{4r + 3}{r} > 2,$$

we might be tempted to multiply both sides by r, but we cannot do this since we do not know if r is positive or ________. Instead, we must add ____ to both sides.

$$\frac{4r + 3}{r} - 2 > 0$$

$$\_\_\_\_\_\_\_\_ > 0$$

$$\_\_\_\_\_\_\_\_ > 0$$

Set the numerator and denominator equal to 0.

____ = 0 or ____ = 0

r = ____ or r = ____

Use these points to divide the regions of the sign graph. Make a sign graph showing where 2r + 3 is positive or negative and where r is positive or negative.

[-3, 3]

$(-\infty, \infty)$

negative

-2

$$\frac{4r + 3 - 2r}{r}$$

$$\frac{2r + 3}{r}$$

2r + 3; r

$-\frac{3}{2}$; 0

As the sign graph shows, the quotient is positive for either of two intervals: ________________. Graph the solution set.

$(-\infty, -3/2) \cup (0, \infty)$

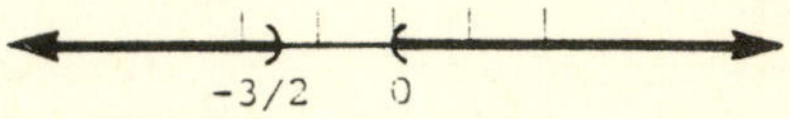

-3/2 0

Solve the rational inequalities of Frames 20–23. Use a sign graph to find the solution.

20. $\dfrac{3m + 1}{m} < -1$

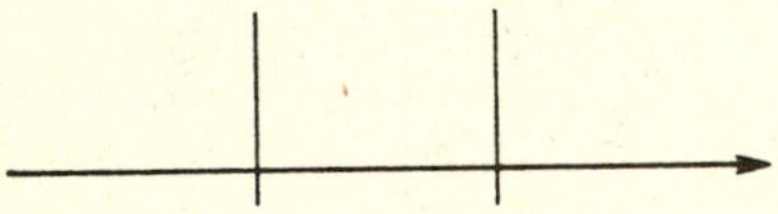

Solution set: ________

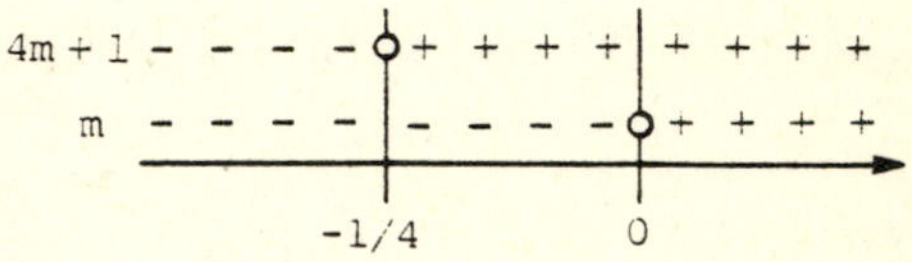

4m + 1 - - - - o+ + + + + + + +
m - - - - - - - - o+ + + +
-1/4 0

$(-1/4, 0)$

21. $\dfrac{3}{x + 2} \geq 1$

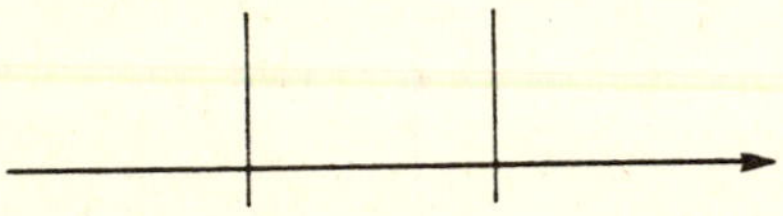

Solution set: ________

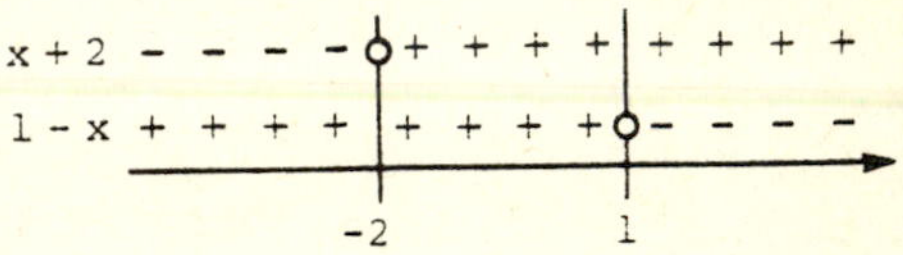

x + 2 - - - - o+ + + + + + + +
1 - x + + + + + + + + o- - - -
-2 1

$(-2, 1]$

Even though the original inequality has ≥, and not >, the number -2 is not part of the solution since it makes the denominator equal ___.

0

22. $\dfrac{3z + 4}{z - 1} < 2$

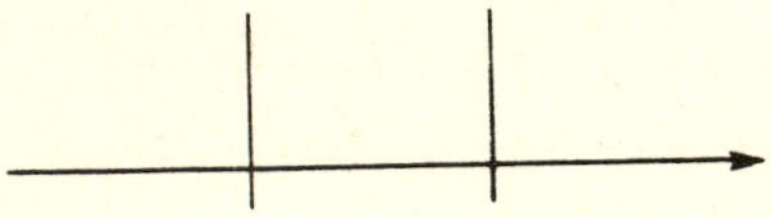

Solution set: ________

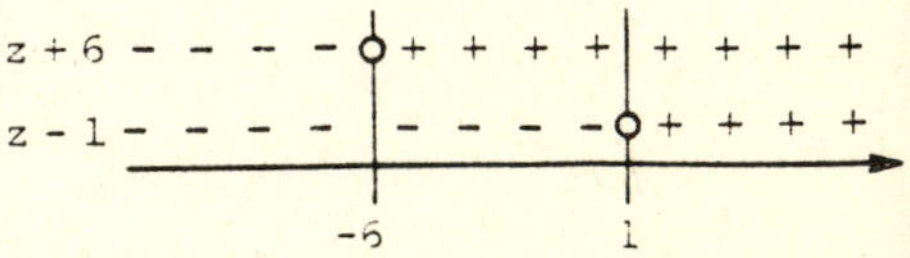

z + 6 - - - - o+ + + + + + + +
z - 1 - - - - - - - - o+ + + +
-6 1

$(-6, 1)$

23. $\frac{a - 2}{3a - 1} < 1$

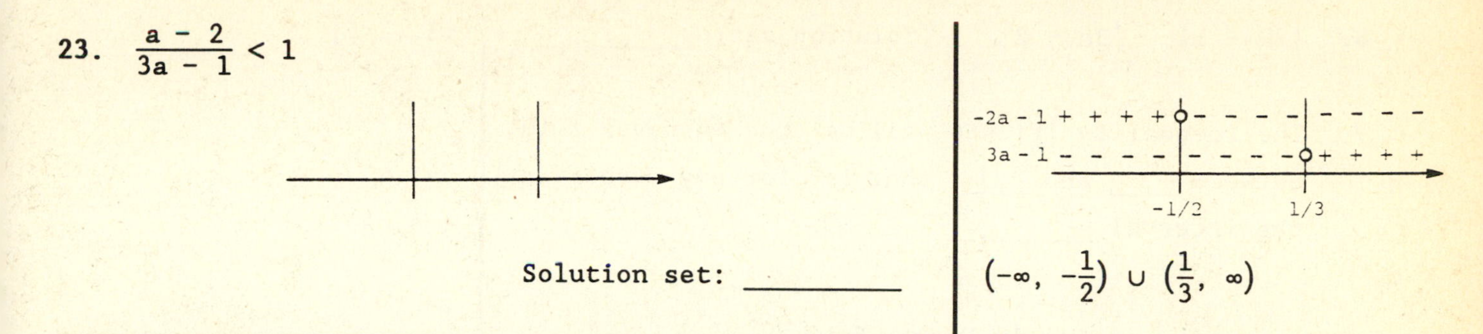

Solution set: ______ | $(-\infty, -\frac{1}{2}) \cup (\frac{1}{3}, \infty)$

2.8 Absolute Value Equations and Inequalities

1. An equation such as $|x| = 9$ can be solved by using the definition of ______ value. We know that $|x|$ can equal 9 if either

   $x =$ ____ or $x =$ ____,

   giving the solution set ______.

   Answers: absolute; 9; -9; $\{9, -9\}$

Solve each equation in Frames 2–6.

2. $|y + 1| = 7$

   Write two equations:

   $y + 1 = 7$ or $y + 1 =$ ____.

   Solve each equation to get the solution set ____.

   Answers: -7; $\{6, -8\}$

3. $|-2 + 3r| = 11$ Solution set: ______ | $\{13/3, -3\}$

4. $|8 - 7x| = -2$ Solution set: ______ | $\emptyset$

5. $\left|\frac{2}{r - 1}\right| = 5$ Solution set: ______ | $\{7/5, 3/5\}$

6. $|2m + 1| = |3m - 2|$ Solution set: ________

$\{3, 1/5\}$

7. The inequality $|r| < 4$ will be true whenever r is between ____ and ____; that is, for any number in the interval ________.

-4; 4

$(-4, 4)$

Solve each inequality in Frames 8–12.

8. $|m - 3| \leq 5$

Start with

$$-5 \leq \_\_\_\_\_\_ \leq 5.$$

The solution set is the interval ______.

$m - 3$

$[-2, 8]$

9. $|2y - 1| < 3$ Solution set: ________

$(-1, 2)$

10. $|y| - 3 \leq 2$

Add ___ to both sides. The solution set is ______.

3

$[-5, 5]$

11. $|5r - 3| \leq -1$ Solution set: ________

$\emptyset$

12. $|3p - 7| + 4 < 9$ Solution set: ________

$(2/3, 4)$

13. The inequality $|a| > 2$ is satisfied if ______ or if ______. The solution set is the interval __________.

$a > 2$

$a < -2$

$(-\infty, -2) \cup (2, \infty)$

Solve each inequality in Frames 14–18.

14. $|m - 5| > 3$

Start with

$$m - 5 < \_\_\_\_ \text{ or } m - 5 > \_\_\_\_$$

$$m < \_\_\_\_ \text{ or } m > \_\_\_\_.$$

The solution set is the interval __________.

-3; 3

2; 8

$(-\infty, 2) \cup (8, \infty)$

15.	$\|x - 1\| \geq 3$	Solution set: ________	$(-\infty, -2] \cup [4, \infty)$
16.	$\|y\| - 3 > 1$	Solution set: ________	$(-\infty, -4) \cup (4, \infty)$
17.	$\|2 - 3x\| + 2 \geq 16$	Solution set: ________	$(-\infty, -4] \cup [16/3, \infty)$
18.	$\|-7r + 10\| - 4 > -10$	Solution set: ________	$(-\infty, \infty)$

CHAPTER 2 TEST

Test answers are at the back of this study guide.

Solve each of the following equations.

1. $3k - 2(k - 4) = 5(2k + 1)$ 1. ______

2. $\frac{2p - 1}{p + 1} - \frac{3p}{p + 1} = 1$ 2. ______

3. Solve $y\left(\frac{r}{2} + z\right) = 4r$ for r. 3. ______

4. How many ounces of pear juice selling at 80¢ per 8-oz glass must be mixed with apple juice selling at 40¢ per 8-oz glass to make a mixture that sells at 50¢ per 8-oz glass? 4. ______

5. Andrew and Helen start at the same time from points 300 miles apart and travel toward each other on a straight road. How many miles will Andrew travel until they meet if he travels at 40 mph and Helen travels at 50 mph? 5. ______

Perform each operation. Write each result in standard form.

6. $(-5 - 4i) + (3 - 9i)$ 6. ______

7. $(7 - 2i)(4 + 3i)$ 7. ______

8. $\frac{8 + 3i}{4 - i}$ 8. ______

9. Is i^{207} equal to i, -1, -i, or 1? 9. ______

Solve each equation.

10. $12z^2 + 5z = 2$

10. ______________________

11. $6 + \frac{5}{m} = \frac{6}{m^2}$

11. ______________________

12. Evaluate the discriminant of $81r^2 - 90r + 25 = 0$. Use it to predict the type of solution for the equation.

12. ______________________

13. The math department is writing a placement exam. It took Don 5 hr less than Art to write a sample test. Working together on a third sample test, they took 6 hr. How long did it take Art to write a sample test?

13. ______________________

Solve each equation.

14. $m^4 - 3m^2 - 10 = 0$

14. ______________________

15. $3(a - 1)^2 + 11(a - 1) - 4 = 0$

15. ______________________

16. $\sqrt{x - 1} = \sqrt{x + 2} - 3$

16. ______________________

17. $(6x - 5)^{1/3} = (5x)^{1/3}$

17. ______________________

Solve each inequality. Write each solution in interval notation.

18. $2r - 4 > 3(r + 1)$

18. ______________________

19. $-4 \le 5a - 1 \le 3$

19. ______________________

20. $x^2 - x - 6 \ge 0$

20. ______________________

21. $\dfrac{y - 4}{y + 1} \leq 0$ 21. ____________

22. Solve $|r - 6| = 2$. 22. ____________

23. Without actually solving the inequality, explain why 1/2 cannot be in the solution set of

$$\frac{3x + 5}{2x - 1} \leq 0?$$

23. ____________

Solve each inequality. Write each solution in interval notation.

24. $|y| \geq 2$ 24. ____________

25. $|2x + 3| < 9$ 25. ____________

CHAPTER 3 PRETEST

Pretest answers are at the back of this study guide.

1. For the points (-2, 5) and (1, 8):

   (a) Find the distance between them.

   (b) Find the midpoint of the segment joining them.

1. (a)________________

   (b)________________

Find the slope of each of the following lines.

2. Through (1, 0) and (2, 3) 2. ________________

3. $4x - 3y = 2$ 3. ________________

4. $y + x = 0$ 4. ________________

For each of the following lines, write the equation in standard form.

5. Through (3, -1) with slope 2 5. ________________

6. Through (-7, -4) and perpendicular to $x = -3$ 6. ________________

7. Through (0, 0) and parallel to $y = 2x + 4$ 7. ________________

Graph the following lines.

8. $3x + 2y = 12$

9. Through $(-1, 4)$ with slope $-2/3$

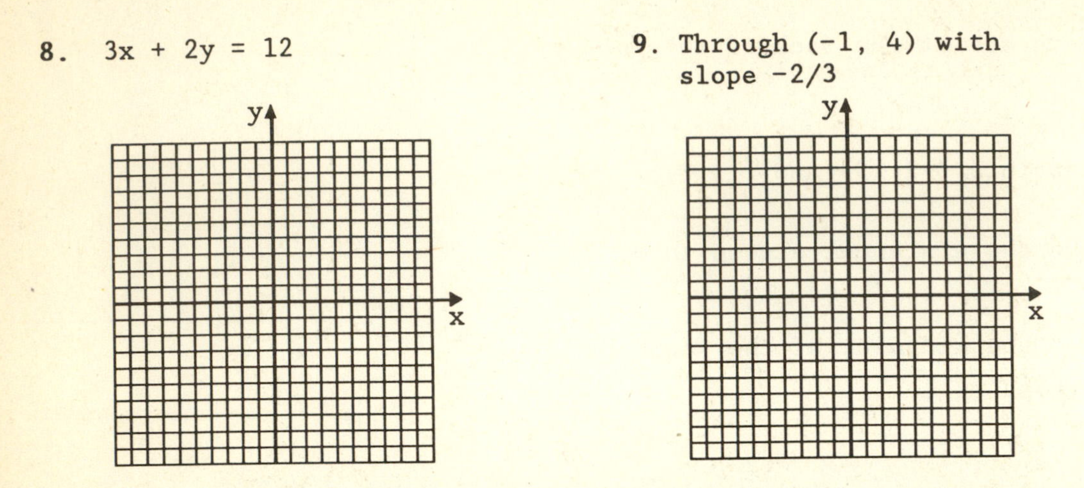

Graph each of the following. Give the axis, vertex, domain, and range of each.

10. $y = (x + 2)^2 - 3$

11. $x = (y + 3)^2 + 2$

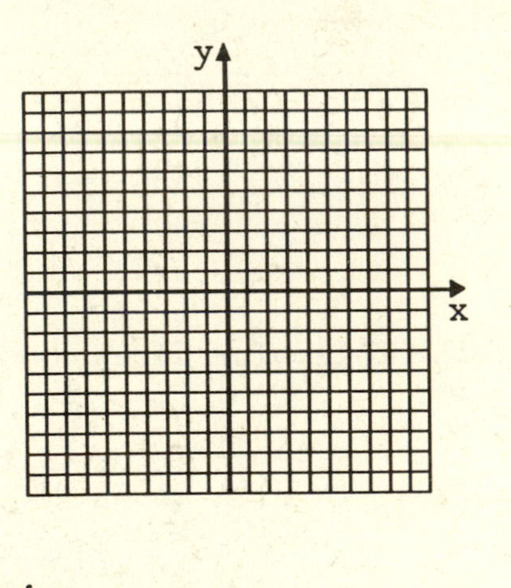

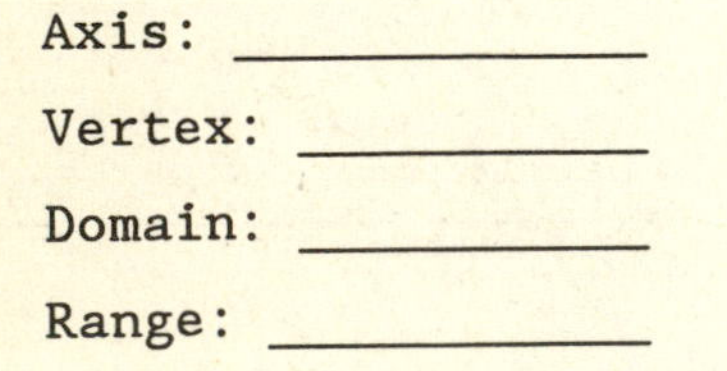

Axis: ____________

Vertex: ____________

Domain: ____________

Range: ____________

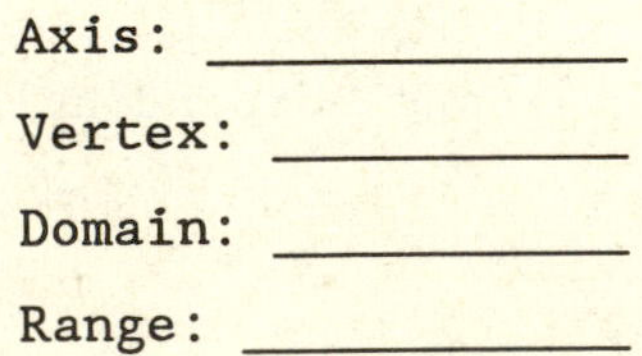

Axis: ____________

Vertex: ____________

Domain: ____________

Range: ____________

12. Give the vertex and axis of the graph of $y = x^2 - 6x + 11$.

12. ____________________

____

13. Family Food Co-op operates a natural food restaurant. By studying data concerning their past costs, they have found that the cost in dollars of operating the restaurant is given by

$$C(x) = x^2 - 80x + 3300,$$

where C(x) is the daily cost to make x meals. Find the number of meals they must sell to minimize the cost. What is the minimum cost? (Hint: Solve by finding the vertex of a parabola.)

13. ____________________

14. Give the equation of the circle having center (-1, 5) and radius $\sqrt{3}$.

14. ____________________

15. Give the center and radius of the circle $x^2 - 4x + y^2 + 6y = 18$.

15. ____________________

____

16. Decide whether the equation $3y + 4x = 0$ is symmetric with respect to the x-axis, y-axis, or origin.

16. ____________________

Graph each of the following. Give the domain and range of each equation.

17. $\frac{x^2}{9} + \frac{y^2}{25} = 1$

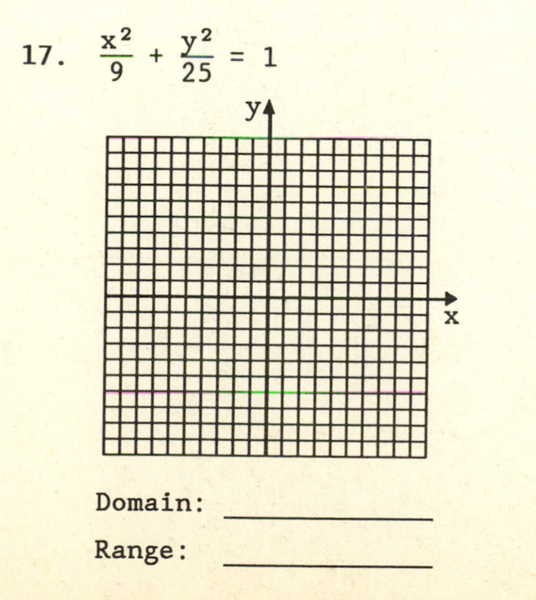

Domain: ____________

Range: ____________

18. $\frac{x^2}{36} - \frac{y^2}{16} = 1$

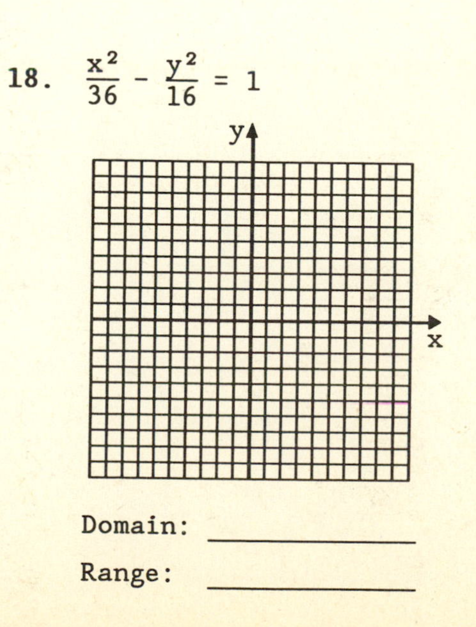

Domain: ____________

Range: ____________

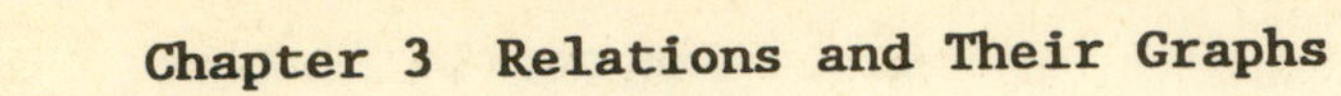

19. $\frac{(x - 2)^2}{4} + \frac{(y + 1)^2}{9} = 1$

20. $2x + y < 4$ where $x \geq 0$

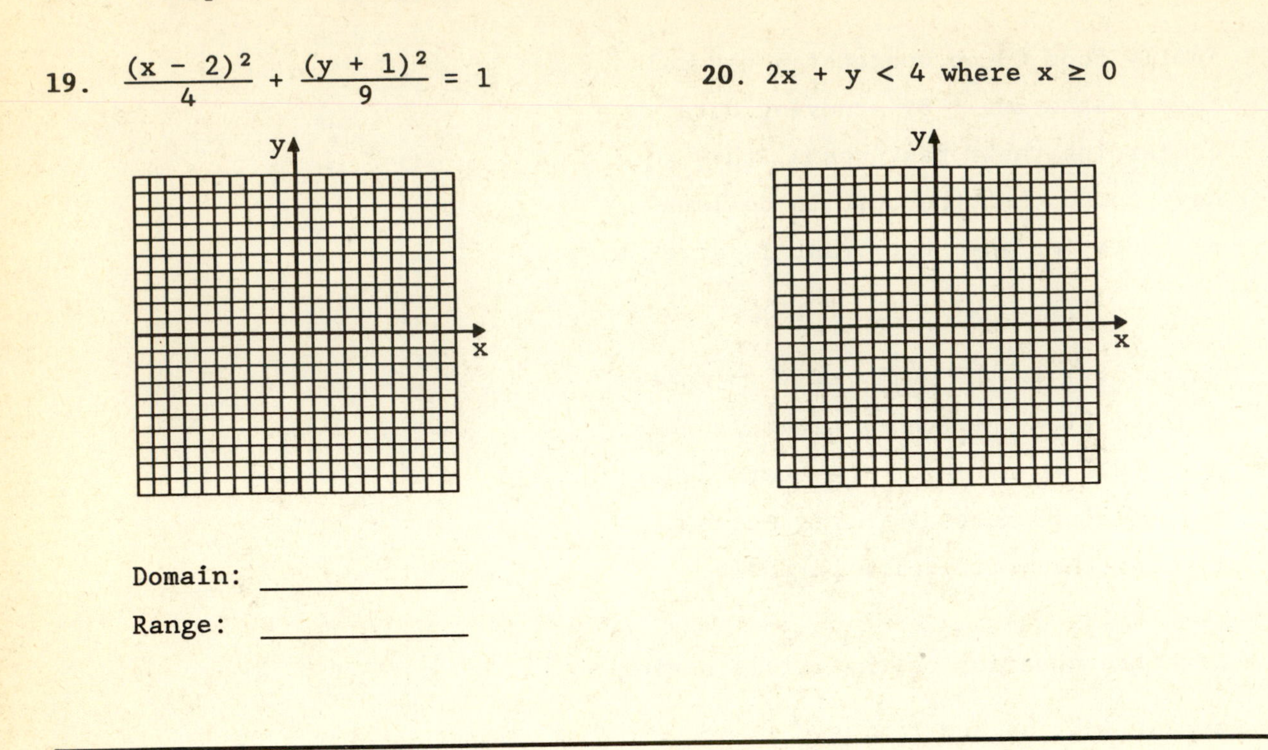

Domain: ____________

Range: ____________

CHAPTER 3 RELATIONS AND THEIR GRAPHS

3.1 Relations and the Rectangular Coordinate System

1. Pairs of related numbers such as (4, 2) are called _____________ pairs.	ordered
2. A set of ordered pairs, for example, $\{(3, 1), (5, 2)\}$, is called a _________.	relation
The domain, or the set of first elements, in this relation is $\{$____, ____$\}$; the range, or	3; 5
the set of second elements, is $\{$____, ____$\}$.	1, 2
3. The solutions for equations such as $2x + 3y = 18$ are written as _________ pairs, such as (12, −2).	ordered
In the ordered pair (12, −2), x = ___ and y = ___.	12; −2
Since both x and y can take any real-number values, both the domain and range are __________.	$(-\infty, \infty)$
4. Find the domain and range of $y = x^2$. For any real number value of x, y is a ______________	nonnegative
number so the domain is ________ and the range	$(-\infty, \infty)$
is ________.	$[0, \infty)$
5. Relations can be graphed using two number lines. The horizontal number line is the _________, and	x-axis
the vertical number line is the _________. The	y-axis
point where the number lines cross is the ______.	origin
The x-axis and y-axis set up a _________________	rectangular
________________ system with the xy-plane divided	coordinate
into four _______________.	quadrants.

6. To graph the ordered pair (-3, -1), start at the ________. Go ____ units in the negative x-direction, then turn and go ____ units in the ________ y-direction. Locate (-3, -1) on the grid below.

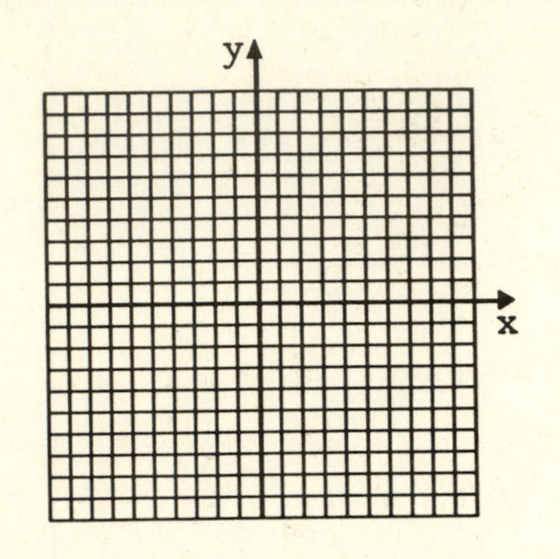

origin; 3

1

negative

(-3, -1)

7. On the grid below, locate (2, 5), (-2, -6), (-1, 5), (0, 2), and (-3, 0).

(-1, 5) (2, 5)

(0, 2)

(-3, 0)

(-2, -6)

8. The distance between the points (x_1, y_1) and (x_2, y_2) is found by the ________ formula:

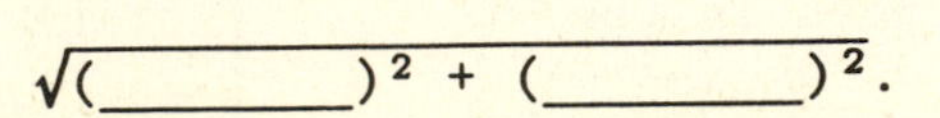

$\sqrt{(\_\_\_\_\_\_)^2 + (\_\_\_\_\_\_)^2}$.

distance

$x_2 - x_1$; $y_2 - y_1$

Find the distance between each pair of points in Frames 9–12.

9. (-6, 3) and (-1, -9)

Let $x_1 = -6$, with $y_1 =$ ____, $x_2 =$ ____, and $y_2 =$ ____. — 3; -1; -9

The distance is

$\sqrt{(-1 - \_\_\_)^2 + (-9 - \_\_\_)^2}$ — -6; 3

$= \sqrt{(\_\_\_\_)^2 + (\_\_\_\_)^2}$ — 5; -12

$= \sqrt{\_\_\_ + \_\_\_\_}$ — 25; 144

$= \sqrt{\_\_\_\_} =$ ____. — 169; 13

10. (3, -5), (-4, 7) Distance: __________ — $\sqrt{193}$

11. (5, -12), (3, 0) Distance: __________ — $\sqrt{148}$; or $2\sqrt{37}$

12. (-2, 9), (-2, 15) Distance: __________ — 6

13. Are the points (-5, 5), (8, -4), and (6, -6) the vertices of a right triangle? Use the ________ formula to find the length of each side. — distance

Distance between (-5, 5) and (8, -4): ______ — $5\sqrt{10}$

Distance between (8, -4) and (6, -6): ______ — $2\sqrt{2}$

Distance between (6, -6) and (-5, 5): ______ — $11\sqrt{2}$

Use the __________ theorem to decide whether or not the sum of the squares of two sides equals the square of the length of the longest side. — Pythagorean

$(\_\_\_\_)^2 + (\_\_\_\_)^2 = (\_\_\_\_)^2$ — $2\sqrt{2}$; $11\sqrt{2}$; $5\sqrt{10}$

________ + ________ = ________ — 8; 242; 250

This (*is/is not*) a right triangle. — is

14. Points that lie on a line are said to be ____________. — collinear

15. Three points are collinear if the sum of the __________ between two pairs of the points is equal to the __________ between the remaining pair.

distances
distance

Decide whether or not the given points are collinear.

16. (4, 7), (8, 12), (-4, -3)

The distance between (4, 7) and (8, 12) is

$$\sqrt{(\_\_\_\_)^2 + (\_\_\_\_)^2} = \sqrt{\_\_\_ + \_\_\_}$$

$$= \_\_\_\_.$$

8 - 4; 12 - 7; 16; 25

$\sqrt{41}$

The distance between (4, 7) and (-4, -3) is

$$\sqrt{\_\_\_ + \_\_\_} = \_\_\_\_.$$

64; 100; $2\sqrt{41}$

The remaining pair of points is __________ and __________.

(8, 12)
(-4, -3)

The distance between them is ______. Therefore, the three points (*are/are not*) collinear.

$3\sqrt{41}$
are

17. (-7, 3), (-1, -1), (8, -7)

The points (*are/are not*) collinear.

are

18. (-3, 6), (1, 4), (13, -1)

The points (*are/are not*) collinear.

are not

19. The midpoint of the line segment connecting (x_1, y_1) and (x_2, y_2) is

(__________, __________).

$\frac{x_1 + x_2}{2}, \frac{y_1 + y_2}{2}$

Find the midpoint of the segment connecting the points in Frames 20 and 21.

20. (-1, 9), (8, 17) __________

(7/2, 13)

21. (0, -15), (19, -8) __________

(19/2, -23/2)

22. A line segment has (2, -7) as one endpoint. The midpoint is (-1, 3/2). Find the other endpoint of the segment.

Let the other endpoint be (x_2, y_2). Then x_1 = ____ and y_1 = ____. — 2; -7

By the midpoint formula,

$$-1 = \frac{2 + x_2}{2},$$

or x_2 = ____. — -4

Find y_2: y_2 = ____. — 10

The other endpoint is (____, ____). — -4; 10

3.2 Linear Relations

1. An equation that can be written in the form

$$y = ax + b,$$

where a and b are real numbers is called a ________ ________. — linear; relation

2. The standard form of a linear relation is

________ = C, — Ax + By

where A, B, and C are real and A and B are not both ___. — 0

3. The domain and range of Ax + By = C are both $(-\infty, \infty)$ since any real number can be used for ___ or ___. — x; y

4. Every linear relation has a graph which is a ________ line. To graph a straight line, we need at least ____ different points of the line.

straight
two

5. Often, two good points are found with the x- ________ and y-intercept. The x-intercept is the x-value of any point where the graph crosses the ________, while the y-intercept is the y-value of any point where the graph crosses the ________.

intercept
x-axis
y-axis

6. To find the x-intercept, let ___ = 0 and to find the y-intercept, let ___ = 0.

y
x

Graph the straight lines of Frames 7–12.

7. $3x - 4y = 12$
Find the x-intercept: if $y = 0$, then $x =$ ____.
If $x = 0$, then $y =$ ____. Use the intercepts to graph $3x - 4y = 12$ on the grid at the left below.

4
-3

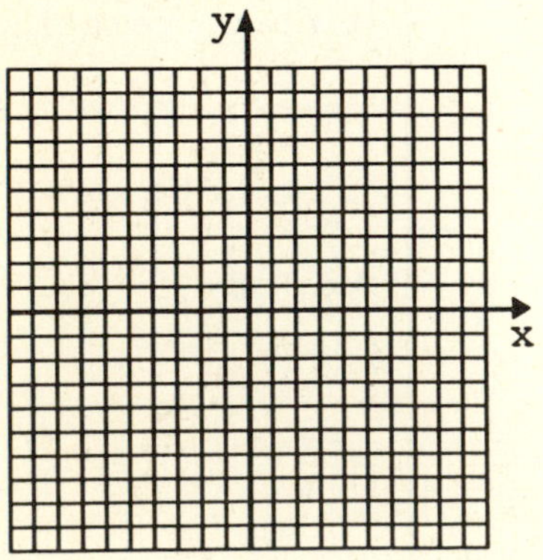

y
x

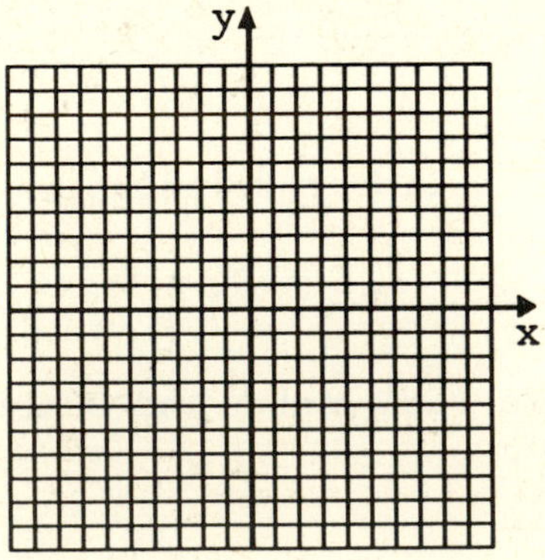

y
x

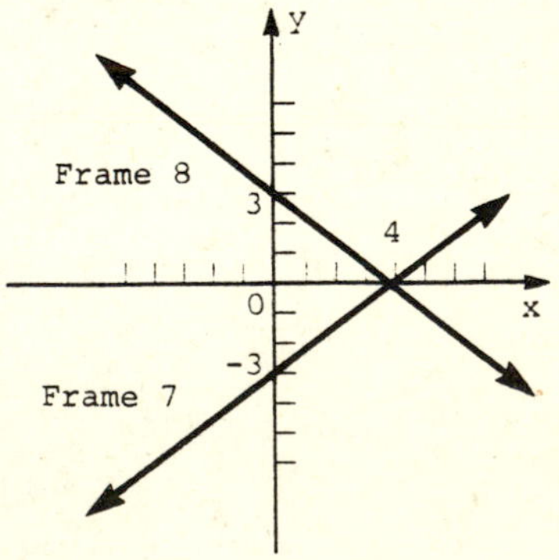

y
x
Frame 8
3
4
0
-3
Frame 7

8. Graph $4x + 5y = 15$ on the right above.

9. Graph $x - 6y = 6$ on the left below.

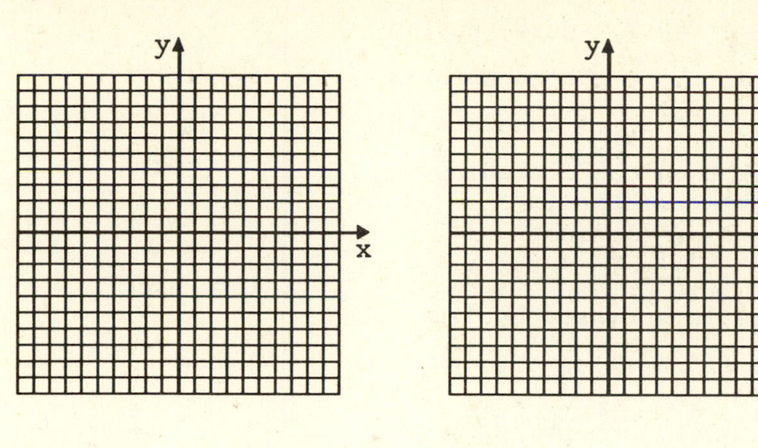

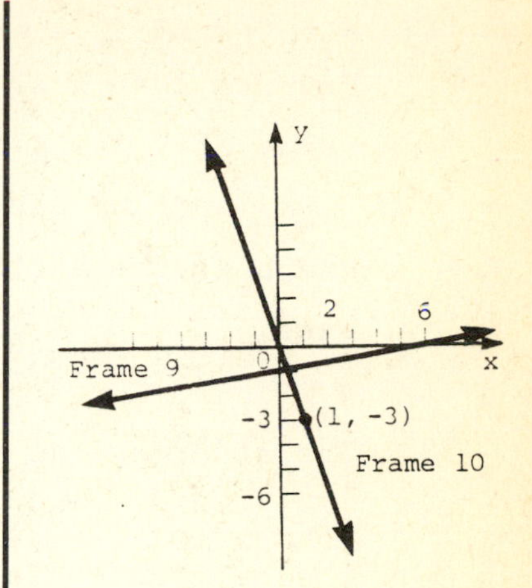

10. To graph $y = -3x$, let $y = 0$, getting $x =$ ____. Also, if $x = 0$, then $y =$ ____. Here both intercepts lead to the same point, _______. To get a second point for the graph, choose a different value for x (or y): say $x = 2$. If $x = 2$, then $y =$ ____, giving ______. Use (0, 0) and (2, -6) to complete the graph at the right above.

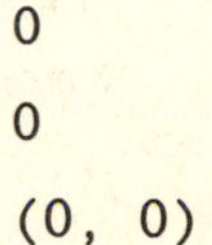

0
0
(0, 0)
-6; (2, -6)

11. $y = 4$

This equation can be rewritten as ____ + $y = 4$, which shows that for any value of ____, y is always _____. The domain of the relation is $(-\infty, \infty)$, but the range is _____. Complete some ordered pairs having a value ____ for y. (The y-intercept is ____; there is ____ x-intercept.) Draw the graph at the left below.

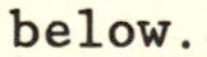

0x
x
4
{4}
4
4
no

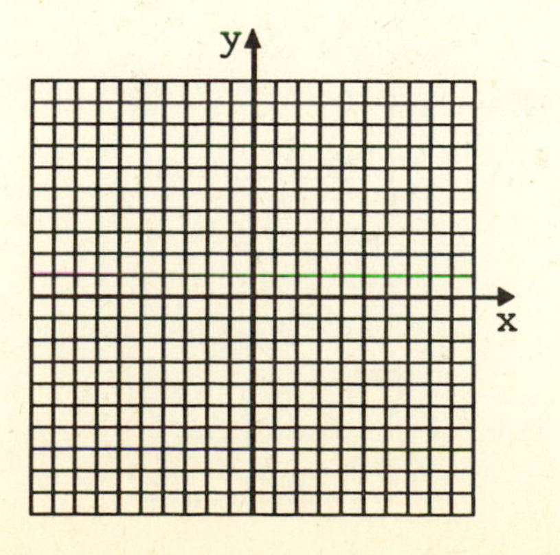

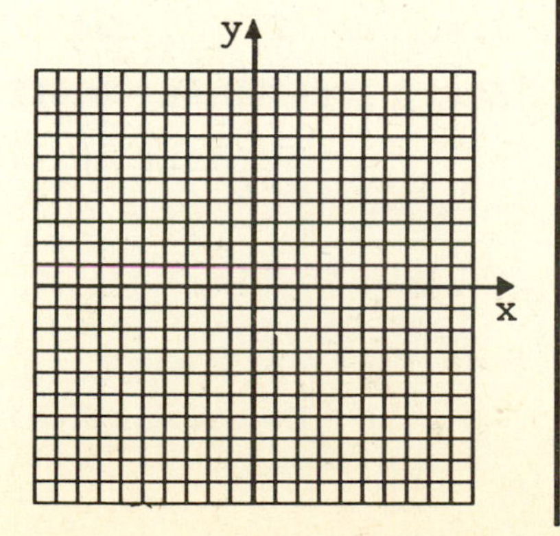

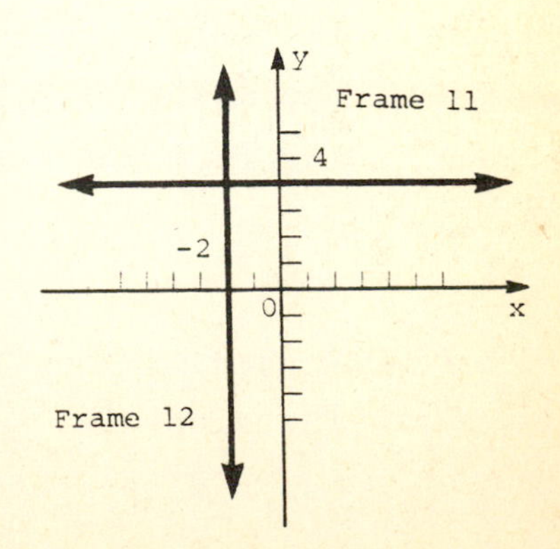

12. $x = -2$

Rewrite $x = -2$ as ________ $= -2$. Here, for any value of y, x is always ______. The domain is ______ and the range is _________. Complete the graph at the bottom right on the preceding page.

$x + 0y$

$\{-2\}$; $(-\infty, \infty)$

13. The steepness of a line is measured by its ___________. The slope of the line through the distinct points (x_1, y_1) and (x_2, y_2) is

$$m = \_\_\_\_\_\_\_\_\_\_.$$

The letter m represents ______. The difference $y_2 - y_1$ is called the _________________.

slope

$\dfrac{y_2 - y_1}{x_2 - x_1}$

slope

change in y

Find the slope of the lines in Frames 14–19.

14. Through (-11, 2) and (4, -5)

Let $x_1 = -11$, with $y_1 =$ _____, $x_2 =$ _____, and $y_2 =$ _____. Use the definition of slope.

$$m = \frac{-5 - \_\_\_\_}{4 - \_\_\_\_}$$

$$m = \_\_\_\_\_\_$$

2; 4

-5

2

-11

$-\dfrac{7}{15}$

15. Through (8, 5) and (9, -7) $m =$ ___________

-12

16. $y = 3$

Two points on this line are (2, __) and (-1, __). The slope is ___. Every line whose equation has the form $y = k$ (a ______________ line) has slope ___.

3; 3

0

horizontal

0

17. $x = -4$

The slope is _____________. Every vertical line has a slope that is _____________.

undefined

undefined

18. $2x - y = 4$

Find two points on the line, such as (___, 0) and (0, ___). The slope is ___.

2

-4; 2

19. $4x - 7y = 8$ $m =$ ________

$\frac{4}{7}$

20. If we know the slope of a line and one ________ that the line passes through, its equation can be found with the ________-slope form of the equation of a line: If a line has slope m and passes through (x_1, y_1), an equation is

$$y - y_1 = \_\_\_\_\_\_\_\_.$$

point

point

$m(x - x_1)$

Find an equation of the lines in Frames 21–27.

21. $m = -2$, line goes through $(-1, 5)$

Here $m = -2$, $x_1 =$ ___, and $y_1 =$ ___. Use the point-slope form.

$$y - \_\_\_\_ = \_\_\_\_ (x - \_\_\_\_)$$

$$y - 5 = -2(\_\_\_\_\_\_)$$

Write the equation in standard form.

____

-1; 5

5; -2; -1

$x + 1$

$2x + y = 3$

22. $m = -3/4$, through $(8, -2)$ ________

$3x + 4y = 16$

23. Through $(-7, 4)$, with slope 0. ________

$y = 4$

24. Through $(-3, -5)$, with undefined slope. The only lines with undefined slope are those of the form ________. Here $x =$ _____, so our line has equation ________.

$x = k$; -3

$x = -3$

25. Through (-1, 5) and (2, 7)
The slope of this line is m = ______. (Use the definition of slope.) Use the slope and either point to find the equation: ____________.

2/3

$2x - 3y = -17$

26. Through (-5, 6) and (2, -4) ____________

$10x + 7y = -8$

27. Through (3, 4) and (-3, -2) ____________

$x - y = -1$

28. The slope-__________ form of the equation of a line says that if ___ is the slope and ___ is the y-intercept of a line, then the equation of the line is ______________.

intercept

m; b

$y = mx + b$

29. To find the slope of the line $3x - 2y = 5$, first write the equation in slope-intercept form, by solving for ___.

$$-2y = \_\_\_\_\_\_$$

$$y = \_\_\_\_\_\_$$

The slope is m = ____ and the y-intercept is _____.

y

$-3x + 5$

$\frac{3}{2}x - \frac{5}{2}$

3/2

-5/2

30. To graph $3x - 2y = 5$, first locate the y-intercept, ______. Go ______ units down on the y-axis. To locate a second point, use the slope ___. From the intercept, go over _____ units in the positive x-direction, and then turn and go up ___ units in the y-direction. This gives the second point (______), which can be used to complete the graph at the left on the following page.

-5/2; 5/2 or $2\frac{1}{2}$

3/2; 2

3

2, 1/2

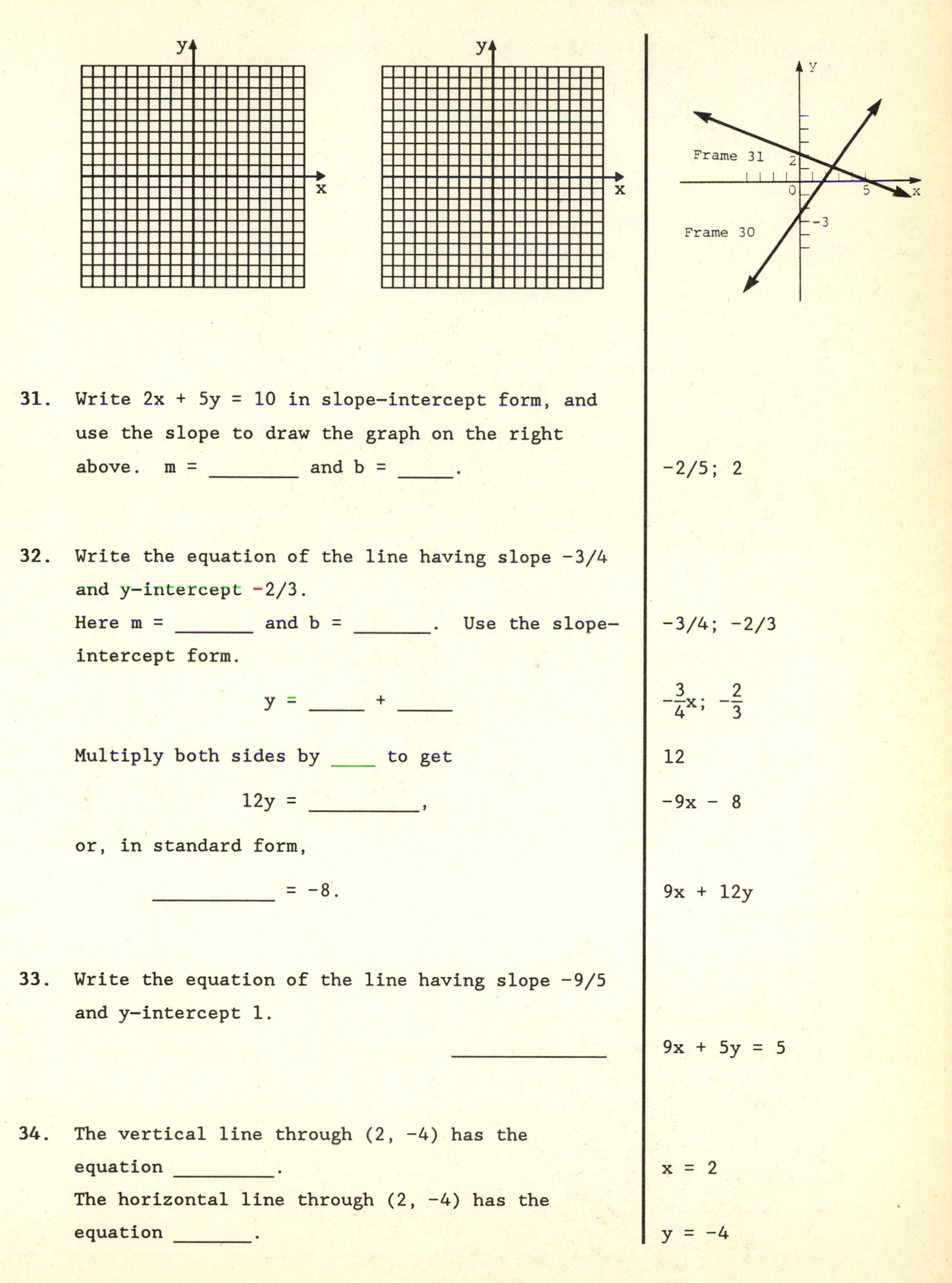

31. Write $2x + 5y = 10$ in slope-intercept form, and use the slope to draw the graph on the right above. m = ________ and b = _____.

$-2/5$; 2

32. Write the equation of the line having slope $-3/4$ and y-intercept $-2/3$.
Here m = _______ and b = _______. Use the slope-intercept form.

$-3/4$; $-2/3$

$$y = \_\_\_\_\_ + \_\_\_\_\_$$

$-\frac{3}{4}x$; $-\frac{2}{3}$

Multiply both sides by ____ to get

12

$$12y = \_\_\_\_\_\_\_\_\_\_,$$

$-9x - 8$

or, in standard form,

$$\_\_\_\_\_\_\_\_\_\_\_ = -8.$$

$9x + 12y$

33. Write the equation of the line having slope $-9/5$ and y-intercept 1.

____

$9x + 5y = 5$

34. The vertical line through $(2, -4)$ has the equation _________.

$x = 2$

The horizontal line through $(2, -4)$ has the equation _______.

$y = -4$

35. Two distinct nonvertical lines with the same slope are ____________.	parallel
36. Find the equation of the line through (2, 6) and parallel to $2x + y = 19$. First, find the slope of $2x + y = 19$; $m =$ _____.	-2
Then use $m =$ _____ and (2, 6) to get the necessary equation: ________________.	-2 $2x + y = 10$
37. Find the equation of the line parallel to $7x - 3y = 5$ and going through $(-1, 3)$. ______________	$7x - 3y = -16$
38. Two lines are perpendicular if the product of their slopes is ____.	-1
39. Find the equation of the line through $(-1, 4)$ and perpendicular to $3x - y = 6$. The slope of $3x - y = 6$ is ____.	3
Since we want the perpendicular line, we need the line with slope ________. (The product of the slopes must be -1).	$-1/3$
Find the equation we need. ______________	$x + 3y = 11$
40. Use slopes to decide whether the points $(-2, 0)$, $(1, 3)$, and $(3, 5)$ lie on the same line. If these points do lie on the same line, the slope of the lines through the first and second points and through the second and third points will be the same. The slope of the line through $(-2, 0)$ and $(1, 3)$ is ___;	1
the slope of the line through $(1, 3)$ and $(3, 5)$ is ____.	1
Therefore, the three points (*do/do not*) lie on the same line.	do

3.3 Parabolas: Translations and Applications

1. Parabolas are the graphs of ________ relations; that is, relations with equations of the form ________________ or ________________, $a \neq 0$.

quadratic

$y = ax^2 + bx + c$; $x = ay^2 + by + c$

2. In this section we see how to graph parabolas. As an example, let us graph $y = x^2 - 2$. Start by completing the following table of ordered pairs.

x	−2	−1	0	1	2	3
y	2	−1	−2	____	____	____
Ordered pairs	(−2, 2)	(−1, −1)	(0, −2)	____	____	____

−1; 2; 7

(1, −1); (2, 2); (3, 7)

Graph $y = x^2 - 2$ at the left below. The lowest point on the graph, ________, is the ________. The line of symmetry, $x =$ ____, is the ______. The domain is ________ since any value may be used for x; the range is ________ because the y-values are always greater than or equal to ____.

(0, −2); vertex

0; axis

$(-\infty, \infty)$

$[-2, \infty)$

−2

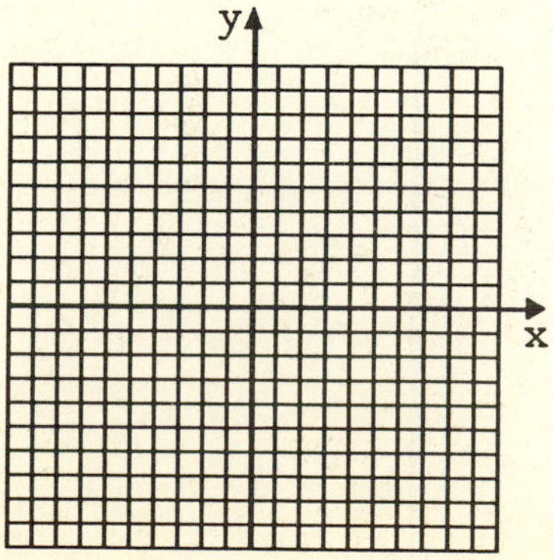

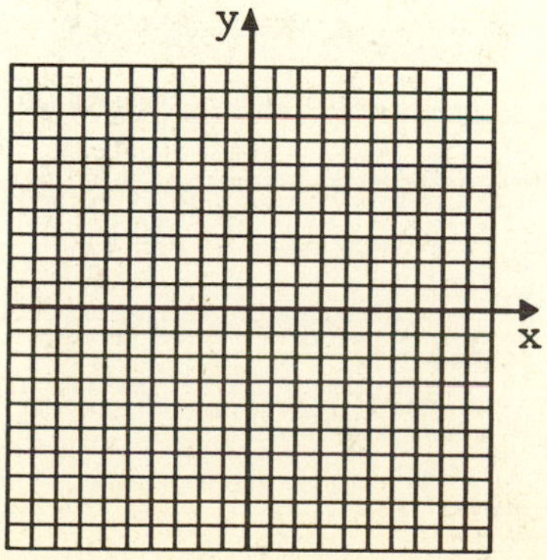

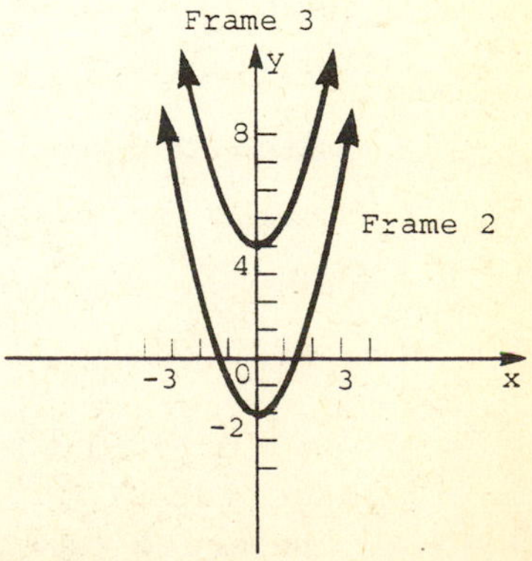

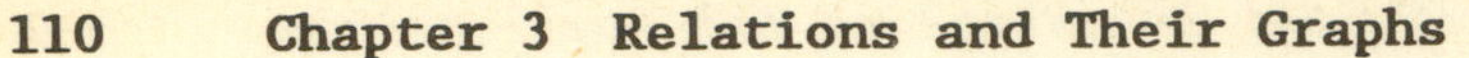

3. Graph $y = x^2 + 4$ on the right on the preceding page. The line of symmetry or ______ is ______.	axis; $x = 0$
4. In the equation $y = x^2 + c$, the number ___ causes the graph to be translated __________, up c units if __________ and down $\lvert c \rvert$ units if _______.	c vertically $c > 0$; $c < 0$
5. For example, the vertex of $y = x^2 - 7$ is _______.	(0, -7)
6. The graph of $y = (x - 2)^2$ is shifted ___ units to the _______. The vertex is _______. Graph $y = (x - 2)^2$ at the left below.	2 right; (2, 0)

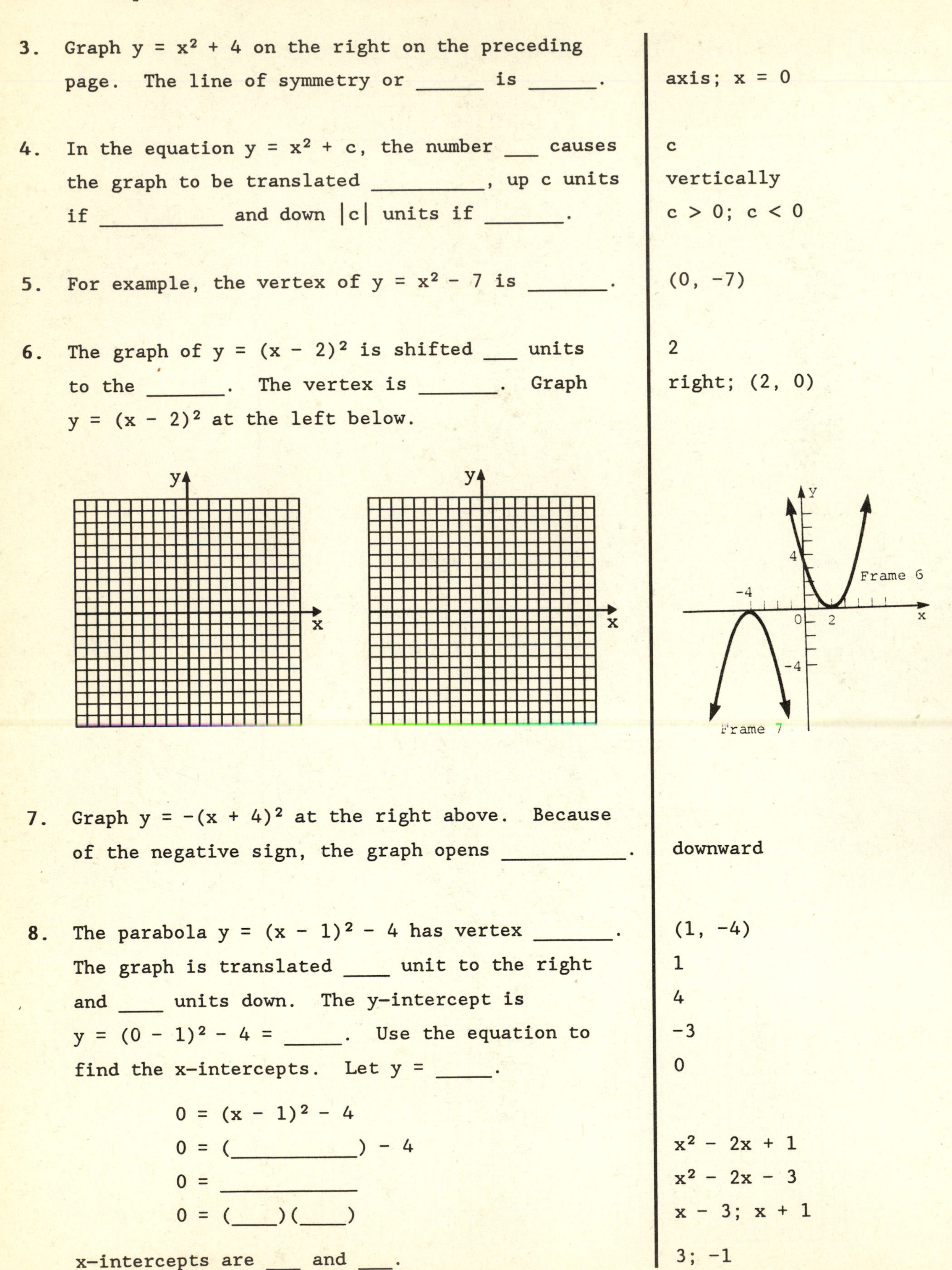

7. Graph $y = -(x + 4)^2$ at the right above. Because of the negative sign, the graph opens __________.	downward
8. The parabola $y = (x - 1)^2 - 4$ has vertex _______. The graph is translated ____ unit to the right and ____ units down. The y-intercept is $y = (0 - 1)^2 - 4 =$ _____. Use the equation to find the x-intercepts. Let y = _____.	(1, -4) 1 4 -3 0
$0 = (x - 1)^2 - 4$	
$0 = ($__________$) - 4$	$x^2 - 2x + 1$
$0 =$ ____________	$x^2 - 2x - 3$
$0 = ($____$)($____$)$	$x - 3$; $x + 1$
x-intercepts are ___ and ___.	3; -1

Graph this parabola at the left below.

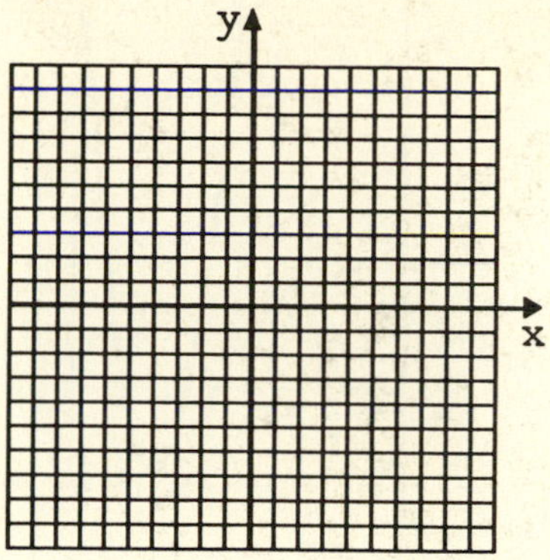

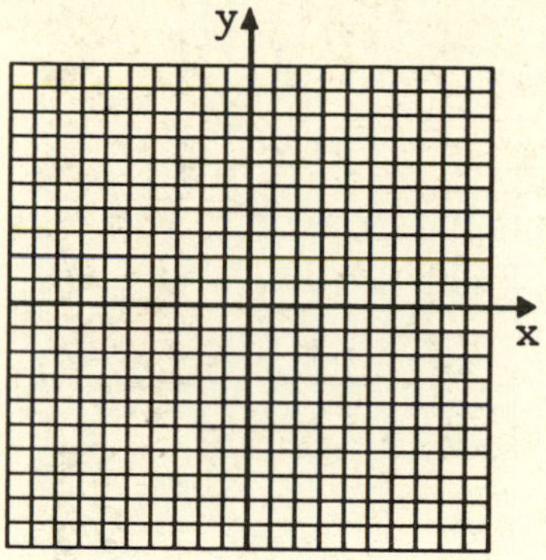

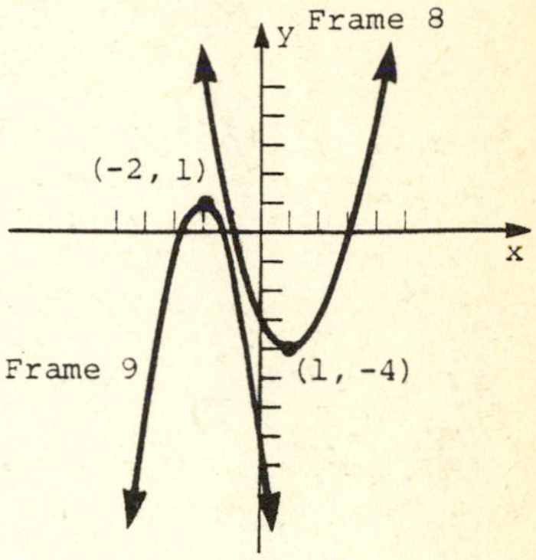

Frame 8
(-2, 1)
Frame 9
(1, -4)

9. Graph $y = -2(x + 2)^2 + 1$ at the right above. The −2 in the equation affects the ______ of the graph. Since $|-2|$ is > 1, the graph is ____________ than the graph of $y = x^2$.

width

narrower

10. Graph $y = 2x^2 + 12x + 5$

Begin by rewriting the equation in the form ________________. First factor ___ out of $2x^2 + 12x$.

$y = a(x - h)^2 + k$; 2

$$y = 2(\_\_\_\_\_\_\_\_) + 5$$

$x^2 + 6x$

Now get a perfect square by dividing _____ by 2 and squaring the result: $6/2 =$ ____; $3^2 = 9$. Add the square, _____, to the quantity inside the parentheses and subtract $2 \cdot$ ____ $=$ ______ outside the parentheses. Note that we have, in effect, added 0 and thus have not changed the value of the equation.

6

3

9

9; 18

This gives

$$y = 2(x^2 + 6x + 9) + 5 - \_\_\_\_\_$$

18

which is $y = 2(\_\_\_\_)^2 - 13$. The vertex is (____, ____). Graph the parabola on the next page.

$x + 3$

−3; −13

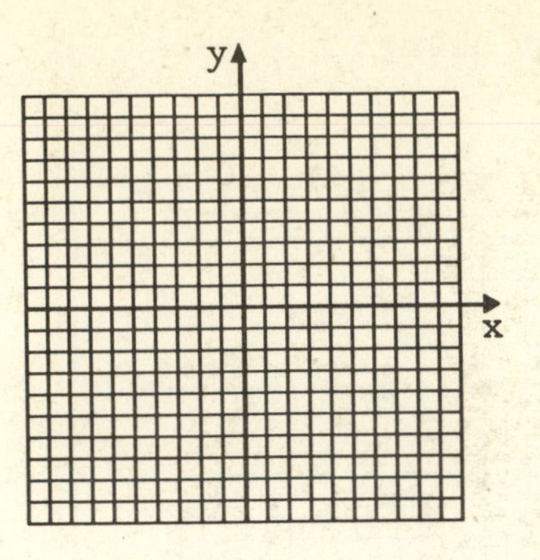

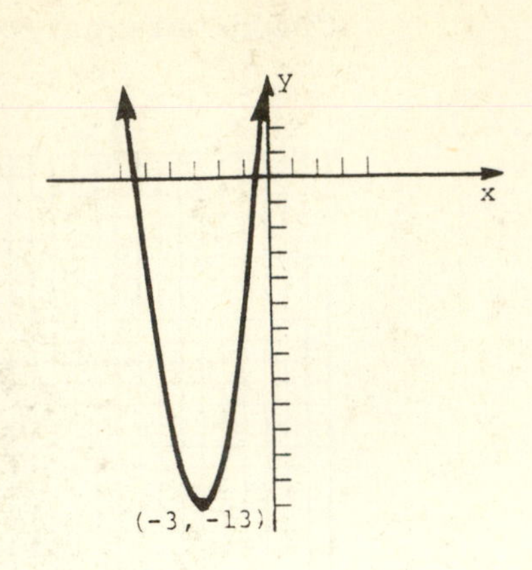

11. To find the x-value of the vertex of the parabola $y = -3x^2 + 9x - 1$, we may use the formula ______. The x-value is _____. Substitute this value into the equation to find the y-value of the vertex, which is ______. Use this information to draw the graph in Frame 12.

$-b/(2a)$

3/2

23/4

12. Graph $y = -3x^2 + 9x - 1$ below.

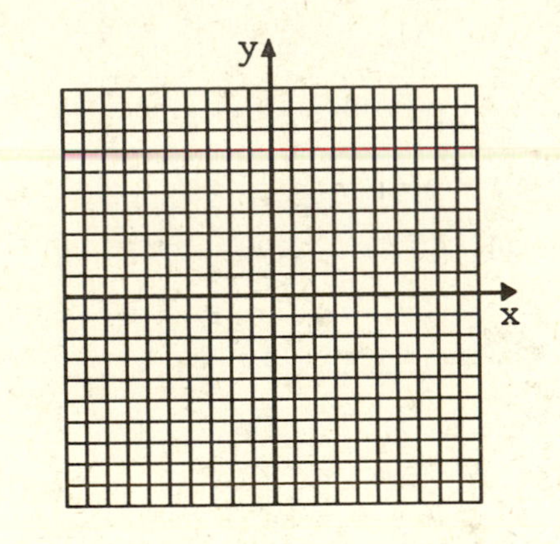

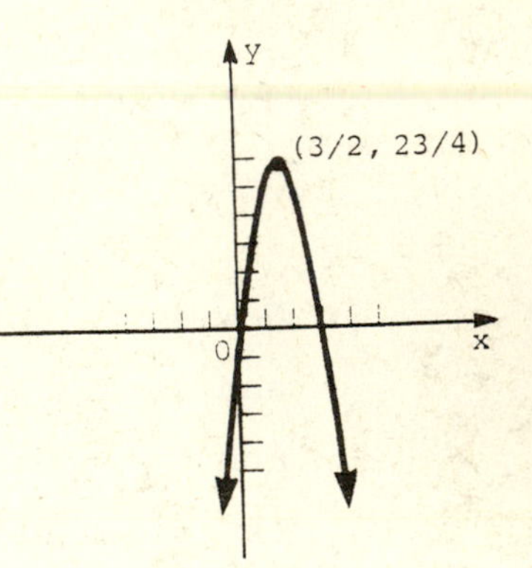

13. Graph $y = x^2 + 6x + 9$ at the left below.

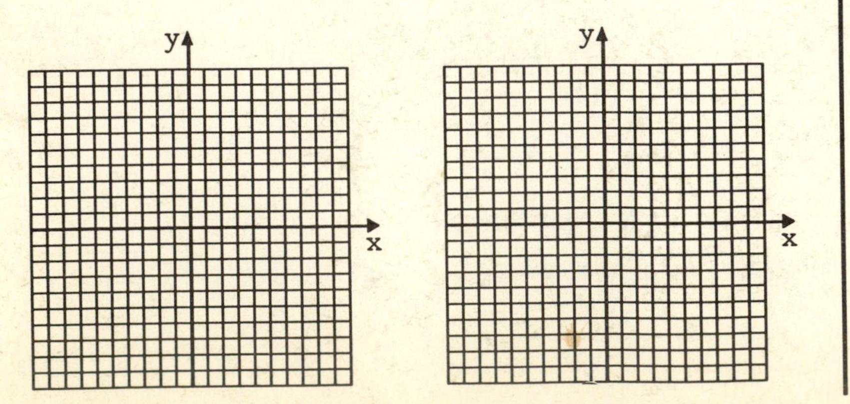

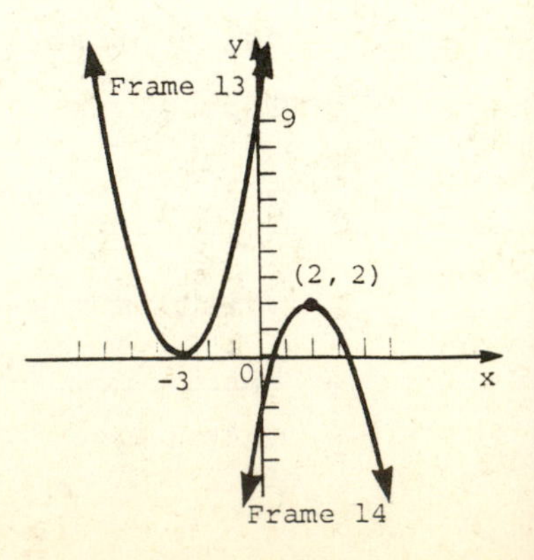

14. Graph $y = -x^2 + 4x - 2$ at the right on the previous page.

15. **The number, in thousands, of ladybugs in a certain area of North Carolina depends on the number of quarts of ladybugs that are stocked. If the number of ladybugs, $T(x)$, is given by $T(x) = 8x - x^2$, where x represents the number of quarts, find the number of quarts that will produce the maximum number of ladybugs. What is the maximum number of ladybugs?**

First we recognize that the graph of the parabola $T(x) = 8x - x^2$ opens __________. Thus, the vertex will be the ________ point of the graph of the equation $T(x) = 8x - x^2$. Rewrite the equation in the form $T(x) = a(x - h)^2 + k$

downward

highest

$$T(x) = -(\_\_\_\_\_\_) = -(x - 4)^2 + 16.$$

$x^2 - 8x$

The vertex is __________. The number of quarts that will produce the maximum number of ladybugs is the x-value of the vertex. Thus, ____ quarts will produce the maximum number of ladybugs. The maximum number of ladybugs in thousands is _____. The maximum number is ____________.

(4, 16)

4

16; 16,000

16. **The profit of a 500-room hotel during a 3-day convention depends on the number of unoccupied rooms. If the profit, $P(x)$, is given by $P(x) = 10{,}000 + 100x - x^2$, where x is the number of unoccupied rooms, find the maximum profit and the number of unoccupied rooms which produce maximum profit.**

____

$12,500 with 50 unoccupied rooms

17. A parabola is defined geometrically as the set of all points in a plane that are ____________ ____________ from a point, called the ________, and a line, called the ______________.

equally distant; focus directrix

Use this definition to write an equation for the parabola with focus F(2, 4) and directrix $y = -2$.

Name a point P(x, y) of the parabola. Then a point D(x, _____) is on the directrix and on the line drawn through P ______________ to the directrix.

-2 perpendicular

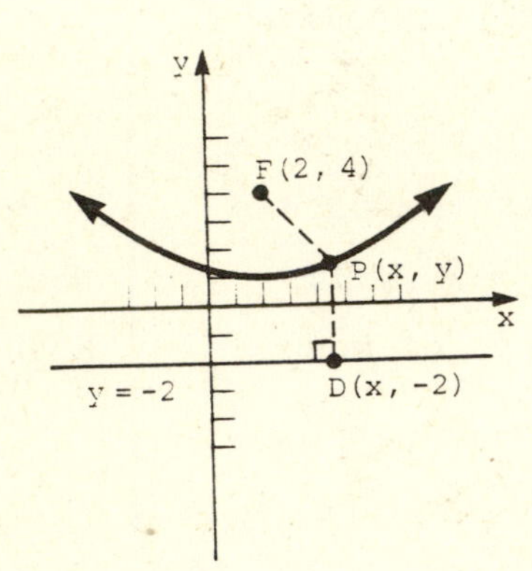

Use the distance formula to write the equation, recalling that the distance of a point to a line is the ____________________ distance.

perpendicular

d(P, F) = d(P, D)

$$\sqrt{(x - \_\_\_)^2 + (y - \_\_\_)^2}$$

$$= \sqrt{(x - \_\_\_)^2 + [y - (\_\_\_)]^2}$$

2; 4

x; -2

Square both sides and simplify.

$$x^2 - \_\_\_ + \_\_\_ + y^2 - \_\_\_ + \_\_\_$$

$$= y^2 + \_\_\_\_ + \_\_\_$$

4x; 4; 8y; 16

4y; 4

Collect the terms involving y on one side.

$$\_\_\_\_\_\_\_\_\_ = 12y$$

$x^2 - 4x + 16$

Write in the form of $y = ax^2 + bx + c$.

$$y = \frac{1}{12}x^2 - \_\_\_\_ + \_\_\_\_$$

$\frac{1}{3}x$; $\frac{4}{3}$

Now we can find an equation in the form $y = a(x - h)^2 + k^2$:

$$y = \frac{1}{12}(x - \_\_\_\_)^2 + \_\_\_\_$$

2; 1

The vertex of the parabola is ________.

(2, 1)

18. The graph of $y = x^2$ is a ______________ opening __________. Exchanging x and y gives ________, a parabola opening to the ________. The domain is ________ and the range is ________.

parabola
upward; $x = y^2$
right
$[0, \infty)$; $(-\infty, \infty)$

19. The vertex of $x = -y^2$ is _________. Graph this parabola on the following grid.

(0, 0)

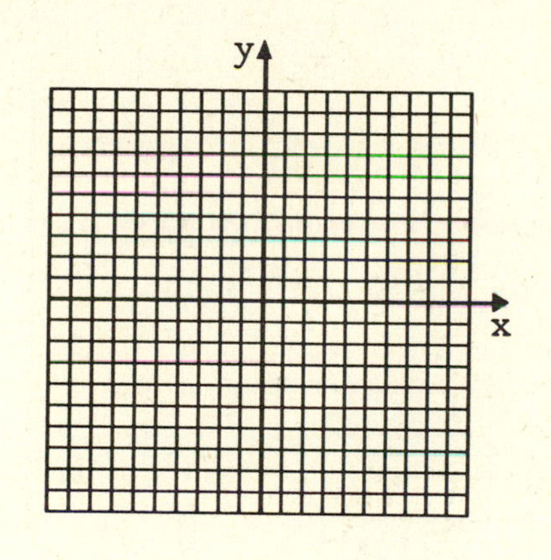

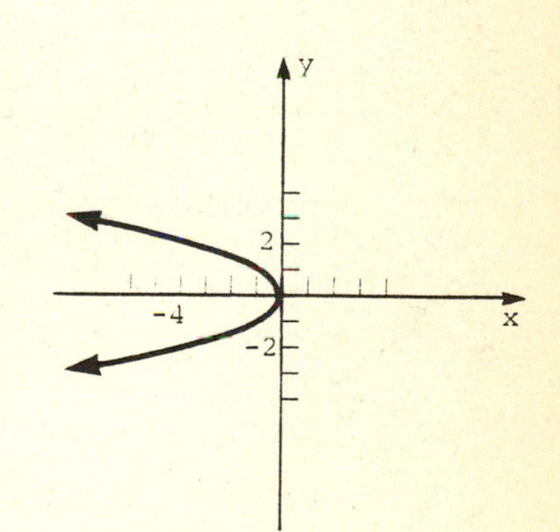

20. Graph $x = (y + 2)^2 - 3$ on the grid below. The vertex is __________.

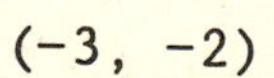
(-3, -2)

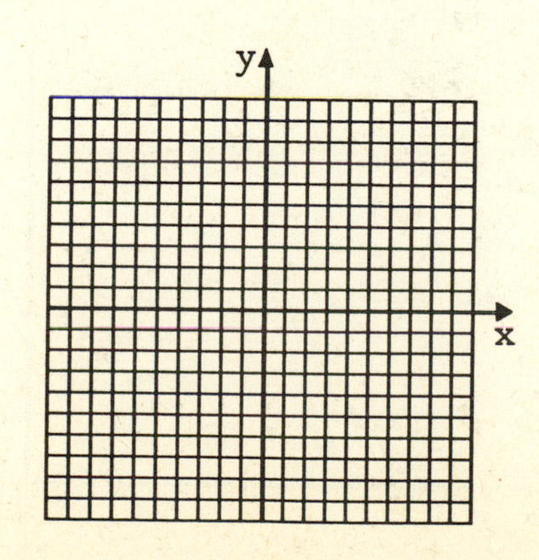

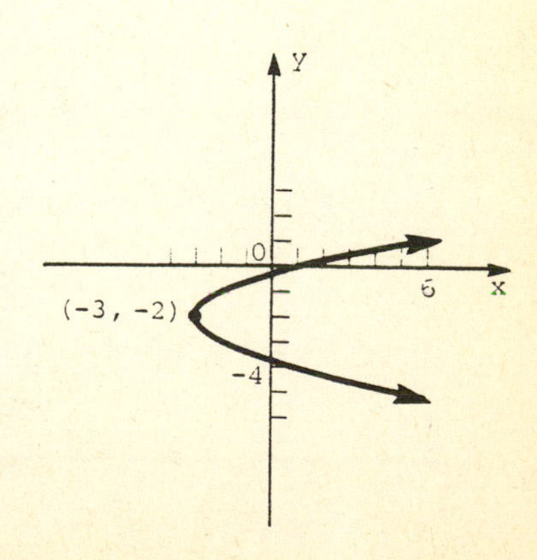

21. Graph $x = 2y^2 + 4y + 2$.

Complete the square on y.

$$x = 2(\_\_\_\_\_\_\_\_) + 2 - 2 = 2(y + 1)^2$$

The vertex is (___, ___). (Remember to reverse the ordered pair.) The parabola opens to the ______ because the 2 is positive. Sketch the graph.

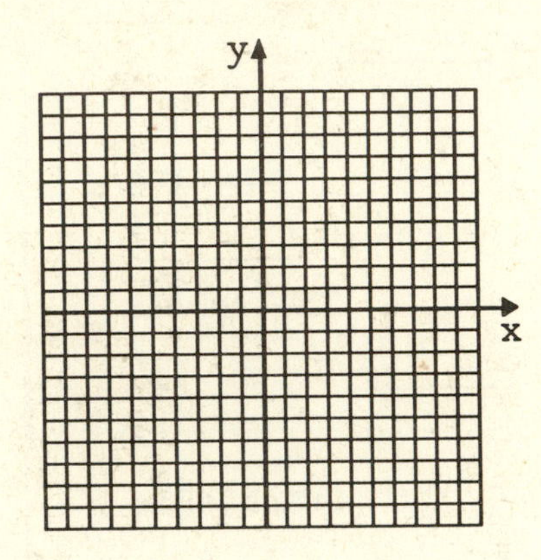

Answers: $y^2 + 2y + 1$; 0; −1; right

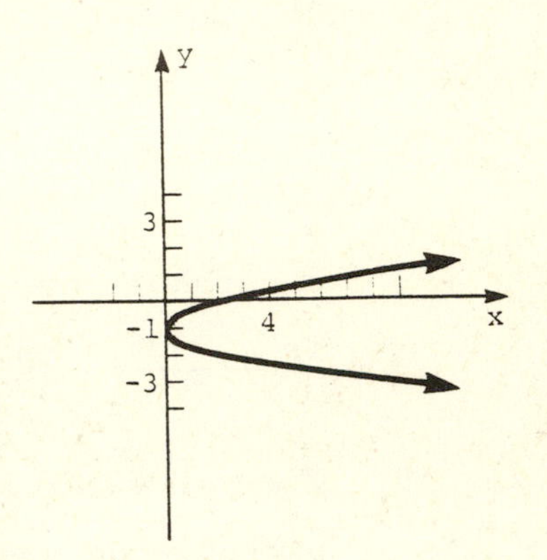

22. Graph $x = -2y^2 + 4y + 3$ on the grid below. The vertex is ______, and the parabola opens to the ______.

Answers: (5, 1); left

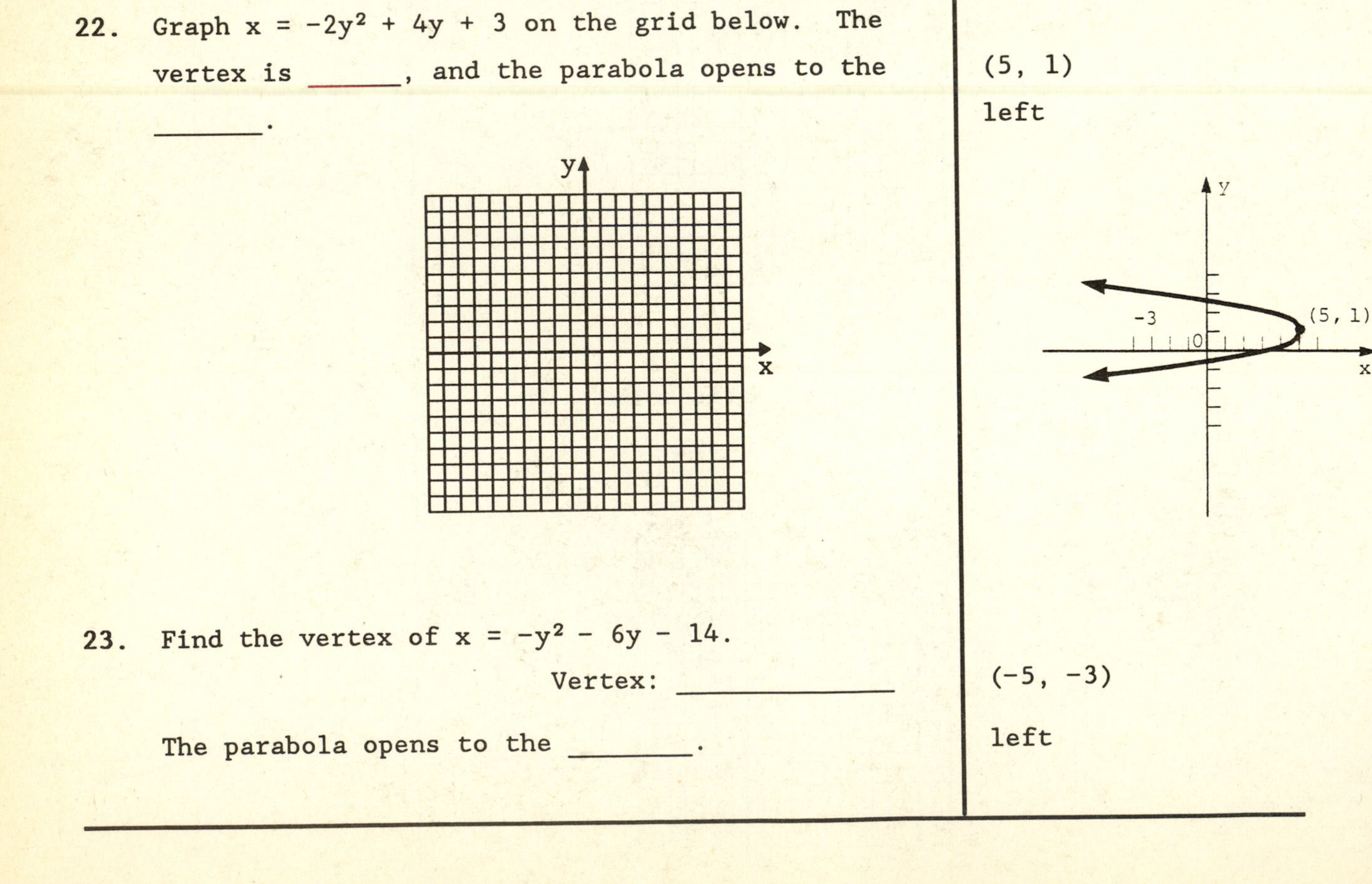

23. Find the vertex of $x = -y^2 - 6y - 14$.

Vertex: __________

The parabola opens to the _______.

Answers: (−5, −3); left

3.4 The Circle and Symmetry

1. A ________ is a relation defined as the set of points in a plane a given __________ from a given _______. The given point is the _______ and the given distance is the ________.	circle distance point; center radius
2. The equation of the circle with center at (h, k) and radius r is ______________.	$(x-h)^2 + (y-k)^2 = r^2$
3. The equation of the circle with center at the origin and radius 4 is ______________.	$x^2 + y^2 = 16$
4. Write the equation of the circle having center at (3, -7) and radius 2. ______________.	$(x-3)^2 + (y+7)^2 = 4$
5. Write the equation of the circle having center at (4, 0) and radius 5. ______________.	$(x-4)^2 + y^2 = 25$
6. To graph $x^2 + 2x + y^2 - 4y = 11$, first ________ the square on x and y separately.	complete
$(x^2 + 2x \quad) + (\_\_\_\_\_\_) = 11$	$y^2 - 4y$
$(x^2 + 2x + \_\_) + (y^2 - 4y + \_\_) = 11 + \_\_ + \_\_$	1; 4; 1; 4
$(\_\_\_\_)^2 + (\_\_\_\_)^2 = 16$	x + 1; y - 2
The center is ________, with radius ___. Draw the graph on the next page.	(-1, 2); 4

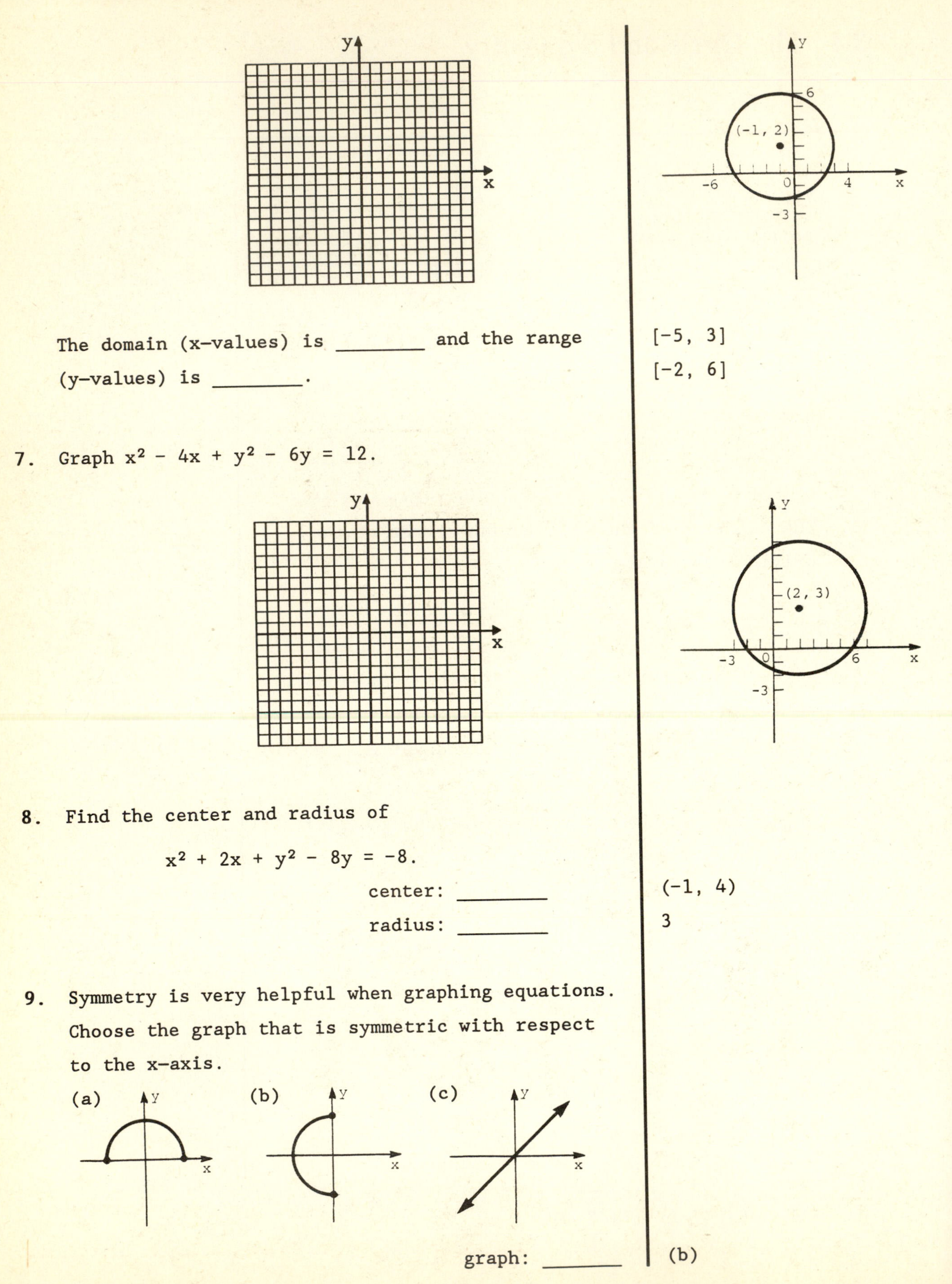

The domain (x–values) is ________ and the range (y–values) is ________.

[−5, 3]
[−2, 6]

7. Graph $x^2 - 4x + y^2 - 6y = 12$.

8. Find the center and radius of

$$x^2 + 2x + y^2 - 8y = -8.$$

center: ________

radius: ________

(−1, 4)
3

9. Symmetry is very helpful when graphing equations. Choose the graph that is symmetric with respect to the x–axis.

(a) (b) (c)

graph: ________

(b)

10. Choose the graph in Frame 9 that is symmetric with respect to the y-axis.

graph: _______ | (a)

11. Finally, graph (c) is symmetric with respect to the _________. | origin

Complete the statements of the tests for symmetry in Frames 12–14.

12. A graph is symmetric with respect to the y-axis if replacing ______ with ______ results in an ____________ equation. | x; −x | equivalent

13. A graph is symmetric with respect to the x-axis if replacing ____ with ____ results in an equivalent equation. | y; −y

14. A graph is symmetric with respect to the origin if replacing x with ____ and y with ____ results in an equivalent equation. | −x; −y

Test for symmetry in Frames 15–18.

15. $y = x^2$

Replacing x with −x gives

$$y = (\quad)^2 = \_\_\_\_.$$

| $-x$; x^2

The graph is symmetric with respect to the _____. | y-axis

16. $x = y^2$ Symmetry: ___________ | x-axis

17. $x^2 + y^2 = 16$ Symmetry: ___________ | origin; x-axis; y-axis

18. $x + y = 10$ Symmetry: ___________ | none

3.5 The Ellipse and the Hyperbola

1. An ________ is the set of all points in the plane the ____ of whose distances from two fixed points is __________.

ellipse
sum
constant

2. The equation of the ellipse centered at the origin with x-intercepts a and -a and y-intercepts b and -b is

__________________,

where $a \neq b$.

$\frac{x^2}{a^2} + \frac{y^2}{b^2} = 1$

Graphs the ellipses in Frames 3 and 4.

3. $\frac{x^2}{16} + \frac{y^2}{4} = 1$

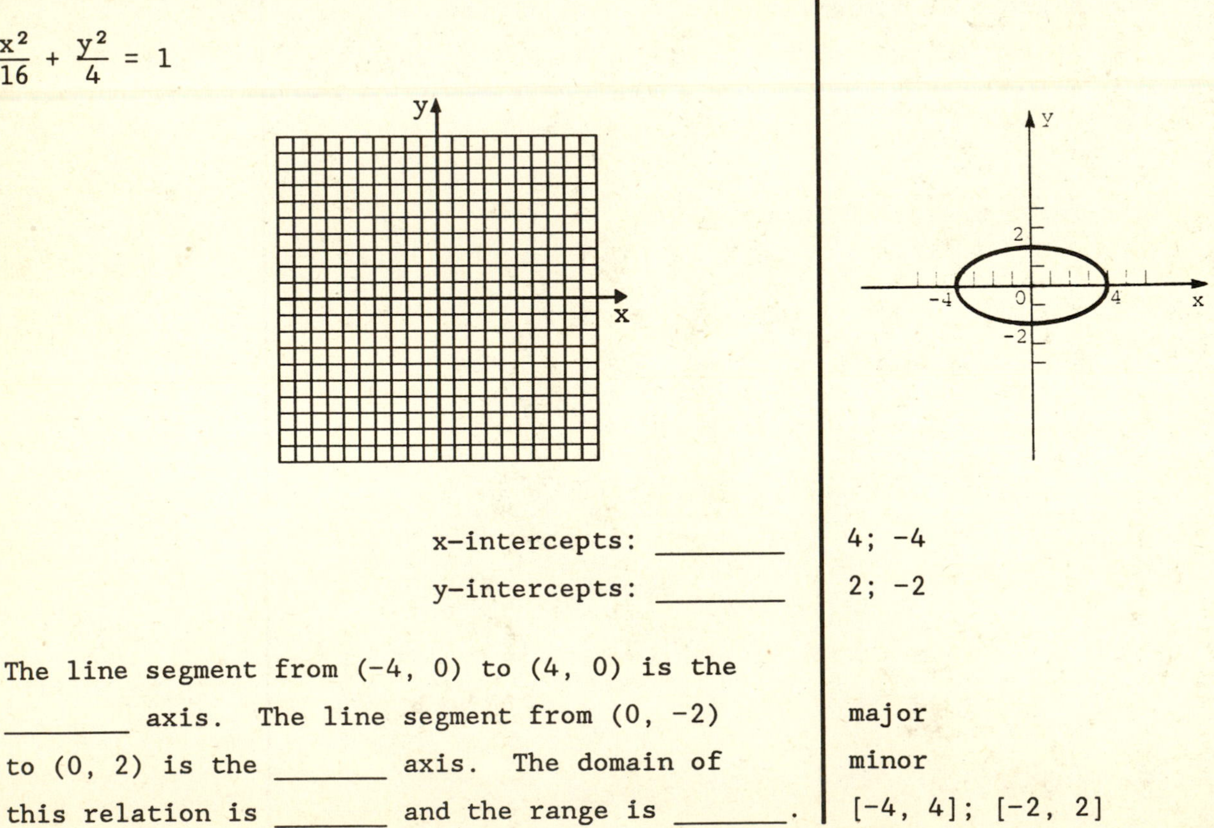

x-intercepts: ________

y-intercepts: ________

4; -4
2; -2

The line segment from (-4, 0) to (4, 0) is the ________ axis. The line segment from (0, -2) to (0, 2) is the _______ axis. The domain of this relation is _______ and the range is _______.

major
minor
[-4, 4]; [-2, 2]

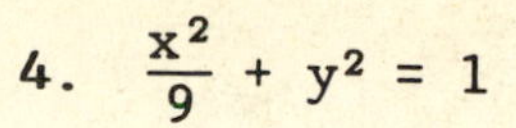

4. $\frac{x^2}{9} + y^2 = 1$

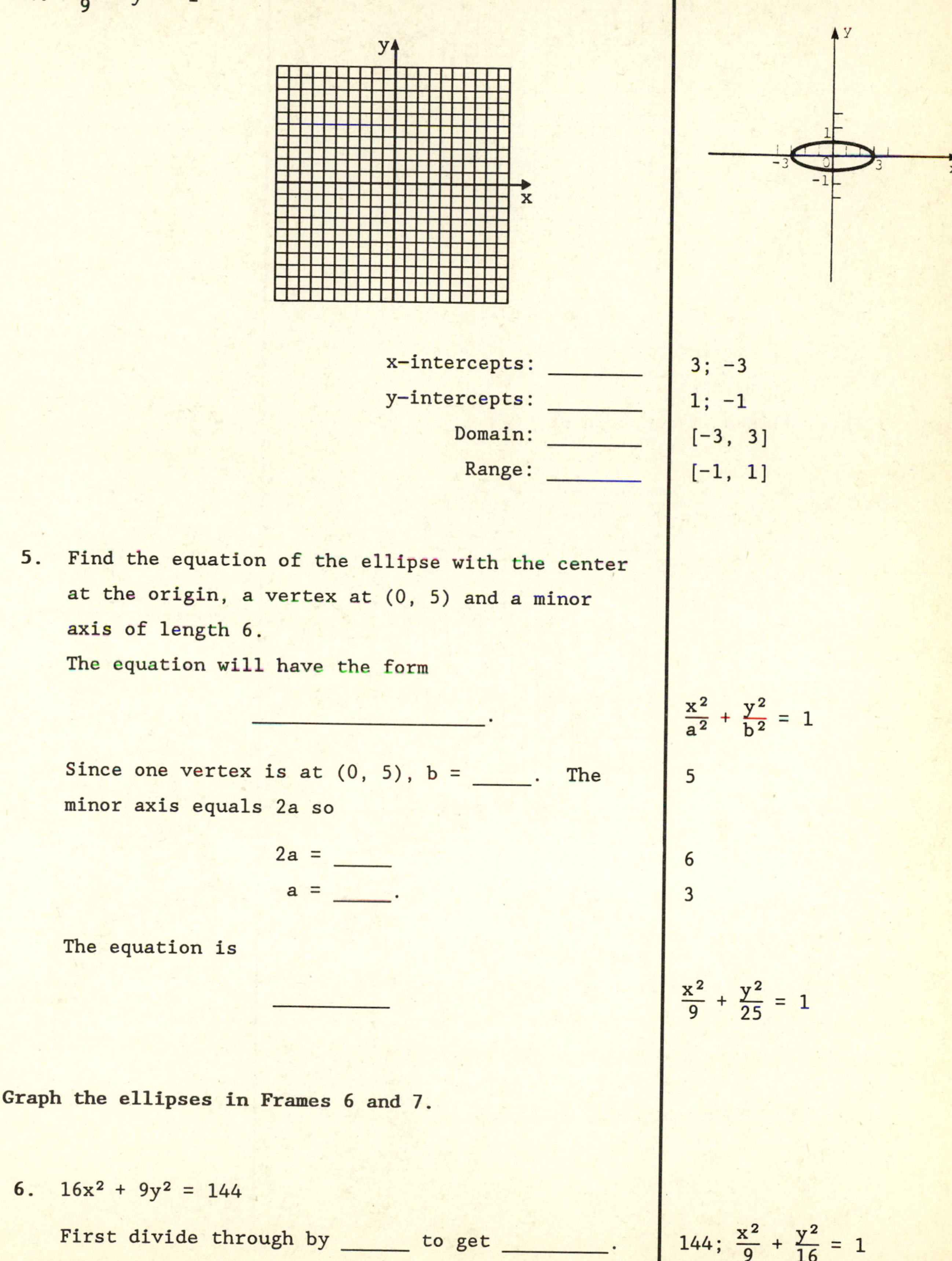

x-intercepts: ________ 3; −3

y-intercepts: ________ 1; −1

Domain: ________ [−3, 3]

Range: ________ [−1, 1]

5. Find the equation of the ellipse with the center at the origin, a vertex at (0, 5) and a minor axis of length 6.

The equation will have the form

________________. $\frac{x^2}{a^2} + \frac{y^2}{b^2} = 1$

Since one vertex is at (0, 5), b = _____. The minor axis equals 2a so 5

2a = _____ 6

a = _____. 3

The equation is

__________ $\frac{x^2}{9} + \frac{y^2}{25} = 1$

Graph the ellipses in Frames 6 and 7.

6. $16x^2 + 9y^2 = 144$

First divide through by _____ to get _________. 144; $\frac{x^2}{9} + \frac{y^2}{16} = 1$

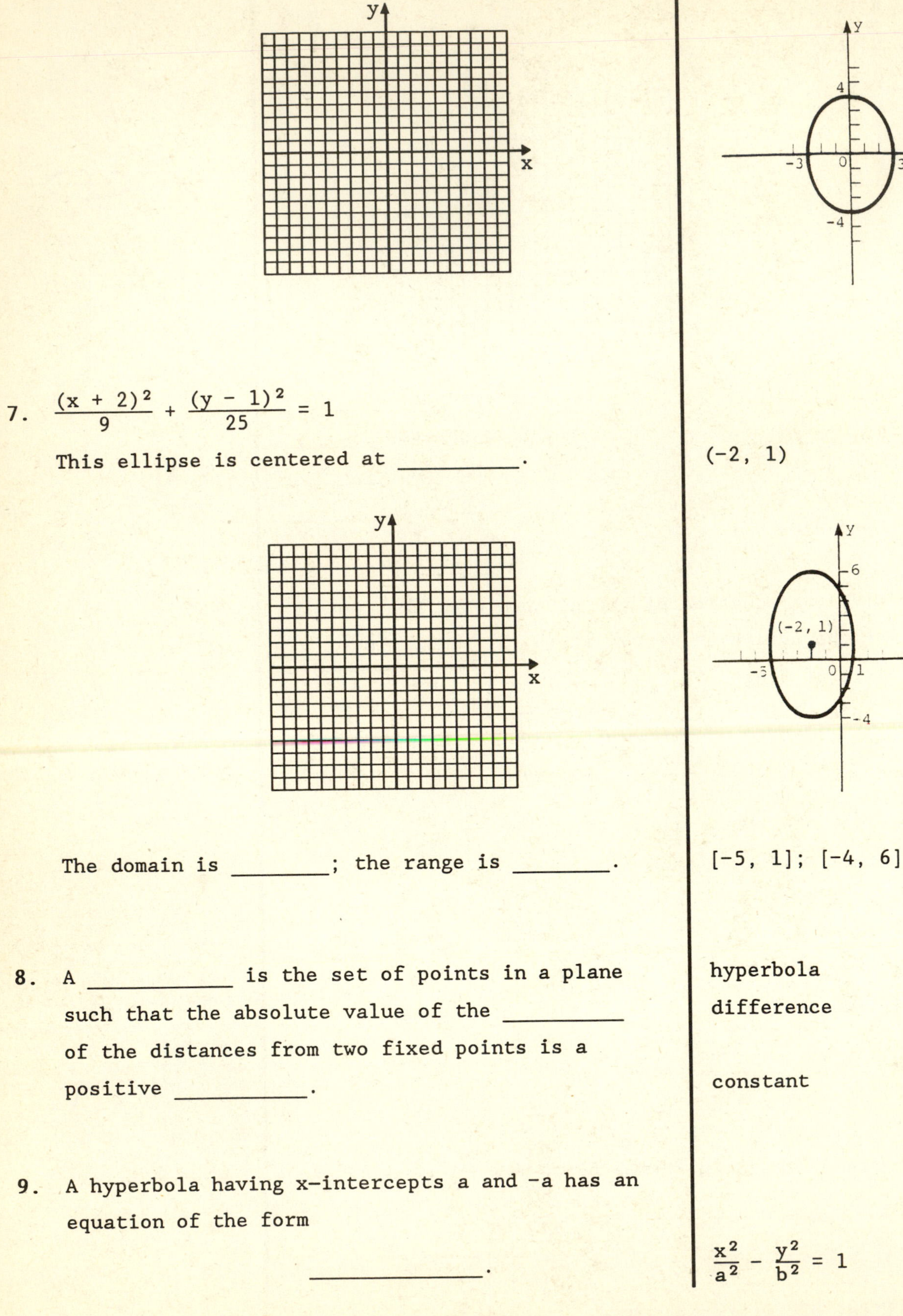

7. $\dfrac{(x + 2)^2}{9} + \dfrac{(y - 1)^2}{25} = 1$

This ellipse is centered at ________.

(-2, 1)

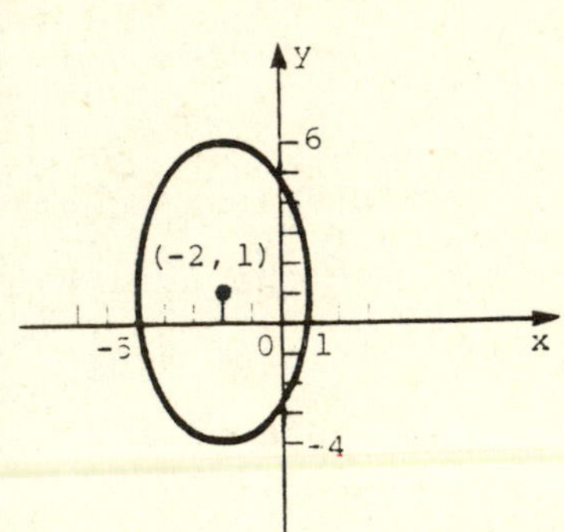

The domain is ________; the range is ________.

[-5, 1]; [-4, 6]

8. A ____________ is the set of points in a plane such that the absolute value of the __________ of the distances from two fixed points is a positive __________.

hyperbola
difference
constant

9. A hyperbola having x-intercepts a and -a has an equation of the form

______________.

$\dfrac{x^2}{a^2} - \dfrac{y^2}{b^2} = 1$

10. A hyperbola having y-intercepts b and -b has an equation of the form

_______________.

$\frac{y^2}{b^2} - \frac{x^2}{a^2} = 1$

11. To graph hyperbolas, we can find the ___________. The asymptotes of either

$$\frac{x^2}{a^2} - \frac{y^2}{b^2} = 1 \quad \text{or} \quad \frac{y^2}{b^2} - \frac{x^2}{a^2} = 1$$

are y = ______ and y = ______.

The asymptotes are the extended diagonals of the _________________ rectangle.

asymptotes

$\frac{b}{a}x$; $-\frac{b}{a}x$

fundamental

Graph the hyperbolas of Frames 12–15.

12. $\frac{x^2}{16} - \frac{y^2}{9} = 1$

Asymptotes are y = ____ and y = ____.

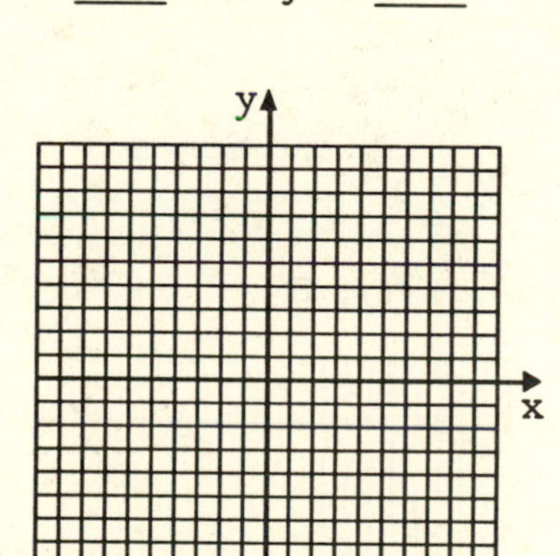

$\frac{3}{4}x$; $-\frac{3}{4}x$

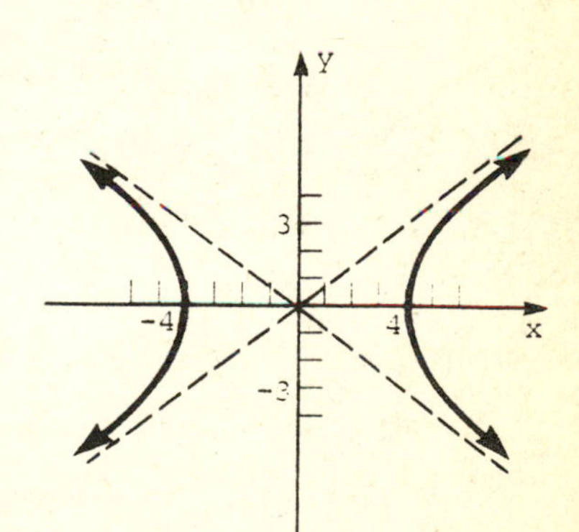

The vertices of this hyperbola are ________ and ________. The transverse axis is the line from ________ to ________. The midpoint of the transverse axis (___, ___) is the ___________ of the hyperbola. The domain of this relation is _____________________ and the range is _________.

(-4, 0)

(4, 0)

(-4, 0); (4, 0)

0; 0; center

$(-\infty, -4] \cup [4, \infty)$; $(-\infty, \infty)$

13. $\frac{y^2}{25} - \frac{x^2}{16} = 1$

Asymptotes are y = ______ and y = ______.

$\frac{5}{4}x$; $-\frac{5}{4}x$

The vertices of the fundamental rectangle are ________, ________, ________, and ________.

(-4, 5); (4, 5)
(-4, -5); (4, -5)

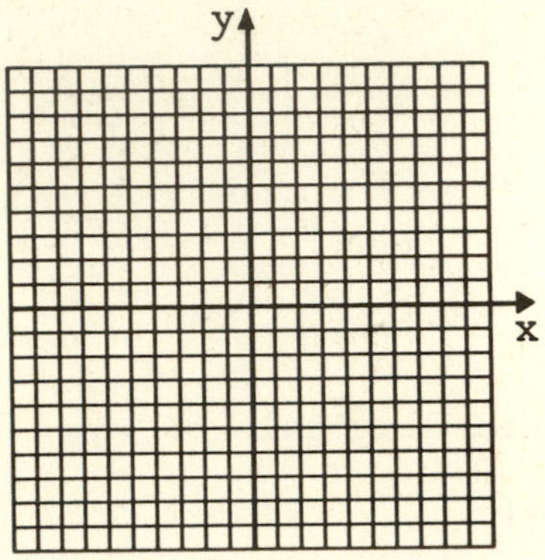

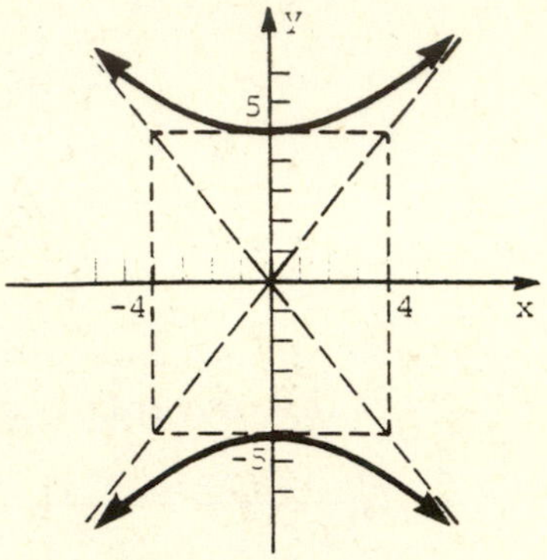

The domain of this relation is __________ and the range is ______________.

$(-\infty, \infty)$

$(-\infty, -5] \cup [5, \infty)$

14. $16y^2 - x^2 = 16$

Divide through by ____. Use the points ________, ________, ________, and ________ for the fundamental rectangle.

16; (4, 1)

(4, -1); (-4, 1);
(-4, -1)

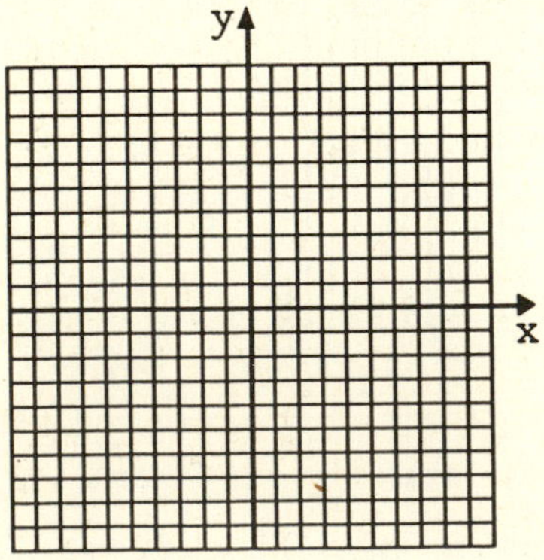

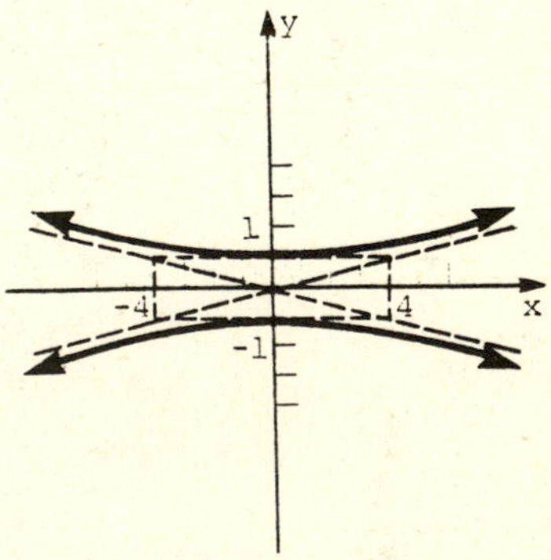

15. 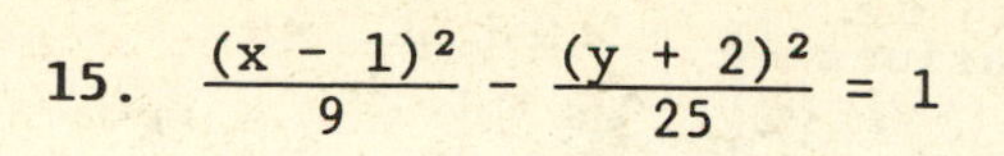

$$\frac{(x-1)^2}{9} - \frac{(y+2)^2}{25} = 1$$

This hyperbola is centered at ________.

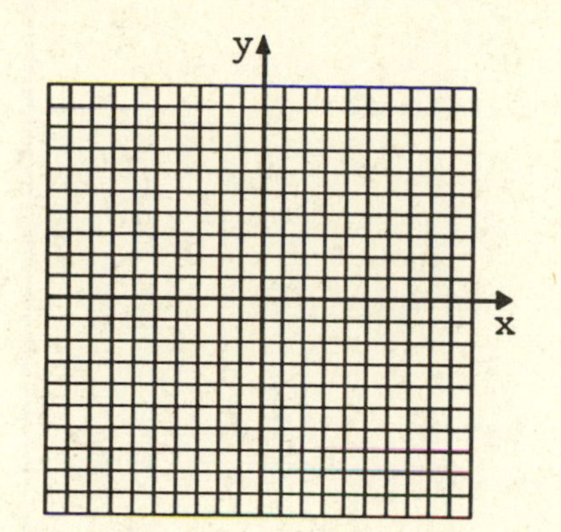

(1, −2)

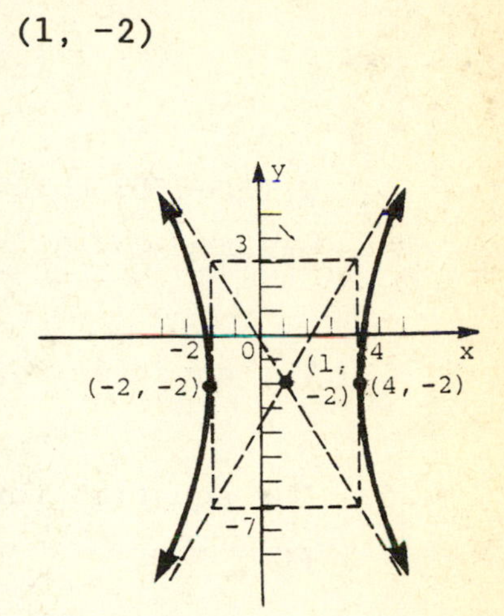

16. Write an equation for the hyperbola, center at the origin, a vertex at (4, 0), and $b = 2a$, where the asymptotes are the lines

$$y = \pm\frac{b}{a}x.$$

The equation will have the form

____

$a =$ ____, $b =$ ____

The equation is

____

$\frac{x^2}{a^2} - \frac{y^2}{b^2} = 1$

4; 8

$\frac{x^2}{16} - \frac{y^2}{64} = 1$

3.6 The Conic Sections

1. Hyperbolas, parabolas, ellipses, and ________ are all conic ________.

circles

sections

2. All conic sections in this chapter have equations of the form

$$Ax^2 + \_\_\_\_\_\_\_\_\_\_ = 0.$$

$Bx + Cy^2 + Dy + E$

3. Written in this form, the equation for a ________ has $A = C \neq 0$.

circle

4. The equation for a hyperbola has AC ___ 0.

$<$

5. The equation for an ellipse has A ____ C and AC ___ 0.

$\neq$

$>$

6. The equation for a ____________ has either $A = 0$ or $C = 0$, but not both.

parabola

Without actually graphing, identify the conic sections of Frames 7–15.

7. $x^2 - y^2 = 25$ ______________ hyperbola

8. $x^2 + y^2 = 25$ ______________ circle

9. $x^2 + y = 25$ ______________ parabola

10. $4x^2 + 9y^2 = 36$ ______________ ellipse

11. $4x^2 - 9y^2 = 36$ ______________ hyperbola

12. $9x^2 - 36x + 4y^2 + 32y + 64 = 0$ ______________ ellipse

13. $2x^2 + 4x + y - 1 = 0$ ______________ parabola

14. $5x^2 + 36y = 30x + 9y^2 + 36$ ______________ hyperbola

15. $4x^2 + 4y^2 - 9 = 0$ ______________ circle

For each of Frames 16–20 that has a graph, identify the corresponding graph. Transform the equation if necessary.

16. $4x^2 - 4x = 4y^2 + 4y + 7$ ____________ | hyperbola

17. $3x^2 + 2x = 3x^2 + 6y + 7$ ____________ | line

18. $8(x - 2)^2 - 6(y + 1)^2 = 0$ ____________ | point

19. $(x - 3)^2 + (y - 4)^2 = 1$ ____________ | circle

20. $x^2 - 4x + 9y^2 + 18y + 25 = 0$ ____________ | no graph

3.7 Inequalities

---

1. To graph $3x - 4y \le 12$, first graph the boundary ____________. The graph includes the line because the coordinates of the points of the line ____________ the inequality. To decide which side of the line is included in the graph, solve $3x - 4y \le 12$ for ____.

$$-4y \le \_\_\_\_\_\_\_\_$$

$$y \ge \_\_\_\_\_\_\_\_$$

(Do not forget to reverse the inequality symbol.) Because of the ____, graph the region ______ the line. Complete the following graph.

Answers: $3x - 4y = 12$; satisfy; y; $-3x + 12$; $\frac{3}{4}x - 3$; $\ge$; above

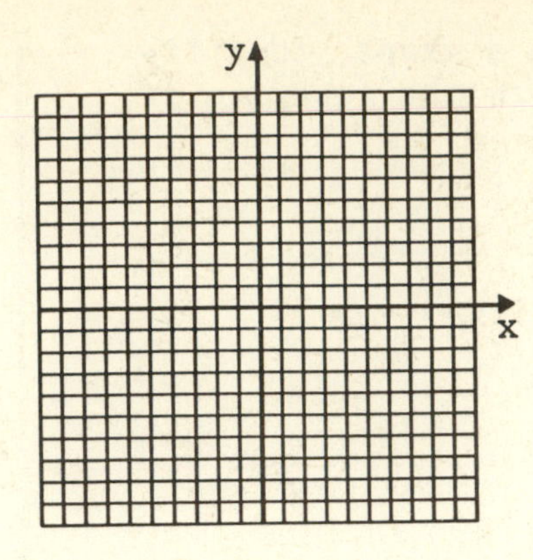

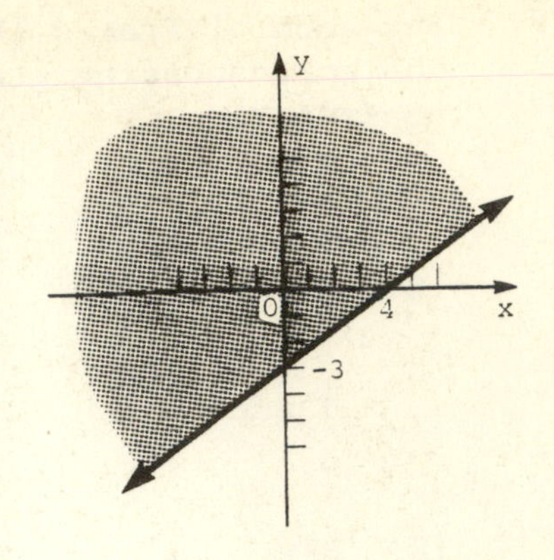

An alternative method for deciding which side of the boundary line to shade, choose a test point not on the ________. Choose (0, 0). Substituting ____ for x and ____ for y give

boundary

0; 0

$$3(0) - 4(0) \le 12$$
$$0 \le 12,$$

which is a (*true/false*) statement.

true

Since (0, 0) leads to a (*true/false*) result, shade the side of the boundary that (*does/does not*) contain (0, 0).

true

does

2. Graph $x - y > -5$ below. Make the line ________ because of the ____.

dashed

>

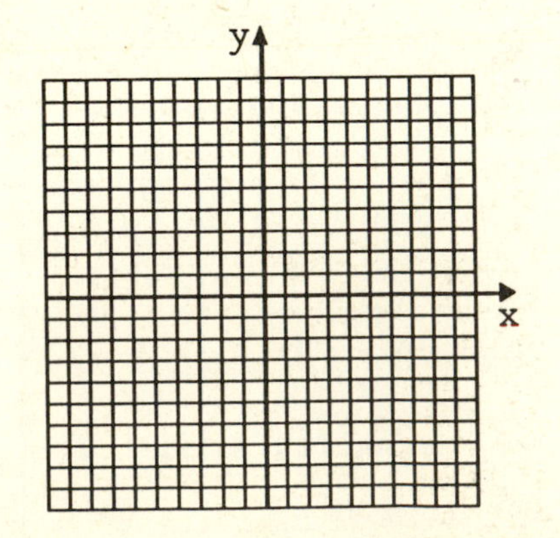

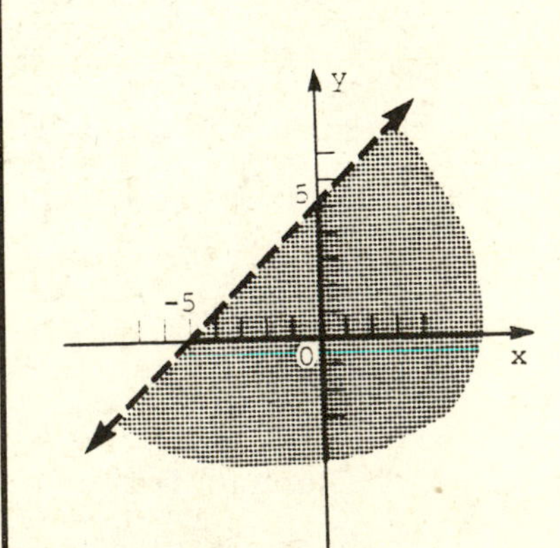

Graph the inequalities of Frames 3–6.

3. $y \geq x - 2$

4. $y < x^2 - 3$

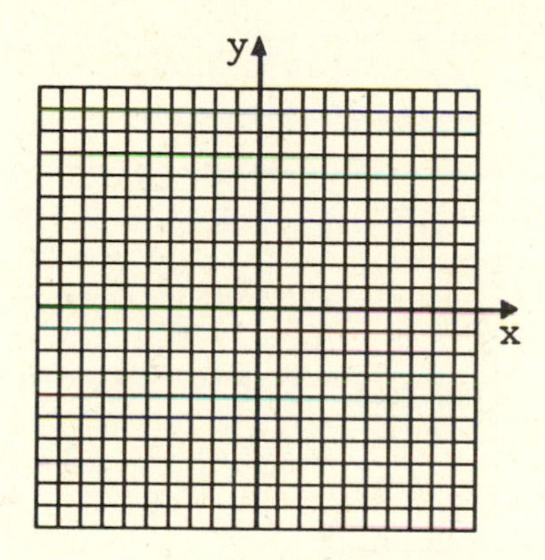

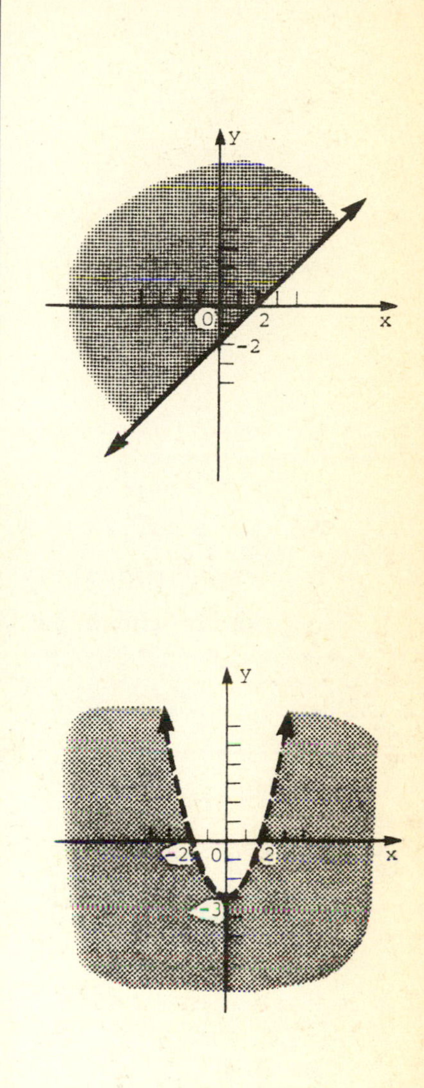

5. $x^2 - y^2 \geq 9$

Since it is not easy to solve for ____, choose a test point, such as (0, 0), to decide on the region to shade.

y

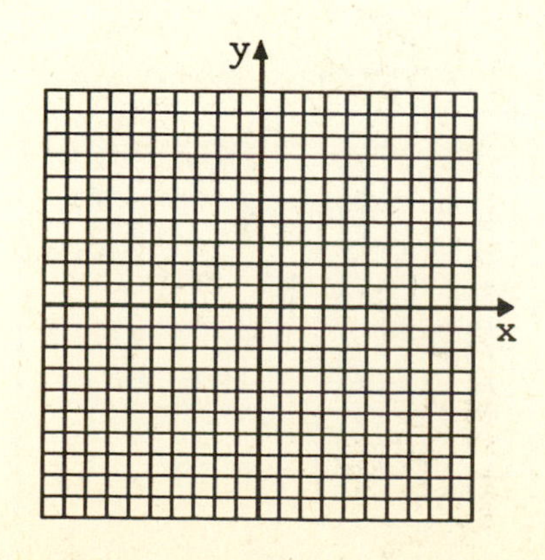

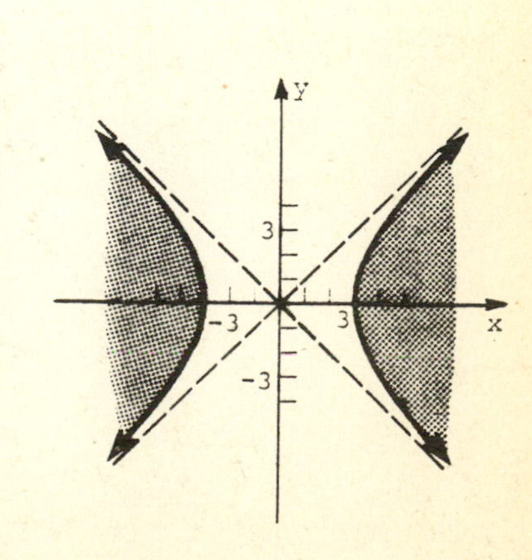

6. $(x - 1)^2 + (y + 2)^2 < 16$

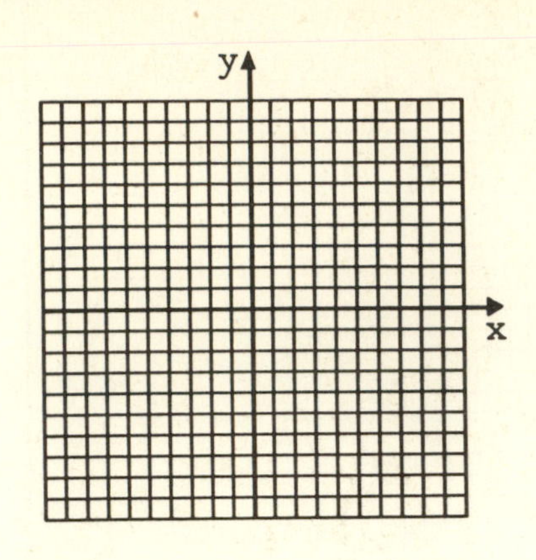

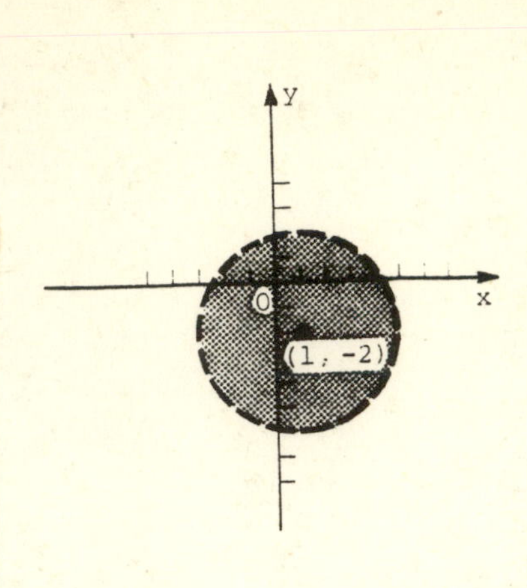

7. To graph $2x - y < 6$, where $3x + 2y > 6$, graph each inequality on the same set of axes and shade the area where they ________________.

intersect or overlap

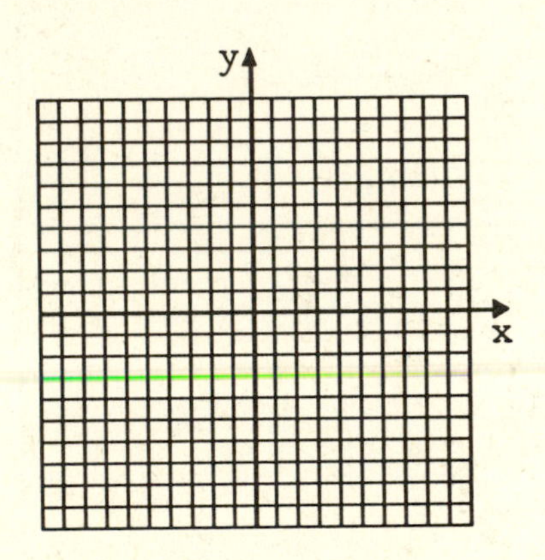

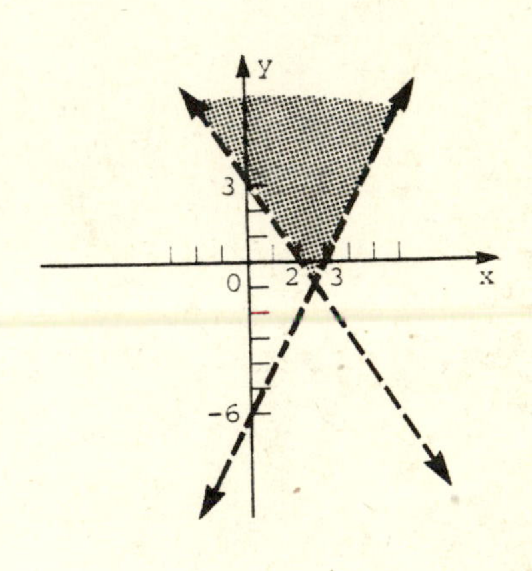

Graph each of the following inequalities in Frames 8–9.

8. $5x - 2y \le 10$, where $x \le 3y$

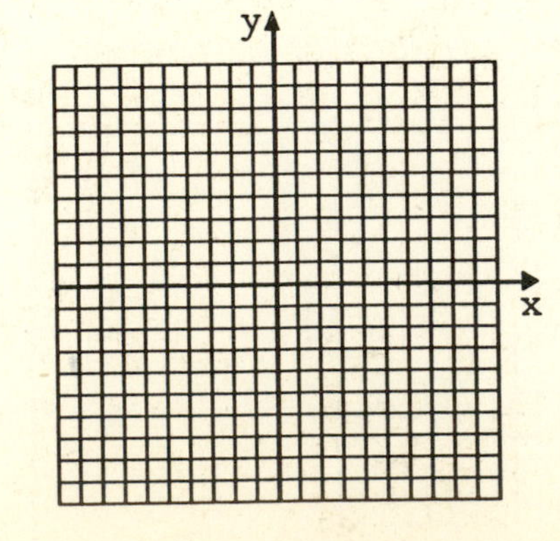

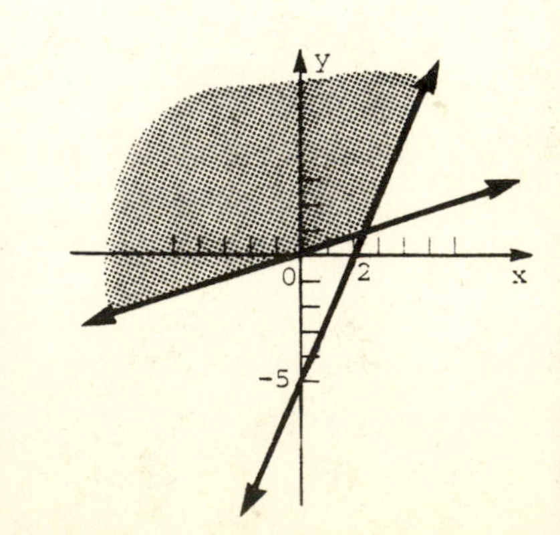

9. $x^2 + y^2 \geq 16$, where
$x^2 + y^2 \leq 25$

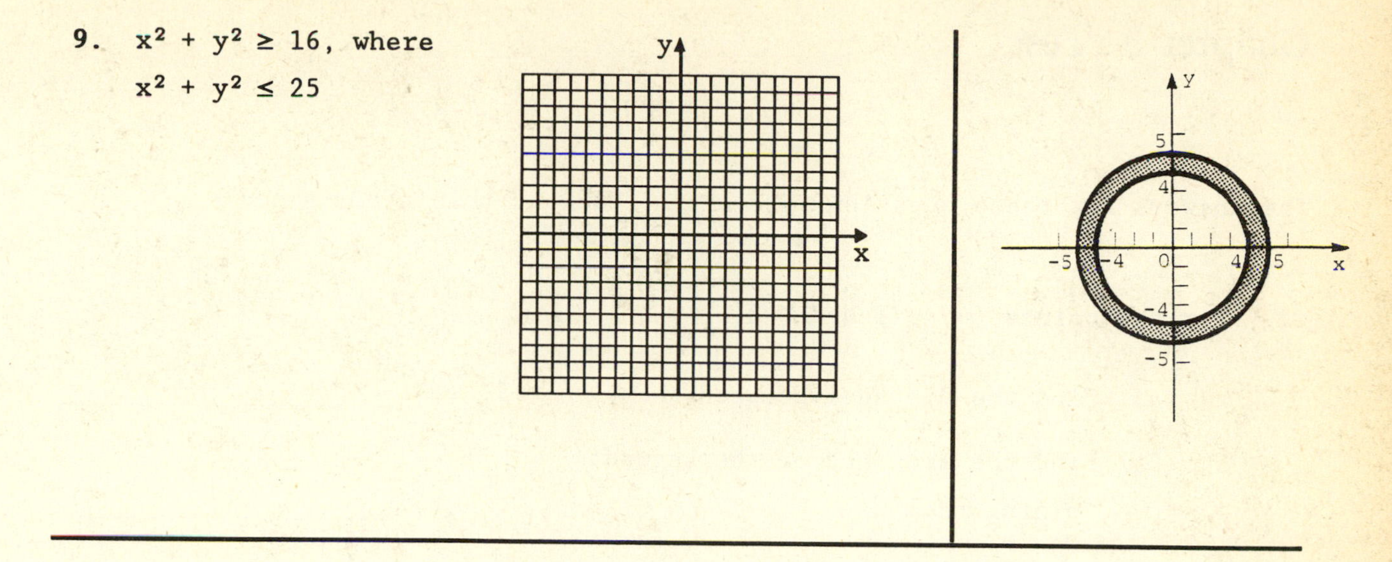

Chapter 3 Test

The answers for these questions are at the back of this study guide.

1. For the points $(4, -7)$ and $(-2, 3)$:

   (a) Find the distance between them. 1. (a) ____________

   (b) Find the midpoint of the segment joining them. (b) ____________

Find the slope of each of the following lines.

2. Through $(2, -5)$ and $(-3, 5)$ 2. ____________

3. $2x - 7y = 11$ 3. ____________

4. Explain the difference between a line with zero slope and undefined slope. 4. ____________

For each of the following lines, write the equation in standard form.

5. Through $(-8, 7)$ and $(1, -4)$ 5. ____________

6. Through $(3, -2)$ and perpendicular to $y = 7$. 6. ____________

7. Through $(-3, 7)$ and parallel to $7x - 2y = 9$. 7. ____________

Graph the following lines.

8. $2x + 7y = 14$

9. Through $(3, -2)$, with $m = -3/4$

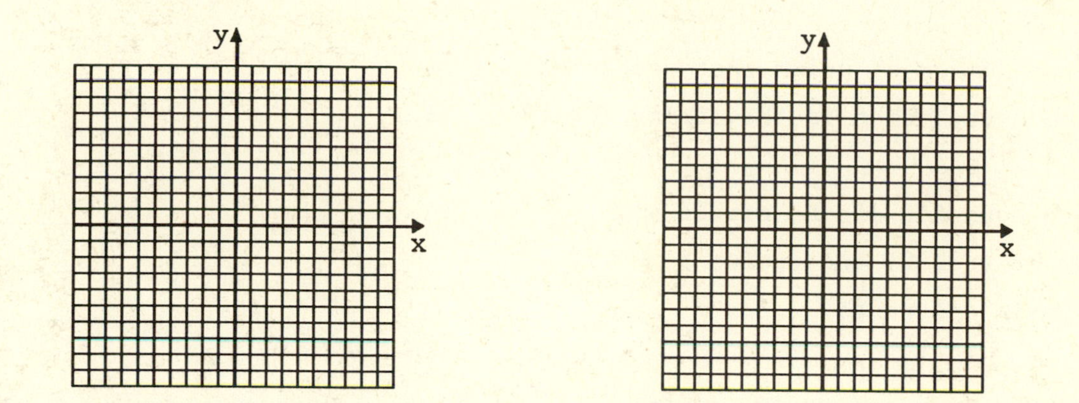

Graph each of the following. Give the axis, vertex, domain, and range of each.

10. $y = -(x - 1)^2 - 2$

11. $x = (y + 1)^2 - 7$

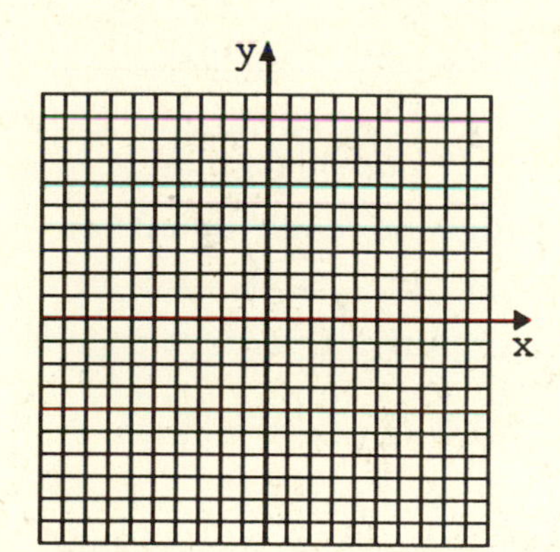

Axis: ______________

Vertex: ______________

Domain: ______________

Range: ______________

Axis: ______________

Vertex: ______________

Domain: ______________

Range: ______________

12. Give the vertex and axis of the graph of $y = x^2 - 4x + 2$.

12. ______________________________

____

13. The cost in dollars to make x units of a product is

$$C(x) = x^2 - 4x + 12.$$

Find the number of units that produce minimum cost. What is the minimum cost?

13. ______________________

14. Find an equation of the circle with center at (−3, 2) and radius 5. Give the domain and range.

14. ______________________

____

____

15. Find the center and radius of the circle with equation

$$x^2 - 4x + y^2 + 10y + 20 = 0.$$

15. ______________________

____

16. Decide whether the equation $y = 4x^4 - x^2$ has a graph that is symmetric with respect to the x-axis, y-axis, or origin.

16. ______________________

Graph each of the following. Give the domain and range of each equation.

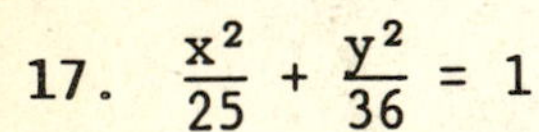

17. $\frac{x^2}{25} + \frac{y^2}{36} = 1$

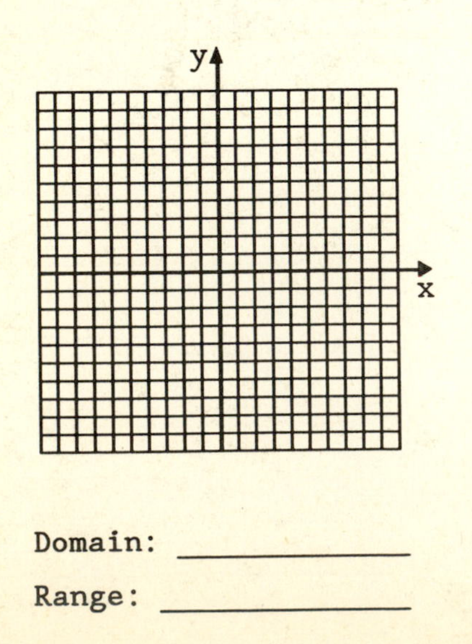

Domain: ______________

Range: ______________

18. $y^2 - x^2 = 4$

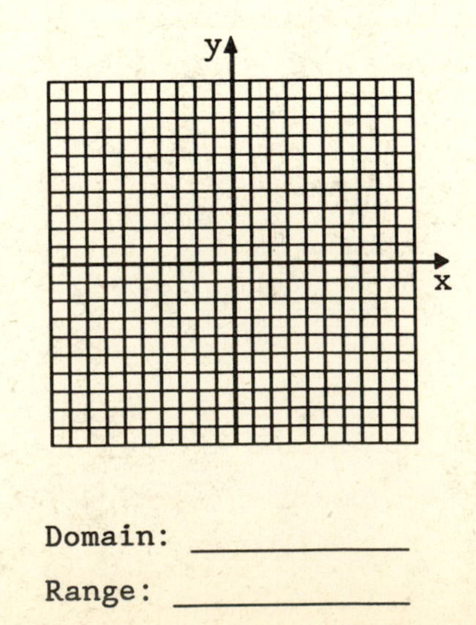

Domain: ______________

Range: ______________

19. $\frac{(x + 2)^2}{9} + \frac{(y - 1)^2}{4} = 1$

20. $3x + y \le 6$, where $2x - y \ge 4$.

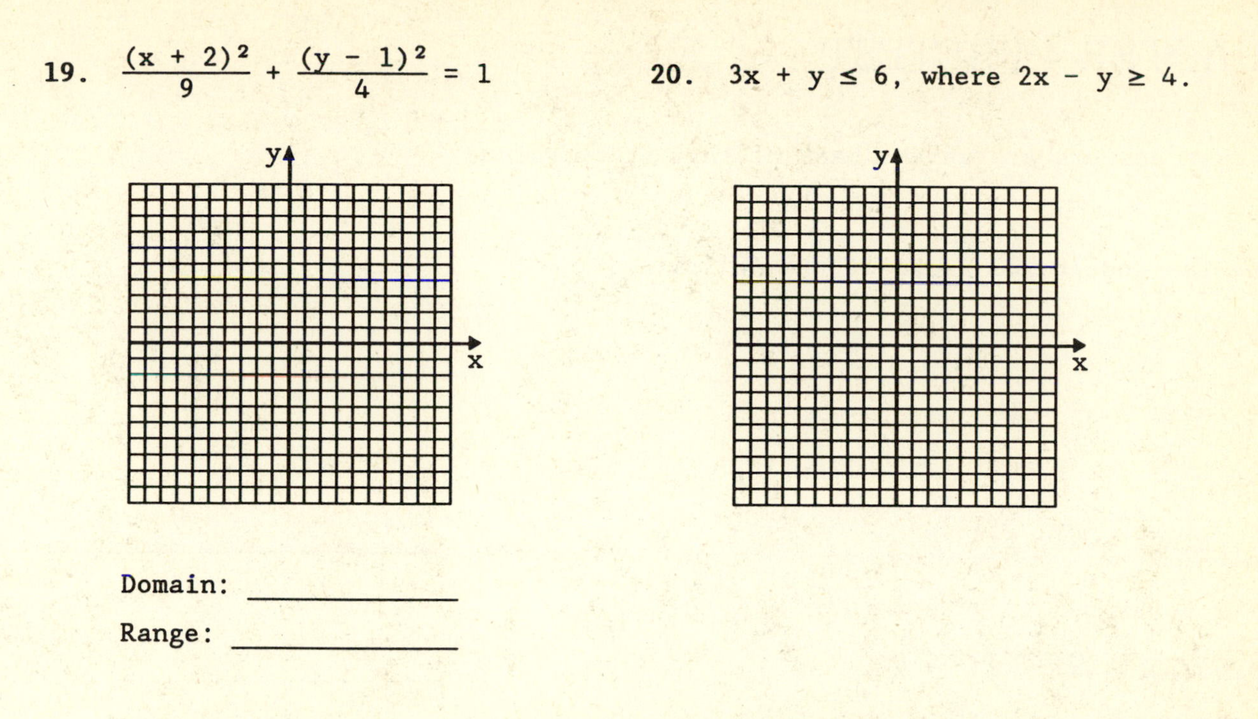

Domain: ______________

Range: ______________

CHAPTER 4 PRETEST

Pretest answers are at the back of this study guide.

Give the domain and range of the following.

1. $h(x) = x^2 + 1$ 1. ____________________

2. $p(x) = \sqrt{x - 5}$ 2. ____________________

3. $r(x) = \sqrt{x^2 - 9}$ 3. ____________________

Let $f(x) = 3x - 5$ and $g(x) = 4x^2 - 3$. Find each of the following.

4. $(f + g)(x)$ 4. ____________________

5. $(f - g)(-2)$ 5. ____________________

6. $(fg)(0)$ 6. ____________________

7. $\left(\frac{f}{g}\right)(-1)$ 7. ____________________

8. $(f \circ g)(x)$ 8. ____________________

9. $(g \circ f)(x)$ 9. ____________________

Graph each function defined as follows.

10. $f(x) = -|x - 2|$

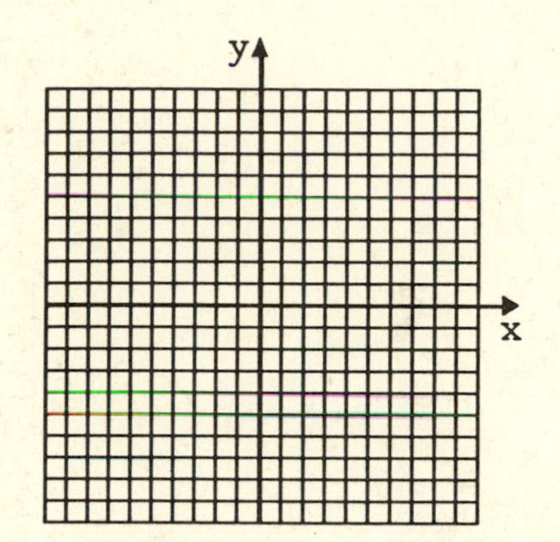

11. $f(x) = [\![x - 2]\!]$

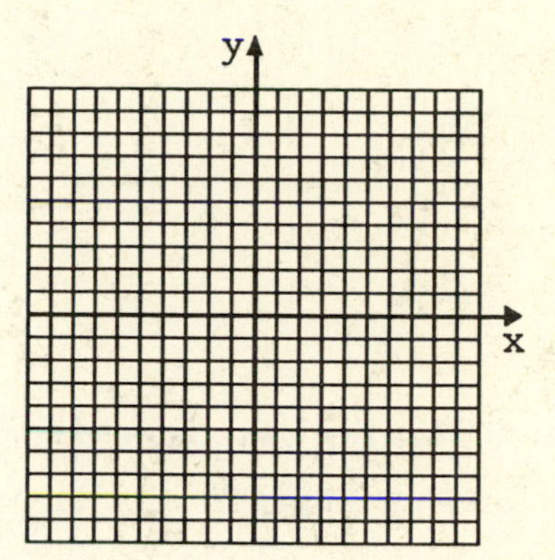

12. $f(x) = \begin{cases} x - 3 & \text{if } x \geq 1 \\ 2 - 4x & \text{if } x < 1 \end{cases}$

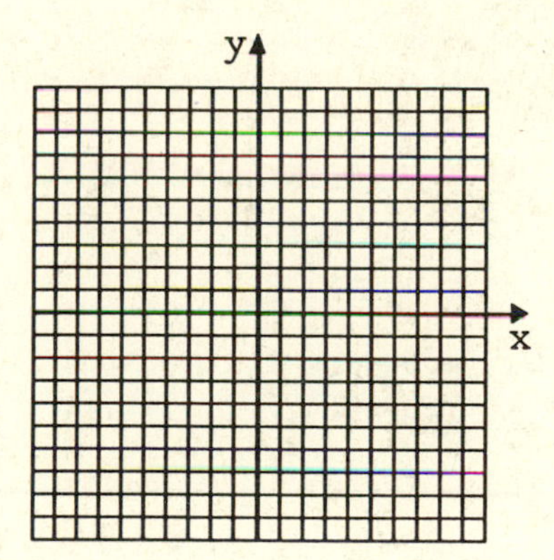

Which of the functions defined as follows are one-to-one?

13. $f(x) = 3x - 2$ 13. ________________

14. $f(x) = \sqrt{x + 3}$, domain $[-3, \infty)$ 14. ________________

15. $f(x) = \sqrt{x^2 - 25}$ 15. ________________

For each of the following that defines a one-to-one function, write an equation for the inverse in the form $y = f^{-1}(x)$.

16. $f(x) = 5 - x^2$; domain $(-\infty, 0]$ 16. ____________

17. $f(x) = \frac{-4x}{2x - 5}$ 17. ____________

18. $f(x) = -\sqrt{2 - x}$ 18. ____________

19. Are the following two functions inverses of each other? Explain why or why not.

    $f(x) = x^2$, domain $(-\infty, 0]$, and $g(x) = -\sqrt{x}$

    19. ____________

20. The illumination produced by a light source varies inversely as the square of the distance from the source. The illumination of a light source at 5 m is 70 candela. What is the illumination 10 m from the source?

    20. ____________

CHAPTER 4 FUNCTIONS

4.1 Functions

1. A function is a relation that assigns to each member of one set ____________ one member of another set.	exactly
2. In Frame 1, the first set is the ________ of the function, while the second set is the _________.	domain range
3. If x represents an element of the domain, x is the _______________ variable. If y represents an element of the range, y is the ____________ variable.	independent dependent
4. If we start with the equation $y = 4x - 7$, and choose a value of x, we will be able to find exactly _____ value of y. Because of this, $y = 4x - 7$ (*is/is not*) a function.	one is
5. If we start with $x = y^2 - 1$, and choose $x = 15$, then $15 = y^2 - 1$, or $y^2 =$ _____. From this, $y =$ _____, or $y =$ _____. One value of x leads to _____ values of y, so $x = y^2 - 1$ (*is/is not*) a function.	16 4; -4 two; is not
6. For the set $\{(2, 3), (3, 4), (4, 3)\}$, the domain is {___, ___, ___} while the range is {___, ___}. Each element in the ____________ has one element assigned to it in the __________ so this set (*is/is not*) a function.	2; 3; 4; 3; 4 domain range is

Write *function* or *not a function* for each set in Frames 7–11.

7. $\{(-1, -4), (1, -1), (3, 2), (5, 5)\}$ — function

8. $\{(-2, 7), (-1, 1), (0, -1), (1, 1), (2, 7)\}$ — function

9. $\{(x, y) | y = -x^2\}$ — function

10. $\{(1, -1), (1, 1), (2, -2), (2, 2)\}$ — not a function

11. $\{(x, y) | x^2 + y^2 = 100\}$ — not a function

12. Letters such as f, g, or h are often used to name ____________. For example, instead of $y = 4x - 7$, we could write

$$f(x) = \_\_\_\_\_\_\_\_\_\_,$$

where f(x) is read ________.

functions; $4x - 7$; f of x

13. If $f(x) = 4x - 7$, then $f(-5)$ is found by substituting _____ for x.

$$f(-5) = 4(\_\_\_\_\_) - 7 = \_\_\_\_\_\_$$

The number _____ in the domain corresponds to ______ in the range.

-5; -5; -27; -5; -27

In Frames 14–17, let $f(x) = -x^2 + 4x - 9$, and $g(x) = 3x - 5$. Find the following.

14. $f(-2) = -(\_\_\_\_)^2 + 4(\_\_\_\_) - 9 = \_\_\_\_$ — -2; -2; -21

15. $f(3) = \_\_\_\_$ — -6

16. $g(-5) = \_\_\_\_$ — -20

17. $f(z) = \_\_\_\_\_\_\_\_$ (Replace x with ____.) — $-z^2 + 4z - 9$; z

18. Let $f(x) = x^2$. Find the following.

$$f(x + h) = (\_\_\_\_\_\_)^2 = \_\_\_\_\_\_\_\_\_\_$$

$$f(x + h) - f(x) = \_\_\_\_\_\_\_\_\_\_$$

$$\frac{f(x + h) - f(x)}{h} = \_\_\_\_\_\_\_\_\_\_$$

This expression is important in calculus. It is called the ________________________.

$x + h$; $x^2 + 2xh + h^2$

$2xh + h^2$

$2x + h$

difference-quotient

19. Let $g(x) = \frac{2}{x}$. Find the difference-quotient.

$$g(x + h) = \_\_\_\_\_\_\_\_\_\_$$

$$g(x + h) - g(x) = \frac{2}{x + h} - \frac{2}{x} = \_\_\_\_\_\_\_\_\_\_$$

$$\frac{g(x + h) - g(x)}{h} = \_\_\_\_\_\_\_\_\_\_$$

$\frac{2}{x + h}$

$\frac{-2h}{x(x + h)}$

$\frac{-2}{x(x + h)}$

20. Recall: The set of all possible values of x is called the ________ of a function, while the set of all possible values of y is the ______.

domain

range

Find the domain and range in Frames 21–25.

21. $f(x) = 9 + 4x$

Here both the domain and range are the set of all ________ numbers. In interval notation, both are ___________.

real

$(-\infty, \infty)$

22. $g(x) = x^2 - 2$

Here x can be any number, so the domain is the interval ________. Since x is squared, we always have x^2 ____ 0. Because of this, $x^2 - 2$ ____ -2. Since $g(x) = x^2 - 2$, the range is ________.

$(-\infty, \infty)$

$\geq$; $\geq$

$[-2, \infty)$

23. $r(x) = \sqrt{x + 4}$

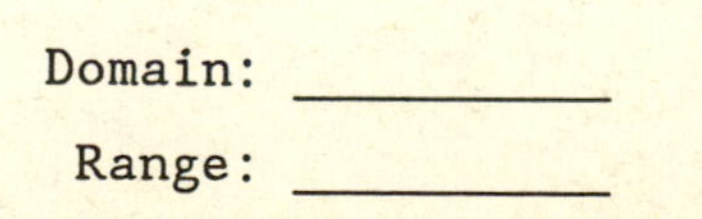

Domain: __________

Range: __________

$[-4, \infty)$

$[0, \infty)$

24. $k(x) = -\sqrt{25 - x^2}$ Domain: ________ Range: ________

[-5, 5]
[-5, 0]

25. $g(x) = |x + 4|$ Domain: ________ Range: ________

$(-\infty, \infty)$
$[0, \infty)$

26. We can tell if a graph is a function by using the ________ line test: a graph is a function if every vertical line cuts the graph in no more than _____ point.

vertical

one

Write *function* or *not a function* in Frames 27–29.

27. ________________

function

28. ________________

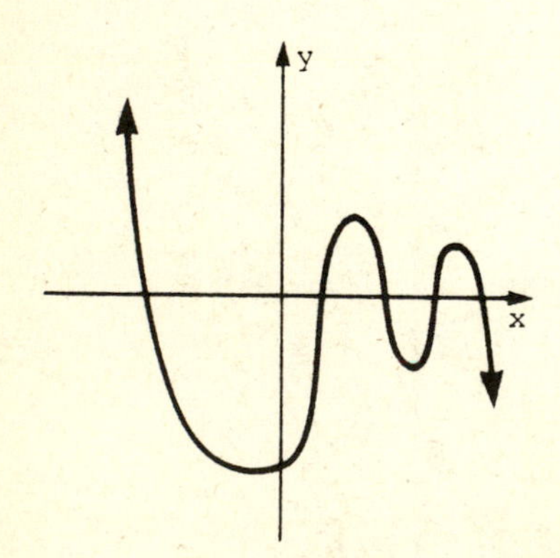

function

29.

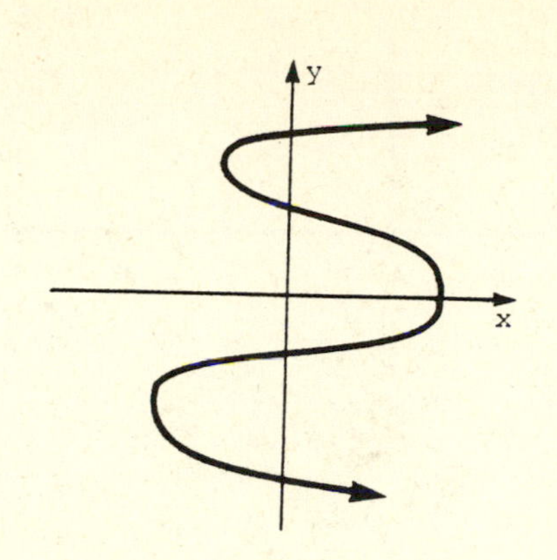

________________ | not a function

30. A linear cost function has the form ______________ where m is the ________ cost and b is the _______ cost.

$C(x) = mx + b$

variable; fixed

The revenue function is __________ where p equals __________.

$R(x) = px$

price

The profit function is _______________. To find the profit, replace x with the number of items _______.

$P(x) = R(x) - C(x)$

sold

31. A toy manufacturer has found that the cost to produce a certain toy is a linear function. If the fixed cost is \$750 and the variable cost is \$20, the cost function is

$$C(x) = \_\_\_\_ x + \_\_\_\_\_.$$

20; 750

Since the toy sells for \$35, the revenue function is

$$R(x) = \_\_\_\_\_ x.$$

35

The profit function is

$$P(x) = \_\_\_\_\_ - (\_\_\_\_\_ + \_\_\_\_\_)$$

35x; 20x; 750

To make a profit, ______ must be positive. Set ___________ ≥ 0 and solve for x.

P(x)

15x - 750

$$\_\_\_\_\_ \geq 750$$

$$x \geq \_\_\_\_$$

15x

50

At least ____ toys must be sold for the toy manufacturer to make a profit.	50

4.2 Variation: Applications of Functions

1. y varies ______________ as x, or y is directly ______________ to x means that a constant ___ exists such that _______.	directly proportional; k $y = kx$
2. **Suppose the garbage of Highland Estates varies directly as the population. If the garbage is 20 tons per year when the population is 100, find the amount of garbage per year when the population is 150.** Since the garbage varies directly as the population,	
$G =$ _____	kP
where G = garbage, P = population, and k is a nonzero constant which we must determine. When $G = 20$ and $P = 100$,	
$20 =$ ______	$100k$
so that	
$k =$ ______.	1/5
Now use $k = 1/5$ and $P = 150$ in the original equation to find the amount of garbage, per year, when the population is 150.	
$G =$ ______.	30
Highland Estates has ____ tons of garbage when the population is 150.	30

3. y varies directly as the nth power of x if a real number k exists such that ________. | $y = kx^n$

y varies inversely as the nth power of x if a real number exists such that

________. | $y = \frac{k}{x^n}$

4. Suppose the cost of manufacturing candles varies inversely as the number of candles produced. If 100 candles are produced, each costs $4. Find the cost per candle if 400 candles are produced.

Let x = the number of candles produced
and y = the cost per candle.

Use the form y = ____. Substitute the given values to find k. | $\frac{k}{x}$

$$\_\_\_\_ = \frac{k}{\_\_\_\_}$$ 4; 100

k = ____ | 400

Use the formula with this value of k.

y = ____ | $\frac{400}{400}$

Cost per candle ____ | $1

5. y varies jointly as the nth power of x and the mth power of z if there exists a real number k such that

y = ________ | $kx^n z^m$

6. Suppose y varies directly as the square of x and inversely as z, and that y = 98 when x = 7 and z = 4. Find y when x = 3 and z = 9.

Start with the equation

y = ____. | $\frac{kx^2}{z}$

Replace y with ____, x with ____, and z with ____, to find k.	98; 7 4
$k =$ ______	8
The equation for y thus becomes	
$y =$ ______.	$\frac{8x^2}{z}$
Now replace x with ____ and z with ____ to find y.	3; 9
$y =$ ______.	8
7. Suppose x varies directly as the square of z and inversely as y. If $x = 6$ when $z = 3$ and $y = 2$, find x if $z = 6$ and $y = 2$. Start with	
$x =$ _____.	$\frac{kz^2}{y}$
Then $6 = \frac{k \cdot 3^2}{2}$	
and $k =$ _____.	$\frac{4}{3}$
Use the values $k = 4/3$, $z = 6$, and $y = 2$ to find x.	
$x =$ _____	24
8. Hiller's Construction Company has won the bid for a new freeway interchange in Kansas City. The company's engineers know that the maximum load a cylindrical column with circular cross section can hold varies directly as the fourth power of the diameter and inversely as the square of the height. If an 11-meter column that is 1 m in diameter will support a load of 10 metric tons, how many metric tons will be supported by a column that is 11 m high and 2 m in diameter?	

Write

$$L = \_\_\_\_\_\_.$$ | $\frac{kd^4}{h^2}$

where L is the load that the column can hold, d is the diameter of the column, h is the height of the column, and k is some nonzero constant. Then

$$10 = \frac{k \cdot (1^4)}{11^2}$$

and $k = \_\_\_\_\_\_.$ | 1210

Now use k = 1210, d = 2, and h = 11 to find the load that the column can hold.

$$L = \_\_\_\_\_\_$$ | 160

The column can hold ______ metric tons. | 160

4.3 Combining Functions: Algebra of Functions

1. We can combine two functions, f and g, to obtain various third functions:

$(f + g)(x) = \_\_\_\_\_\_\_\_\_\_$ | $f(x) + g(x)$

$(f - g)(x) = \_\_\_\_\_\_\_\_\_\_$ | $f(x) - g(x)$

$(fg)(x) = \_\_\_\_\_\_\_\_\_\_$ | $f(x) \cdot g(x)$

$\left(\frac{f}{g}\right)(x) = \_\_\_\_\_\_\_\_\_\_$, where ______. | $\frac{f(x)}{g(x)}$; $g(x) \neq 0$

Find (f + g)(x), (f − g)(x), (fg)(x), and (f/g)(x) in Frames 2–4. Give the domain of each.

2. $f(x) = 9x - 3$, $g(x) = 11x - 10$

$(f + g)(x) = 9x - 3 + \_\_\_\_\_\_ = \_\_\_\_\_\_$ | $11x - 10$; $20x - 13$

$(f - g)(x) = 9x - 3 - (\_\_\_\_\_\_) = \_\_\_\_\_\_$ | $11x - 10$; $-2x + 7$

$(fg)(x) = (9x - 3)(\_\_\_\_\_\_) = \_\_\_\_\_\_$ | $11x - 10$; $99x^2 - 123x + 30$

$\left(\frac{f}{g}\right)(x)$ = ____________	$\frac{9x - 3}{11x - 10}$
The domain for f + g, f − g, and fg is ________.	$(-\infty, \infty)$
For f/g, the denominator must be nonzero, so that $x \neq$ ______. In interval notation, the	$10/11$
domain is ____________________.	$(-\infty, 10/11) \cup (10/11, \infty)$
3. $f(x) = 2x^2 - 7x + 5$, $g(x) = x^2 - 16$	
$(f + g)(x)$ = ____________	$3x^2 - 7x - 11$
$(f - g)(x)$ = ____________	$x^2 - 7x + 21$
$(fg)(x)$ = ____________	$2x^4 - 7x^3 - 27x^2 + 112x - 80$
$\left(\frac{f}{g}\right)(x)$ = ____________	$\frac{2x^2 - 7x + 5}{x^2 - 16}$
The domain for f + g, f − g, and fg is ________.	$(-\infty, \infty)$
The domain for f/g is ________________________.	$(-\infty, -4) \cup (-4, 4) \cup (4, \infty)$
4. $f(x) = 3x - 5$, $g(x) = \sqrt{x + 3}$	
$(f + g)(x)$ = ______________	$3x - 5 + \sqrt{x + 3}$
$(f - g)(x)$ = ______________	$3x - 5 - \sqrt{x + 3}$
$(fg)(x)$ = ______________	$(3x - 5)\sqrt{x + 3}$
$\left(\frac{f}{g}\right)(x)$ = ______________	$\frac{3x - 5}{\sqrt{x + 3}}$
The domain for f + g, f − g, and fg is ________.	$[-3, \infty)$
The domain for f/g is ________.	$(-3, \infty)$
5. Let $f(x) = 9x^2 - 7x - 8$ and $g(x) = -5x + 2$. Find the following.	
$(f + g)(-4)$ = _____ $(g - f)(2)$ = _____	186; -22
$(fg)(1)$ = _____ $\left(\frac{g}{f}\right)(0)$ = _____	18; $-\frac{1}{4}$

6. Let $f(x) = \sqrt{x + 4}$ and $g(x) = \sqrt{3x - 1}$. Find the following.

$(f - g)(2) =$ _____ $(f + g)(5) =$ _____

$(gf)(3) =$ _____ $\left(\frac{f}{g}\right)(1) =$ _____

$\sqrt{6} - \sqrt{5}$; $3 + \sqrt{14}$

$2\sqrt{14}$; $\frac{\sqrt{10}}{2}$

7. The composition of functions f and g is defined as

$(f \circ g)(x) =$ _____ and $(g \circ f)(x) =$ _____.

$f[g(x)]$; $g[f(x)]$

Find $(f \circ g)(x)$ and $(g \circ f)(x)$ in Frames 8–10. Find the domain of each in Frame 10.

8. $f(x) = 7x - 9$, $g(x) = 2x + 1$

$(f \circ g)(x) = f[g(x)]$

$= f($_______$)$ — $2x + 1$

$= 7($_______$) - 9$ — $2x + 1$

$=$ _________ — $14x - 2$

$(g \circ f)(x) = g[f(x)]$

$= g($_______$)$ — $7x - 9$

$= 2($_______$) + 1$ — $7x - 9$

$=$ _________ — $14x - 17$

9. $f(x) = 3x^2 - 2$, $g(x) = x - 5$

$(f \circ g)(x) =$ _________ — $3x^2 - 30x + 73$

$(g \circ f)(x) =$ _________ — $3x^2 - 7$

10. $f(x) = \frac{-2}{x}$, $g(x) = \sqrt{5 - x}$

$(f \circ g)(x) =$ _________ — $\frac{-2}{\sqrt{5 - x}}$

$(g \circ f)(x) =$ _________ — $\sqrt{5 + \frac{2}{x}}$ or $\sqrt{\frac{5x + 2}{x}}$

The domain of $f \circ g$ is the interval ________	$(-\infty, 5)$
since $\sqrt{5 - x}$ is a nonzero real number only when ______ > 0 or $x <$ ______.	$5 - x$; 5
The domain of $g \circ f$ is the set of all real numbers such that x ____ 0 and $5 -$ ____ ≥ 0.	$\neq$; $f(x)$
$5 - f(x) =$ ______	$\dfrac{5x + 2}{x}$
Find the values of x that make the __________	numerator
and the __________ of this fraction each zero.	denominator
These values are ______ and ____. Use a sign	$-2/5$; 0
graph to verify that the domain of $g \circ f$ is the set ______________.	$(-\infty, -2/5] \cup (0, \infty)$
11. Let $f(x) = -x^2 + x - 1$ and $g(x) = 3x - 7$. Find the following.	
$(f \circ g)(2) =$ ____ $(g \circ f)(1) =$ ____	-3; -10
12. Let $f(x) = \sqrt{2x + 1}$ and $g(x) = \sqrt{x + 6}$ Find the following.	
$(f \circ g)(3) =$ ____ $(g \circ f)(12) =$ ____	$\sqrt{7}$; $\sqrt{11}$

4.4 Graphing Basic Functions and Their Variations

1. The function $$f(x) = ax + b$$ is a __________ function, while	linear
$$f(x) = ax^2 + bx + c, \text{ with } a \neq 0,$$ is a __________ function.	quadratic

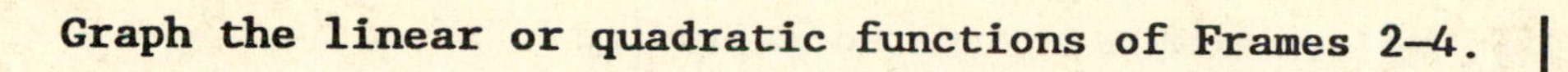

Graph the linear or quadratic functions of Frames 2–4.

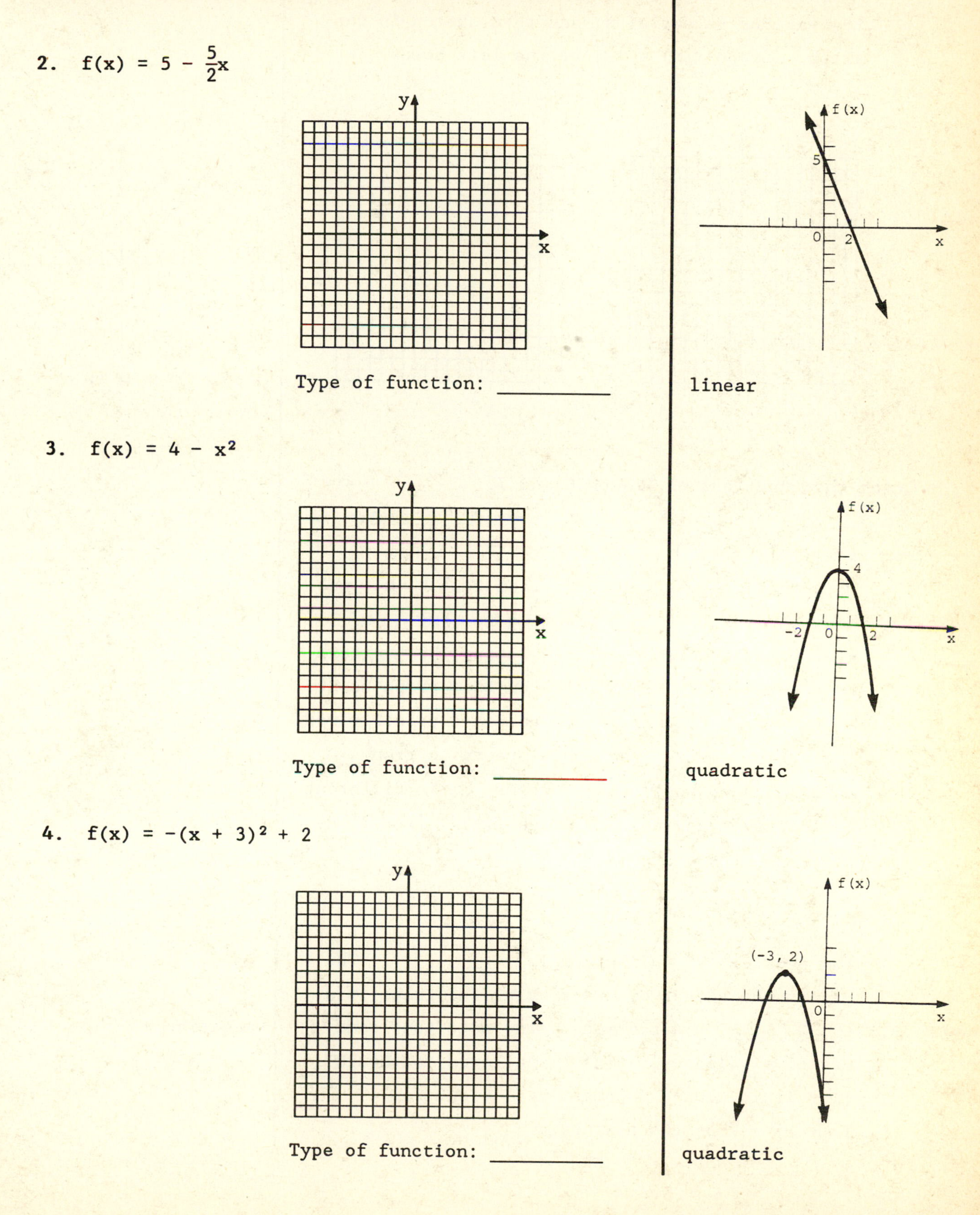

5. To graph $f(x) = |x|$, we can complete ordered pairs for various values of x, and then sketch in the graph. Graph $f(x) = |x|$ at the left below.

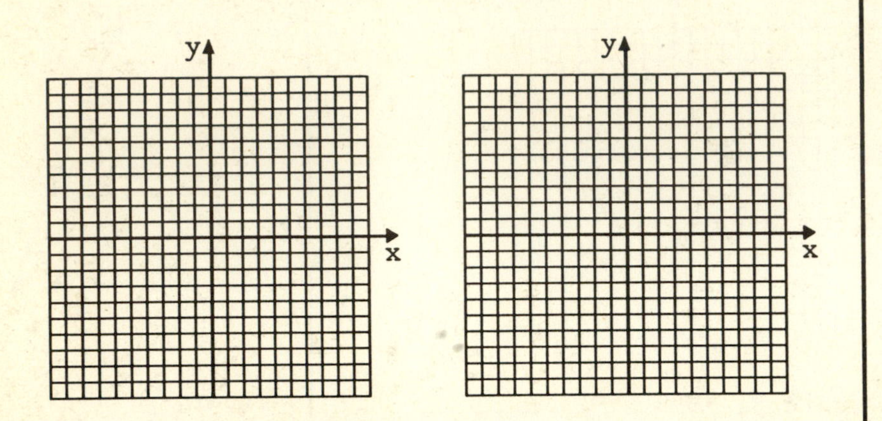

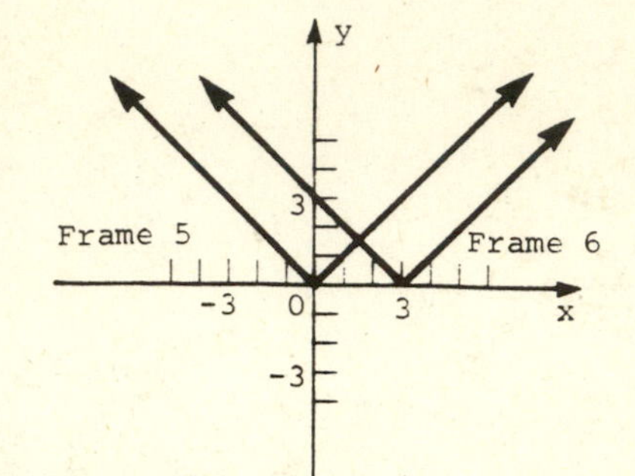

Graph the functions of Frames 6 and 7.

6. $f(x) = |x - 3|$
With the absolute value function $f(x) = |x|$, we know tht $f(x) \geq$ ____. Changing to $f(x) = |x - 3|$ will have the effect of translating the "vertex" on the x-axis. The "vertex" of this graph is at __________. Plotting a few selected points gives the graph. Notice that absolute value graphs have the same basic shape but may be in different positions on the plane. Graph $f(x) = |x - 3|$ at the right above.

0

(3, 0)

7. $f(x) = |x + 1| - 2$
The "vertex" of this graph is __________. The -2 has the effect of translating the graph in a __________ y direction. Also note that some negative values for f(x) are allowed, although $|x + 1|$ is always nonnegative. Graph $f(x) = |x + 1| - 2$ on the next page.

(-1, -2)

negative

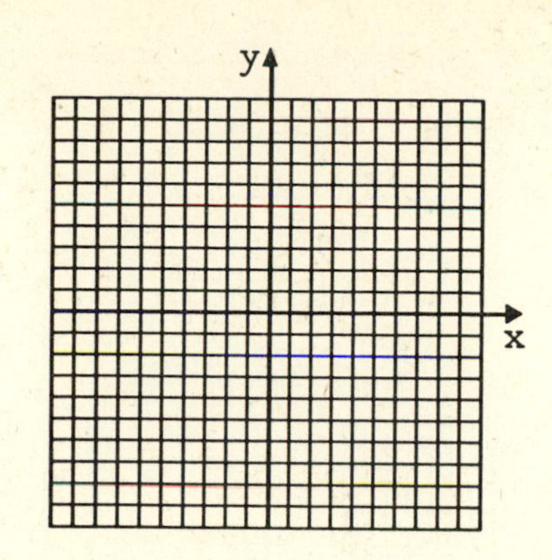

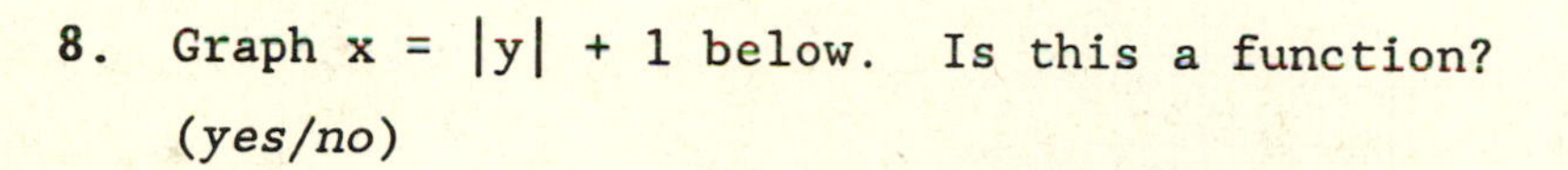

8. Graph $x = |y| + 1$ below. Is this a function? (*yes/no*)

no

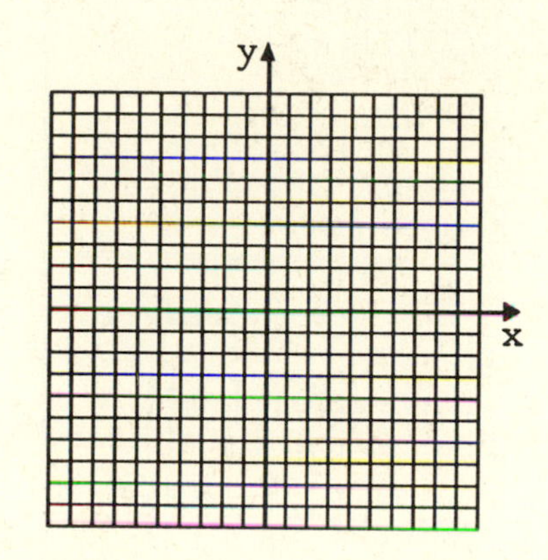

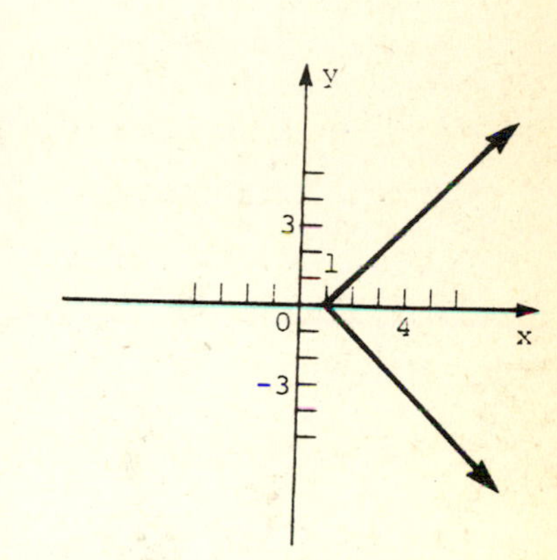

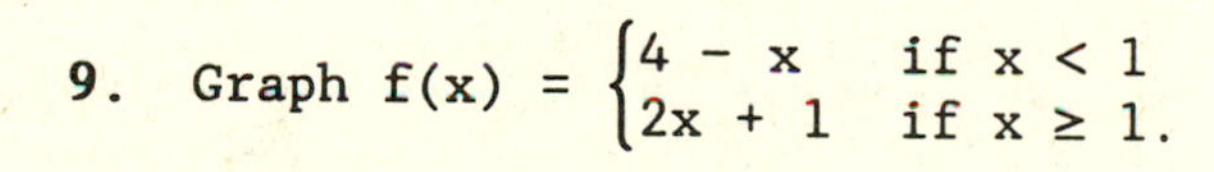

9. Graph $f(x) = \begin{cases} 4 - x & \text{if } x < 1 \\ 2x + 1 & \text{if } x \geq 1. \end{cases}$

Graph $f(x) = 4 - x$ for all values of x less than ____. Graph $f(x) = 2x + 1$ for all ____________.
Complete the graph below.

1; $x \geq 1$

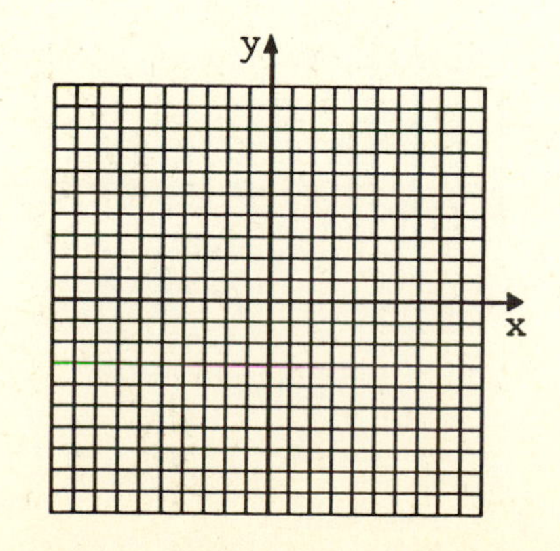

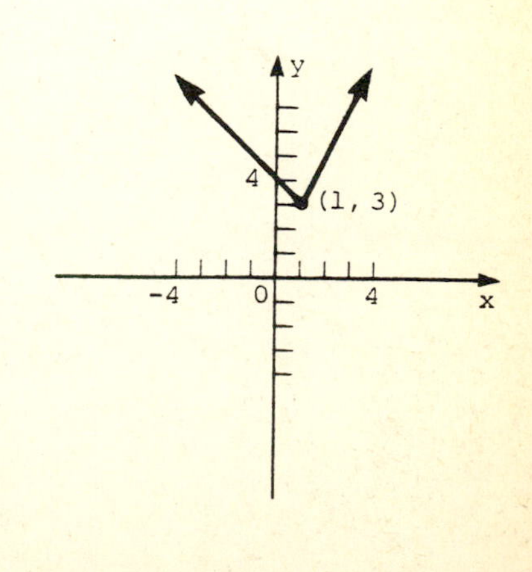

10. Graph $f(x) \begin{cases} x + 1 & \text{if } x \le 2 \\ x - 1 & \text{if } x > 2 \end{cases}$ on the following grid. Notice there is an open endpoint at

____________.

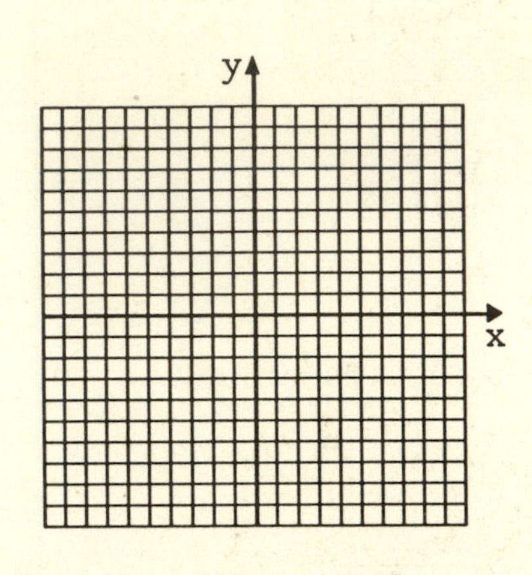

(2, 1)

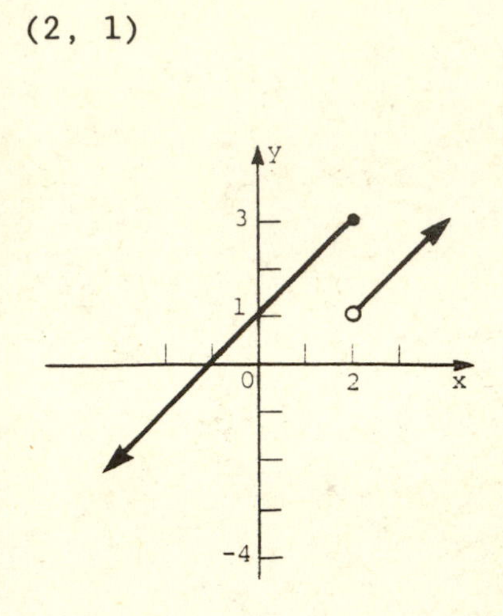

11. $f(x) = [\![x]\!]$ represents the ________________ integer ________ than or equal to x. For example, $[\![-1\frac{1}{4}]\!]$ = ______, $[\![3\frac{1}{8}]\!]$ = ______, and $[\![9]\!]$ = ______. Complete ordered pairs and graph $f(x) = [\![x + 1]\!]$ on the grid below.

greatest

less

-2; 3; 9

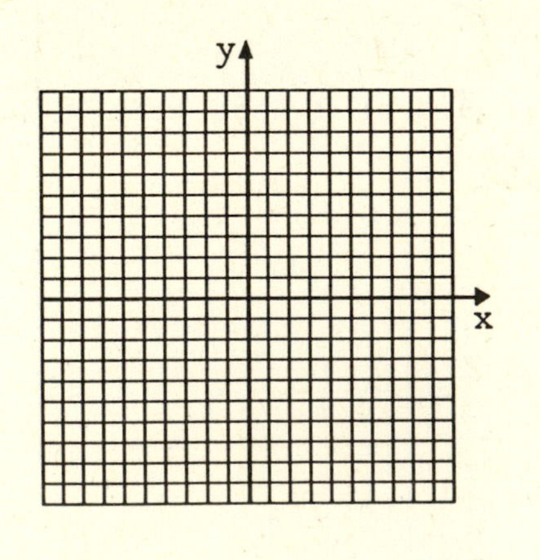

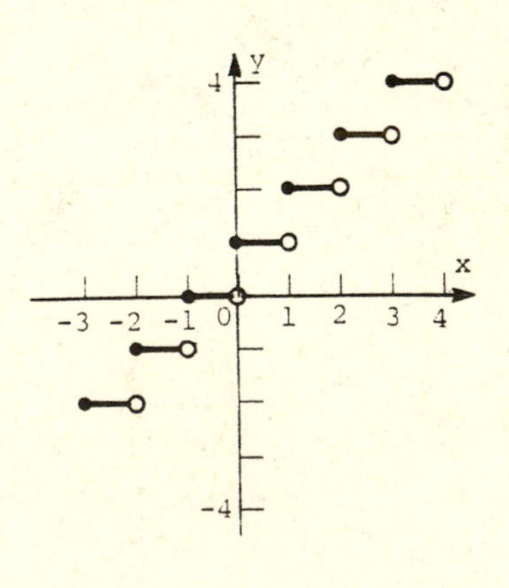

12. Graph $f(x) = 2[\![x]\!] + 1$ on the grid below.

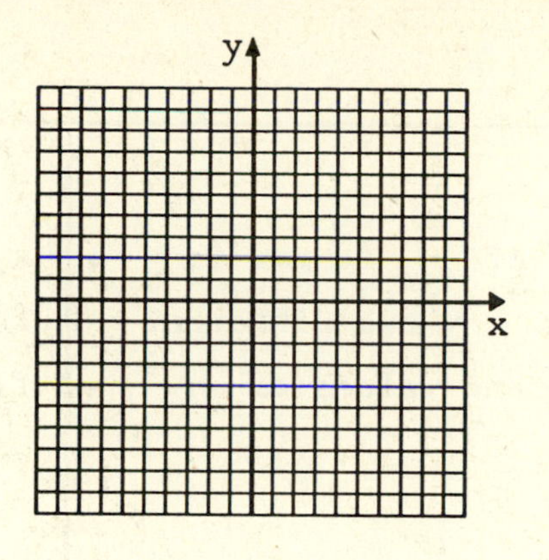

13. To deliver groceries, a grocery store charges \$2 plus \$1 for each block or portion of a block. Complete the graph (distance, cost), started below, for all distances up to five blocks.

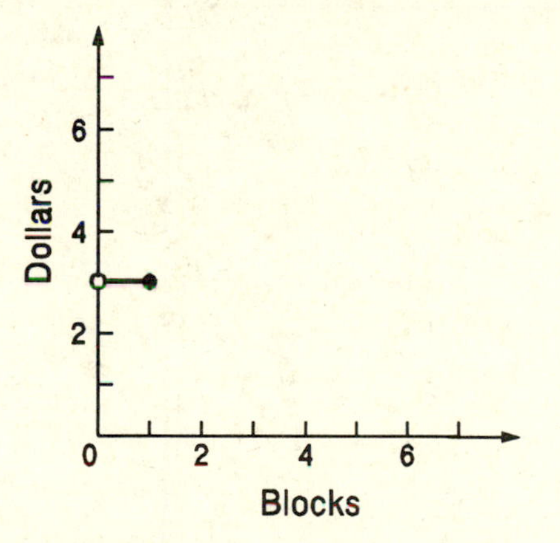

14. A rental firm charges a fixed \$7 for a rental, plus \$4 per day or any fraction of a day. Graph the function (days, cost) started below.

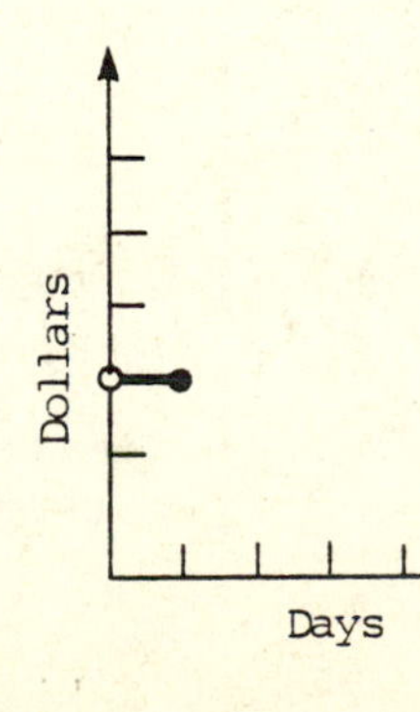

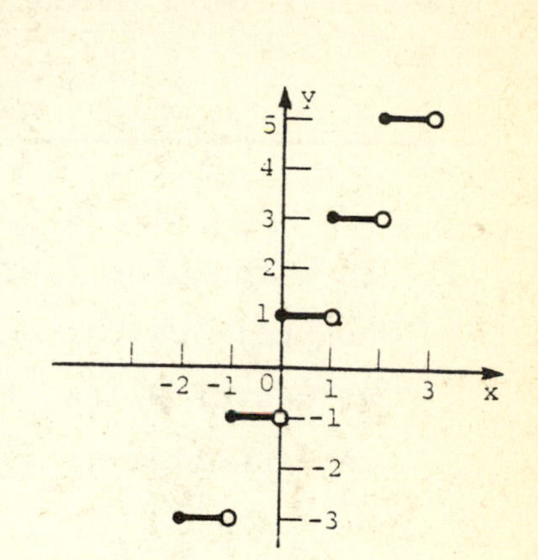

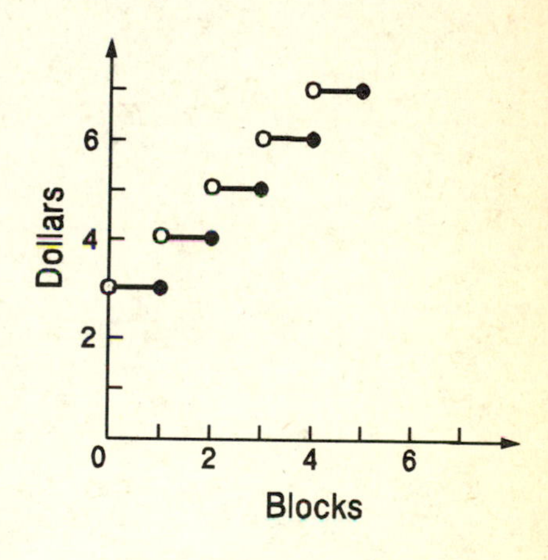

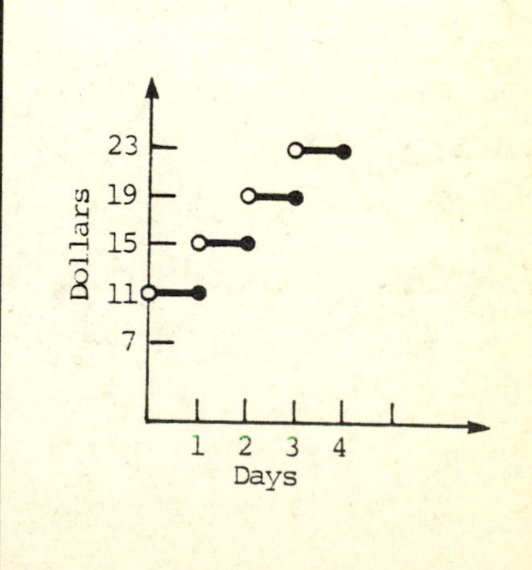

4.5 Inverse Functions

1. In a function, for each value of ____ there is exactly one value of ____. If any two different values of x lead to two different values of ____, the function is ______________.

   Answers: x; f(x); f(x); one-to-one

Identify the functions of Frames 2–4 that are one-to-one.

2. $f(x) = -6x + 5$ ______________

   Answer: one-to-one

3. $f(x) = x^2 - 6$ ______________

   Answer: not one-to-one

4. $f(x) = \sqrt[3]{x + 2}$ ______________

   Answer: one-to-one

5. A function is one-to-one if any ______________ line intersects the graph in no more than ______ point.

   Answers: horizontal; one

6. Which of these functions are one-to-one?

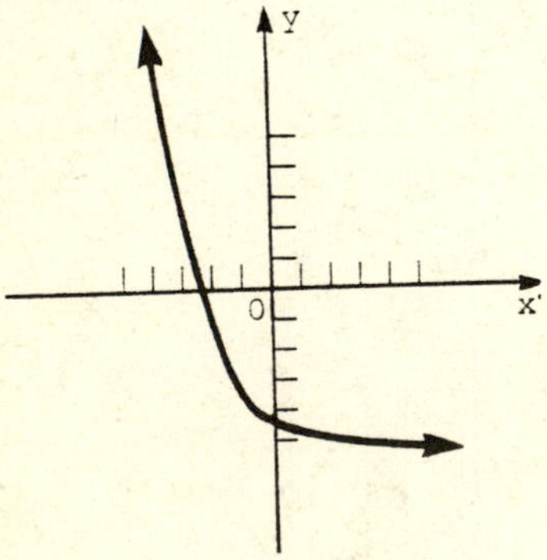

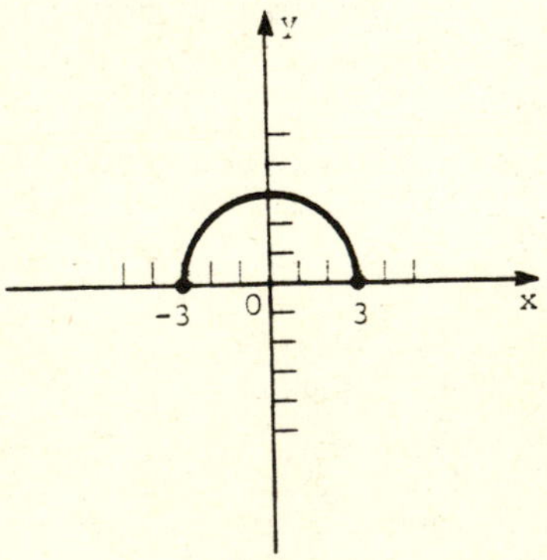

______________ ______________

   Answers: one-to-one; not-one-to-one

7. Recall the definition of the composition of two functions: if f and g are functions, then $(f \circ g)(x) =$ ________ and $(g \circ f)(x) =$ ________.	$f[g(x)]$; $g[f(x)]$
8. Functions f and g are ____________ if	inverses
$(f \circ g)(x) =$ _____ and $(g \circ f)(x) =$ _____.	x; x
9. A function f can have an inverse only if it is ________________.	one-to-one
10. Are $f(x) = 5x - 9$ and $g(x) = \frac{x + 9}{5}$ inverses of each other?	
To find out, find $(f \circ g)(x)$ and __________.	$(g \circ f)(x)$
$(f \circ g)(x) = f[g(x)] = 5(\_\_\_\_\_\_) - 9 =$ ____	$\frac{x + 9}{5}$; x
$(g \circ f)(x) = g[f(x)] = \frac{(\_\_\_\_\_\_) + 9}{5} =$ ____	$5x - 9$; x
Since both compositions equal x, the functions (*are/are not*) inverses.	are
11. Are $f(x) = \sqrt[3]{x + 6}$ and $g(x) = x^3 - 6$ inverses of each other? _______	yes
12. The function $f(x) = 5x - 17$ (*is/is not*)	is
one-to-one, and thus has an ________ function.	inverse
To find the equation of the inverse, replace f(x) with _____. Then, exchange ____ and ____:	y; x; y
Write $y = 5x - 17$ as ________________	$x = 5y - 17$
Solve for y: $y =$ _________	$\frac{x + 17}{5}$
Thus, $f^{-1}(x) =$ __________.	$\frac{x + 17}{5}$

Verify this as follows.	
$(f \circ f^{-1})(x) = 5(\_\_\_\_\_\_) - 17 = \_\_\_\_$	$\frac{x + 17}{5}$; x
$(f^{-1} \circ f)(x) = \frac{(\_\_\_\_\_\_) + 17}{5} = \_\_\_\_$	$5x - 17$; x
We have verified that $f^{-1}(x) = \_\_\_\_\_\_$	$\frac{x + 17}{5}$

Find the inverses of each one-to-one function in Frames 13–17. Be sure to restrict the domain of the inverse if necessary.

13. $f(x) = 3x + 10$ ______	$f^{-1}(x) = \frac{x - 10}{3}$
14. $f(x) = x^3 - 1$ ______	$f^{-1}(x) = \sqrt[3]{x + 1}$
15. $f(x) = x^2 - 4$ ______	not one-to-one
16. $f(x) = \sqrt{3 - x}$ ______	$f^{-1}(x) = 3 - x^2$
To find the domain of f, determine the values of x that will make ____ nonnegative. The domain	$3 - x$
is ______. The range of f is ______.	$(-\infty, 3]$; $[0, \infty)$
Since the range of f equals the domain of ____, f^{-1} must be given as	f^{-1}
$f^{-1}(x) = \_\_\_\_\_\_$, domain ______.	$3 - x^2$; $[0, \infty)$
17. $f(x) = -\sqrt{8 + x}$	
The function is ______ so it has an inverse.	one-to-one
$f^{-1}(x) = \_\_\_\_\_\_$, domain ______	$x^2 - 8$; $(-\infty, 0]$

CHAPTER 4 TEST

Test answers are at the back of this study guide.

Give the domain and range of each of the following.

1. $y = 2 - 3x$ 1. ______
2. $y = -|x - 7|$ 2. ______
3. $y = 2x^2 - 7$ 3. ______

Let $f(x) = 11 - 5x$ and $g(x) = 2x^2 - 7$. Find each of the following.

4. $(f + g)(x)$ 4. ______
5. $(f - g)(-2)$ 5. ______
6. $(fg)(0)$ 6. ______
7. $\left(\frac{f}{g}\right)(-1)$ 7. ______
8. $(f \circ g)(x)$ 8. ______
9. $(g \circ f)(x)$ 9. ______

Graph each function defined as follows.

10. $f(x) = |3 + x|$

11. $f(x) = [\![1 - x]\!]$

12. $f(x) = \begin{cases} 5 - x & \text{if } x \geq 2 \\ 4x - 5 & \text{if } x < 2 \end{cases}$

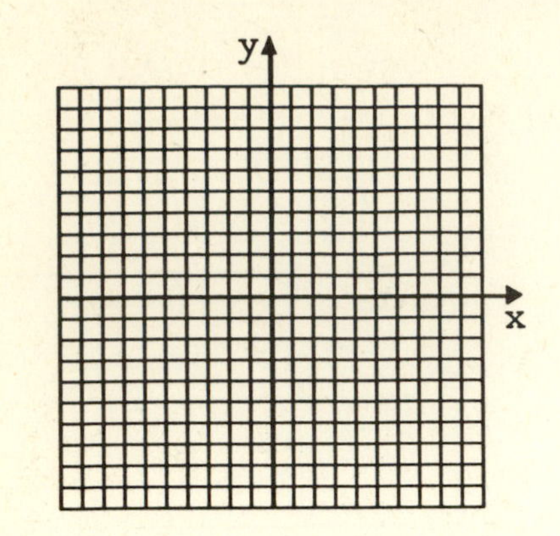

Which of the functions defined as follows are one-to-one?

13. $f(x) = -5 + 7x$ 13. ____________

14. $f(x) = \sqrt[3]{x - 3}$ 14. ____________

15. $f(x) = \sqrt{x^2 - 4}$ 15. ____________

For each of the following that defines a one-to-one function, write an equation for the inverse in the form $y = f^{-1}(x)$.

16. $f(x) = x^2 - 3$; domain $[0, \infty)$ 16. ____________

17. $f(x) = \dfrac{-3x + 1}{x - 2}$ 17. ____________

18. $f(x) = -\sqrt{x + 4}$ 18. ____________

19. Are the following two functions inverses of each other? Explain why or why not.

$f(x) = x^2 - 1$ $g(x) = \sqrt{x + 1}$ 19. ____________

20. The force needed to keep a motorcycle from skidding on a curve varies inversely as the radius of the curve and jointly as the weight of the motorcycle and the square of the speed. It takes 1500 lb of force to keep a 1000-pound motorcycle from skidding on a curve of radius 300 ft at 30 mph. What force is needed to keep the motorcycle from skidding on a curve of radius 300 ft at 50 mph?

20. ____________________

CHAPTER 5 PRETEST

Pretest answers are at the back of this study guide.

Solve each of the following equations.

1. $\left(\frac{1}{2}\right)^x = 4$ 1. ______________________

2. $\frac{1}{27} = x^{-3}$ 2. ______________________

Write each of the following in logarithmic form.

3. $4^{-3} = \frac{1}{64}$ 3. ______________________

4. $e^m = 59.1$ 4. ______________________

Write each of the following in exponential form.

5. $\log_8 \sqrt{32} = \frac{5}{6}$ 5. ______________________

6. $\ln 17 = v$ 6. ______________________

7. What is the y-intercept of the graph of $f(x) = a^x$? 7. ______________________

Graph each of the following functions.

8. $f(x) = 2^{-x}$

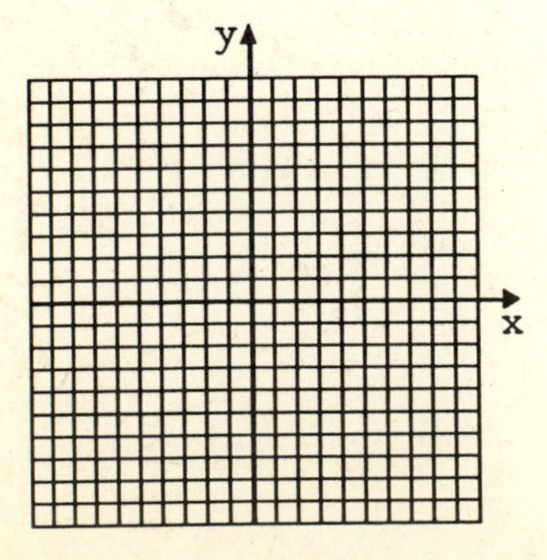

9. $f(x) = \log_3 (1 + x)$

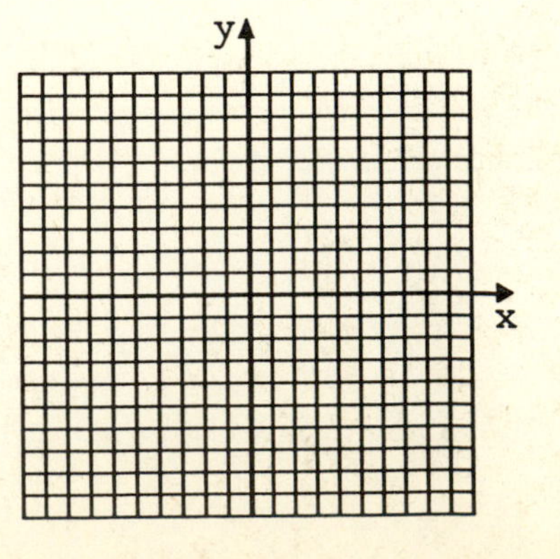

10. Use the properties of logarithms to write the following as a sum, difference, or product of logarithms.

$$\log_5 \frac{p^3\sqrt{q^5}}{3r^2}$$

10. ______________

Find each of the following logarithms. Round to the nearest thousandth.

11. $\ln 5809$

11. ______________

12. $\log_5 17$

12. ______________

13. Without solving the equation

$$\log (3x + 7) = -1,$$

give values of x that cannot be solutions of the equation.

13. ______________

Solve the following equations. Round answers to the nearest thousandths.

14. $.01 = 3^{2w+1}$

14. ______________

15. $\log_4 (1 - x) = 3$

15. ______________

16. $\ln 2x = \ln (x - 3) + \ln 5$

16. ______________

17. An area of Vermont finds its population growing according to the equation $P = 250e^{.1t}$, where t is the time in years.

(a) Find the initial population for the study.

17. (a) ______________

(b) Find the population in 5 yr.

(b) ______________

18. How much will $2750 amount to at 7% compounded continuously for 3 yr?

18. ____________________

19. If a population of insects increases continuously at a rate of 25% per month, how many months, to the nearest tenth, will it take the population to double in size?

19. ____________________

20. How many years, to the nearest tenth, will be needed for $3000 to increase to $5000 at 6% compounded semiannually?

20. ____________________

CHAPTER 5 EXPONENTIAL AND LOGARITHMIC FUNCTIONS

5.1 Exponential Functions

1. So far we have defined a^m only for __________ exponents. In this section we will extend this definition to include all ______ number exponents.	rational real
2. (a) If $a > 0$, $a \neq 1$, and x is any real number, we defined a^x so that a^x is a ________ _______ number.	unique real
(b) If $a > 0$ and $a \neq 1$, then $a^x = a^y$ if ______. Also, if $x = y$, then _________.	$x = y$ $a^x = a^y$
(c) If $a > 1$, then $a^m < a^n$ if ________.	$m < n$
(d) If $0 < a < 1$, then $a^m > a^n$ if ________.	$m < n$
3. Use these properties to solve the equation $2^x = 1/16$. First write 1/16 as $1/16 = 2^{\_\_}$. The equation thus becomes $2^{\_\_} = 2^{\_\_}$, from which x = ____. The solution set is ____.	-4 x; -4 -4; $\{-4\}$

Solve the equations in Frames 4–8.

4. $\left(\frac{2}{3}\right)^m = \frac{27}{8}$ Since $\frac{27}{8} = \left(\frac{2}{3}\right)^{\_\_}$, we have m = ____.	-3; -3
Solution set: _______	$\{-3\}$

5. $64^x = 4$	
Write 64 as ____ so we have the equation ________.	4^3; $4^{3x} = 4$
Use the property in Frame 2:	
$3x =$ ______	1
$x =$ ______	1/3
Solution set: ________	{1/3}
6. $\left(\frac{1}{4}\right)^x = 256$	
Since $\left(\frac{1}{4}\right)^x = 4^{—}$ and $256 = 4^{—}$, $x =$ ___.	$-x$; 4; -4
Solution set: ________	{-4}
7. $\left(\frac{1}{2}\right)^{-r} = 16$	
$\left(\frac{1}{2}\right)^{-r} = 2^{—}$; $16 = 2^{—}$. Thus, $r =$ ___.	r; 4; 4
Solution set: ________	{4}
8. $2^{1-a} = 16$	
$16 = 2^{—}$ so $1 - a =$ ___.	4; 4
$a =$ ________	-3
Solution set: ________	{-3}
9. Another type of equation can be solved using the properties of exponents. Solve $q^{2/3} = 25$.	
$q^{2/3} = 25$	
$(\sqrt[3]{q})^{—} = 25$	2
Take the square root of each side.	
_____ = _____	$\sqrt[3]{q}$; ±5
Cube both sides.	
$q =$ _____	±125
Check both solutions in the original equation. The solution set is ________________.	{-125, 125}

10. We can define a function

$$f(x) = a^x,$$

where $a > 0$, a ___ 1, as a(n) ________________ function.

$\neq$; exponential

11. To graph an exponential function such as

$$f(x) = 3^x,$$

first complete a table of values for the function, and then use these values to sketch the graph. Complete the table below, and then sketch the graph by drawing a smooth ________ through the points.

curve

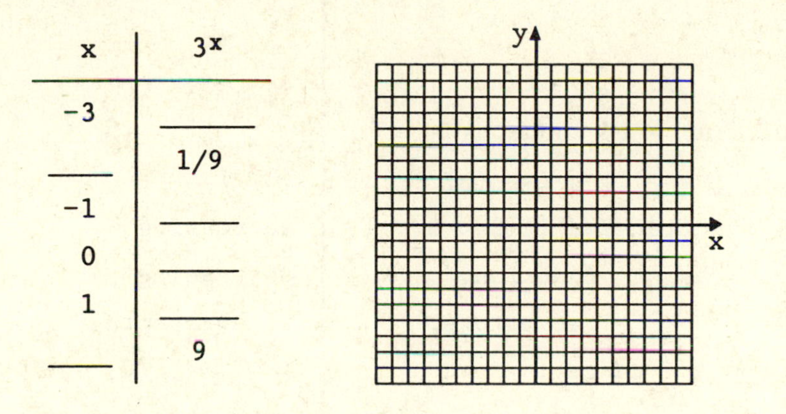

x	3^x
-3	____
____	1/9
-1	____
0	____
1	____
____	9

1/27
-2
1/3
1
3
2

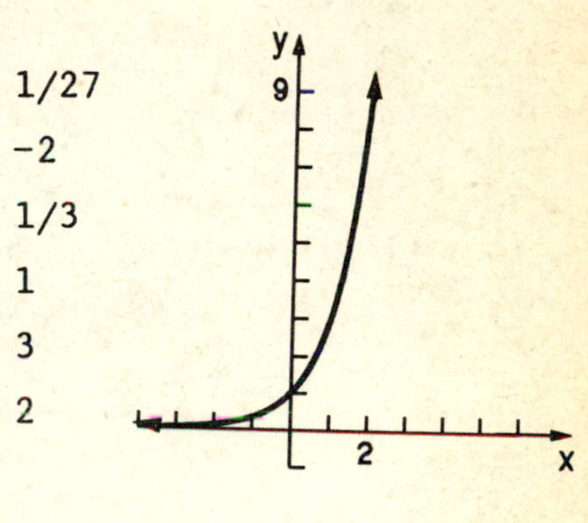

This graph is typical of $f(x) = a^x$ for ________.

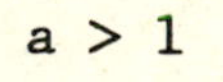
$a > 1$

12. To graph $f(x) = \left(\frac{1}{3}\right)^x$, complete a table of values again, as below. Sketch the graph.

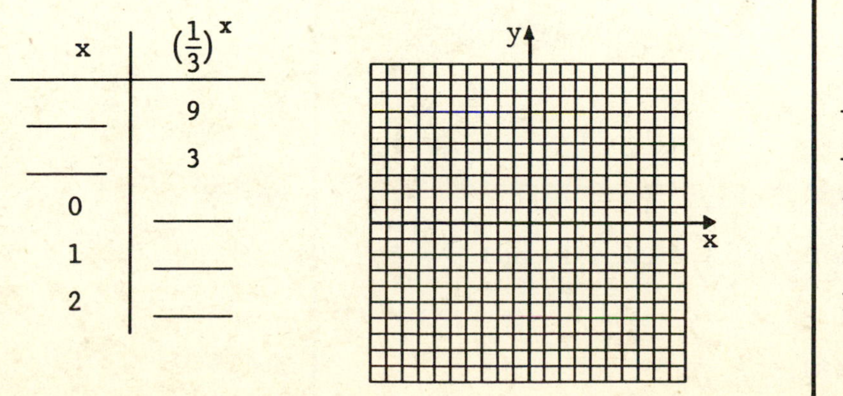

x	$\left(\frac{1}{3}\right)^x$
____	9
____	3
0	____
1	____
2	____

-2
-1
1
1/3
1/9

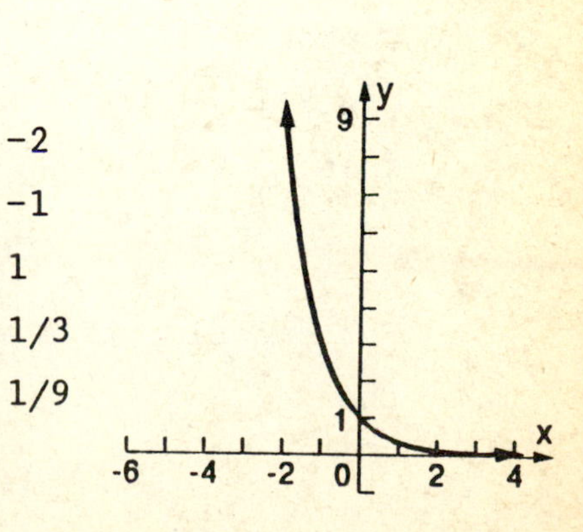

This graph is typical of the graph of $f(x) = a^x$ for ____ < a < ____.

0; 1

13. Graph $y = 4^{-x}$.

x	4^{-x}
-2	____
-1	____
0	____
1	____
2	____

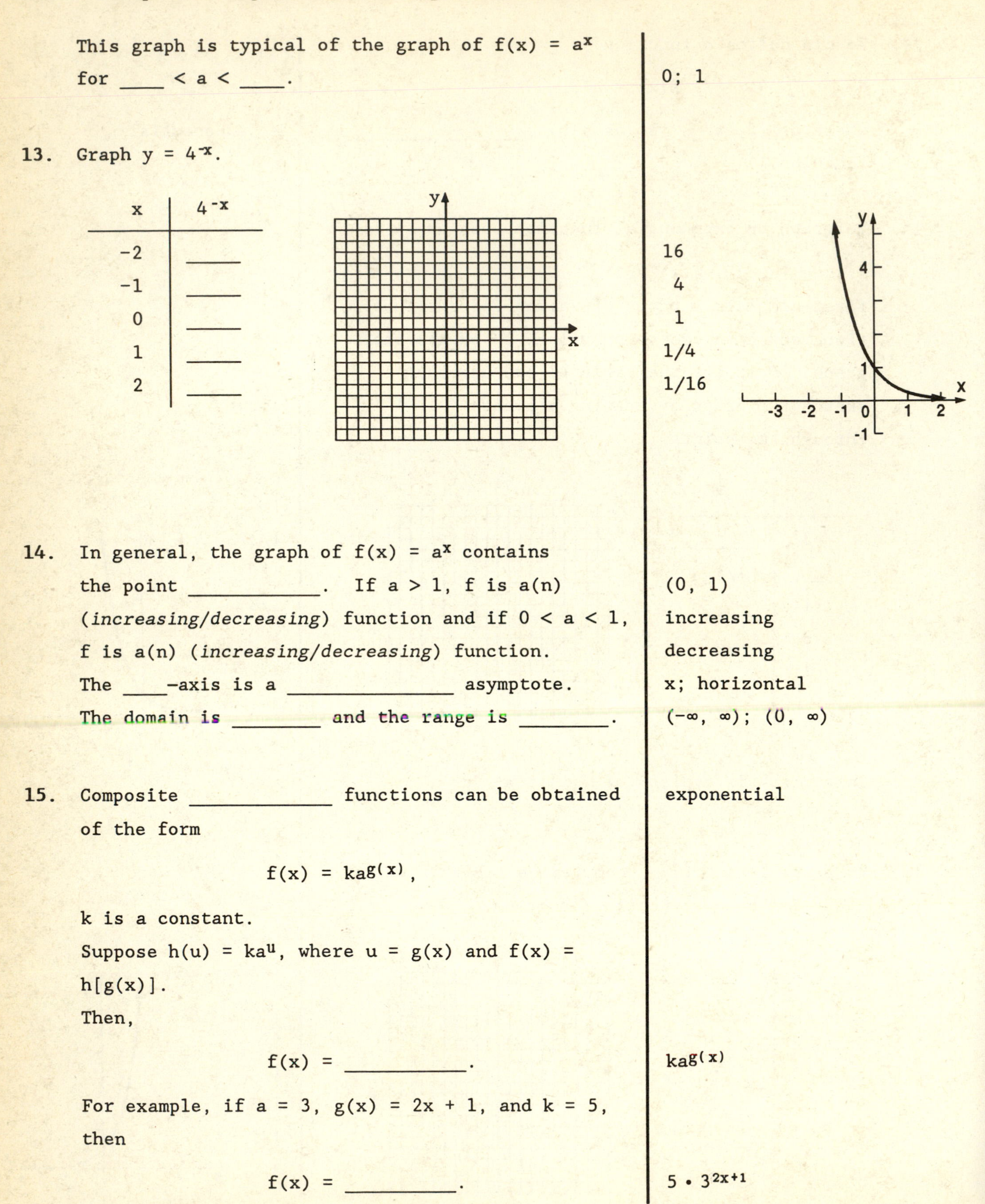

16
4
1
1/4
1/16

14. In general, the graph of $f(x) = a^x$ contains the point ____________. If $a > 1$, f is a(n) (*increasing/decreasing*) function and if $0 < a < 1$, f is a(n) (*increasing/decreasing*) function. The ____-axis is a ______________ asymptote. The domain is ________ and the range is ________.

(0, 1)
increasing
decreasing
x; horizontal
$(-\infty, \infty)$; $(0, \infty)$

15. Composite ____________ functions can be obtained of the form

$$f(x) = ka^{g(x)},$$

k is a constant.

exponential

Suppose $h(u) = ka^u$, where $u = g(x)$ and $f(x) = h[g(x)]$.
Then,

$$f(x) = \text{\_\_\_\_\_\_\_\_\_\_}.$$

$ka^{g(x)}$

For example, if $a = 3$, $g(x) = 2x + 1$, and $k = 5$, then

$$f(x) = \text{\_\_\_\_\_\_\_\_\_\_}.$$

$5 \cdot 3^{2x+1}$

16. Graph $f(x) = 3^{x-1}$.

The graph will have the same shape as the graph of $f(x) =$ _____. Because of the 1 subtracted from x, the graph will be translated 1 unit to the ________ compared to the graph of $f(x) = 3^x$. Thus the point ________ is on the graph instead of (0, 1). Plot a few additional points and draw the graph.

3^x

right

(1, 1)

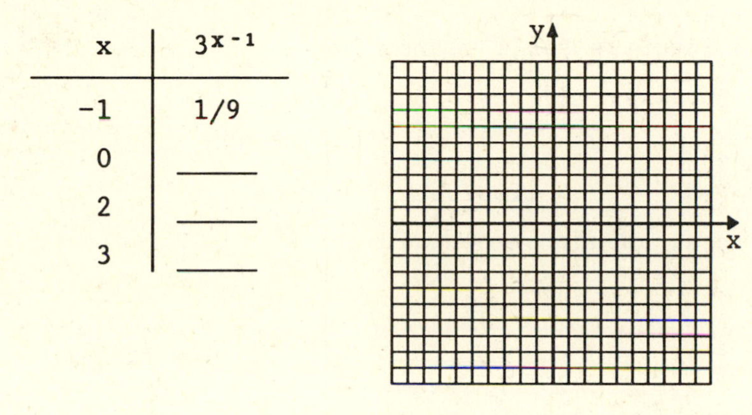

x	3^{x-1}
-1	1/9
0	_____
2	_____
3	_____

1/3

3

9

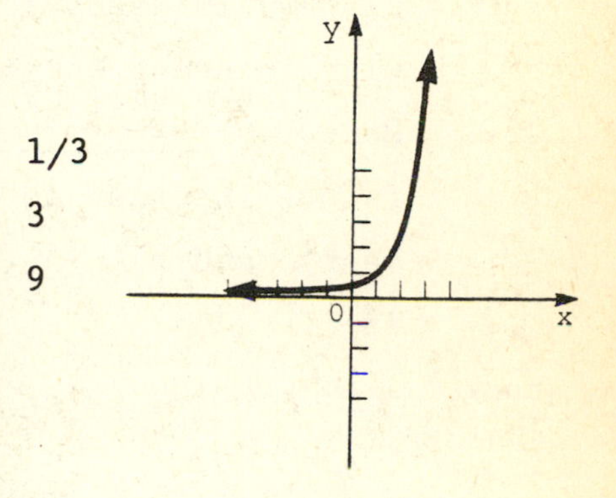

17. Sketch the graph of $f(x) = 2^{-x} - 1$ on the grid below.

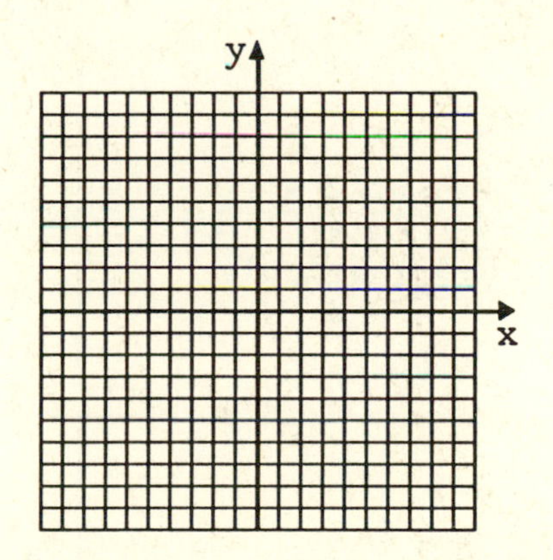

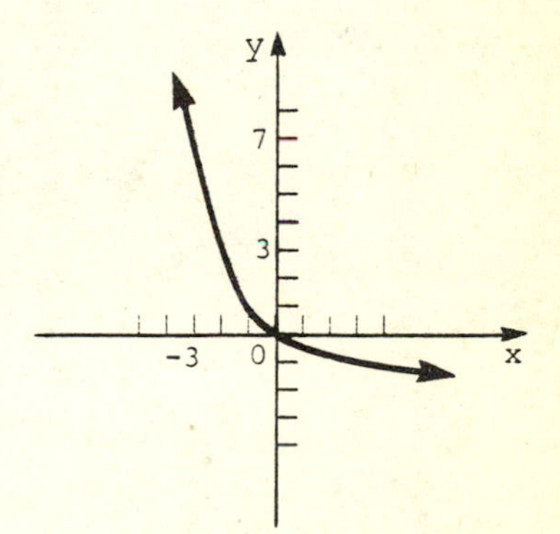

18. Find compound interest with the formula

$$A = P\left(1 + \frac{r}{m}\right)^{\_\_\_},$$

where A is the _______ amount on deposit, or __________ value, P is the _______________ or

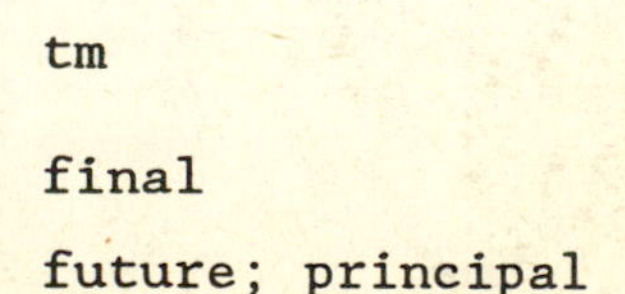

tm

final

future; principal

________ value, r is the interest rate per _______, m is the number of times per year that interest is ____________, and t is the number of ________.	present year compounded years

Find the compound interest in Frames 19–21. (Use a calculator.)

19. \$15,000 is deposited at 7% interest for 6 years, with interest compounded annually.	
Here P = ________, r = _______, m = _______	15,000; .07; 1
(since interest is compounded _______ time per	one
year), and t = ___. The total amount on deposit is	6
A = 15,000(________)$^{\_\_\_}$	1 + .07; 6(1)
= 15,000(1.07)$^{\_\_\_}$	6
Now use a calculator.	
A = 15,000(__________) = __________	1.500730352; 22,510.96
The interest earned is	
___________ - \$15,000 = __________.	\$22,510.96; \$7510.96
20. \$8500 is deposited at 8% compounded quarterly for 3 years.	
The rate of interest is 8% ÷ ____ = _____ per quarter.	4; 2% or .02
Future value (A) = ____________	\$10,780.06
Interest = ____________	\$2280.06
21. \$11,600 is deposited at 6% compounded monthly for 2 1/2 years.	
Future value (A) = ____________	\$13,472.24
Interest = ____________	\$1872.24

22. Find the present value if the future value is $20,000 with interest at 5% compounded semi-annually for 10 years.

A = ________, r = _____, t = _____, and m = ____. — 20,000; .05; 10; 2

Substitute these values into the formula.

$$\_\_\_\_\_\_\_\_ = P(1 + \_\_\_\_)^{\_\_\_\_}$$ — 20,000; .05/2; 20

$$P = \_\_\_\_\_\_\_\_$$ — $12,205.42

23. Find the required annual interest rate to the nearest tenth percent if a $3000 loan compounded quarterly for 3 years yields $5500.

A = ________, P = _____, m = _____, and t = ____. — 5500; 3000; 4; 3

Using the formula, we get

$$\_\_\_\_\_\_ = \_\_\_\_\_\_(1 + \frac{r}{\_\_\_})^{\_\_\_}.$$ — 5500; 3000; 4; 12

Divide both sides by ________. — 3000

$$\_\_\_\_\_\_ = (1 + \frac{r}{4})^{12}$$ — 11/6

Take the _____ root on both sides. — 12th

$$\_\_\_\_\_\_ = 1 + \frac{r}{4}$$ — 1.0518

$$\_\_\_\_\_\_ = \frac{r}{4}$$ — .0518

$$\_\_\_\_\_\_ = r$$ — .207

$$\_\_\_\_\_\_\% = r$$ — 20.7

24. Situations about growth or decay often involve the number ____. To eight significant figures, — e

$$e = \_\_\_\_\_\_\_\_\_\_.$$ — 2.7182818

25. Suppose the population of a city is given by	
$P(t) = 5000e^{.04t}$,	
where t is time in years. Find the population for the following values of t.	
(a) $t = 0$	
$P(0) = 5000e^{\_\_\_}$	$.04(0)$
$= 5000e^{\_\_\_}$	0
$= 5000(\_\_\_)$	1
$= \_\_\_\_\_\_$	5000
The population is ______ when $t = 0$.	5000
(b) $t = 5$	
$P(5) = 5000e^{\_\_\_}$	$.04(5)$
$= 5000e^{.2}$	
Find a value of $e^{.2}$ from a calculator.	
$e^{.2} \approx$ ____________ and	1.22140
$P(5) = 5000(\_\_\_\_\_\_) \approx \_\_\_\_\_\_$.	1.22140; 6107
The population in 5 years will be about ________ people.	6100

5.2 Logarithmic Functions

1. The exponential statement $x = a^y$ can be converted to the ______________ statement	logarithmic
______________,	$y = \log_a x$
read "y is the __________ of x to the base _____."	logarithm; a

Convert each statement in Frames 2–6.

2. $4^3 = 64$	$\log_4$ _____ = _____	64; 3
3. $9^{-2} = \frac{1}{81}$	______________	$\log_9 \frac{1}{81} = -2$
4. $10^3 = 1000$	______________	$\log_{10} 1000 = 3$
5. $\log_2 32 = 5$	______________	$2^5 = 32$
6. $\log_{1/3} 27 = -3$	______________	$\left(\frac{1}{3}\right)^{-3} = 27$

Solve the equations of Frames 7–10.

7. $\log_3 x = 4$

Write the statement in exponential form as _____, — $3^4 = x$

from which x = ____. — 81

Solution set: ______________ — $\{81\}$

8. $\log_x \frac{1}{4} = 2$ Solution set: ______________ — $\{1/2\}$

9. $\log_{25} 125 = m$ Solution set: ______________ — $\{3/2\}$

10. $\log_z 4 = 3$ Solution set: ______________ — $\{\sqrt[3]{4}\}$

Graph the functions of Frames 11 and 12. Complete ordered pairs as needed.

11. $f(x) = \log_2 x$

Write $f(x) = \log_2 x$ in exponential form.

__________________ — $2^y = x$

Choose values for f(x) and find x for ordered pairs:

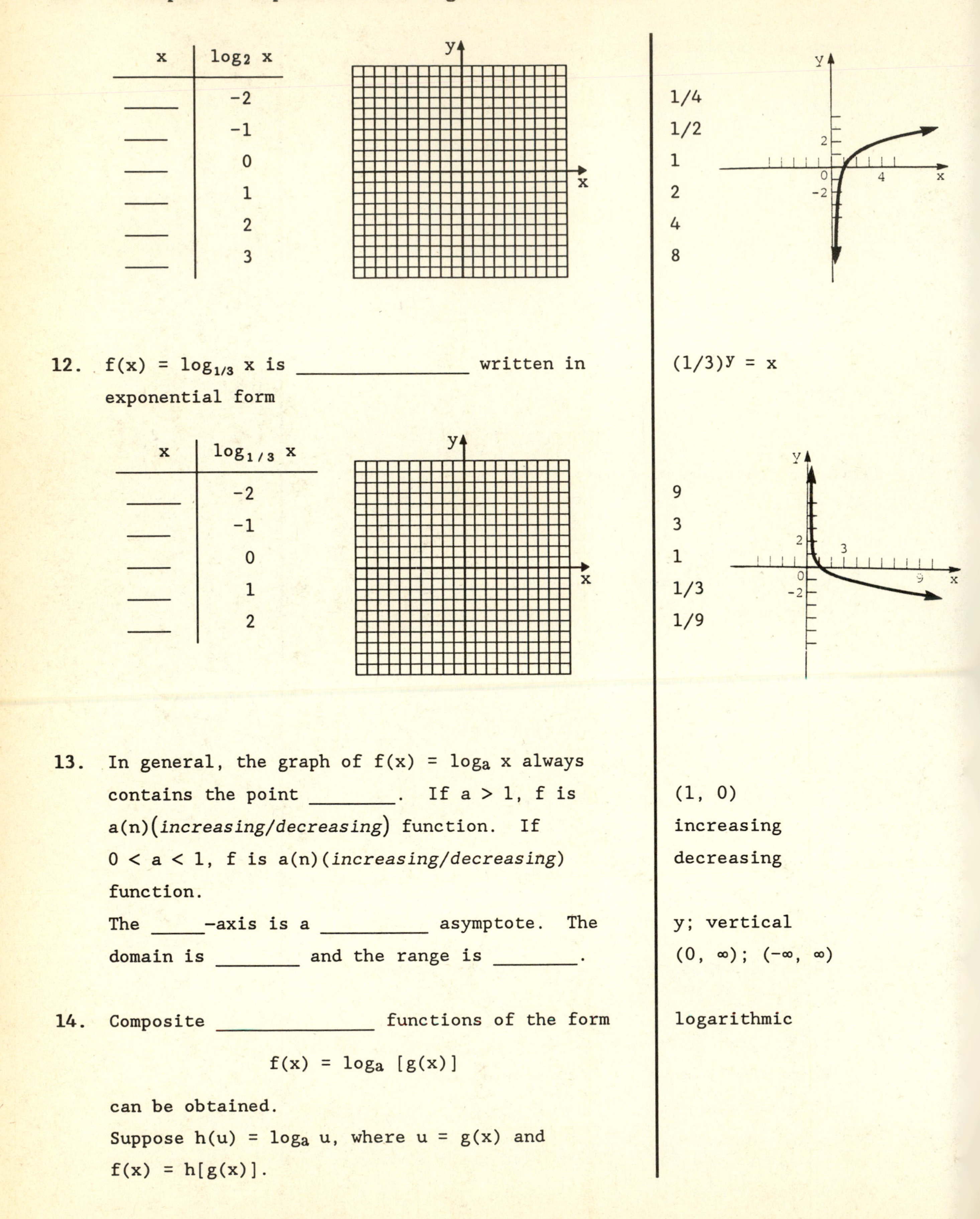

x	$\log_2 x$
____	-2
____	-1
____	0
____	1
____	2
____	3

1/4
1/2
1
2
4
8

12. $f(x) = \log_{1/3} x$ is ______________ written in exponential form

$(1/3)^y = x$

x	$\log_{1/3} x$
____	-2
____	-1
____	0
____	1
____	2

9
3
1
1/3
1/9

13. In general, the graph of $f(x) = \log_a x$ always contains the point ________. If $a > 1$, f is a(n)(*increasing/decreasing*) function. If $0 < a < 1$, f is a(n)(*increasing/decreasing*) function.
The _____-axis is a __________ asymptote. The domain is ________ and the range is ________.

(1, 0)
increasing
decreasing

y; vertical
$(0, \infty)$; $(-\infty, \infty)$

14. Composite ______________ functions of the form

$$f(x) = \log_a [g(x)]$$

can be obtained.
Suppose $h(u) = \log_a u$, where $u = g(x)$ and $f(x) = h[g(x)]$.

logarithmic

Then

$f(x) = $ ____________. | $\log_a [g(x)]$

For example, if $a = 2$ and $g(x) = 3x - 1$, then

$f(x) = $ ____________. | $\log_2 (3x - 1)$

15. Graph $f(x) = \log_3 (x + 1)$.
This graph has the same shape as the graph of ____________ but shifted one unit to the ______. | $f(x) = \log_3 x$; left

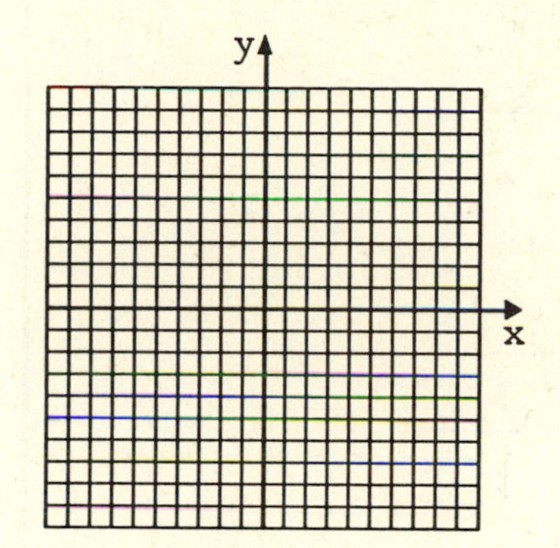

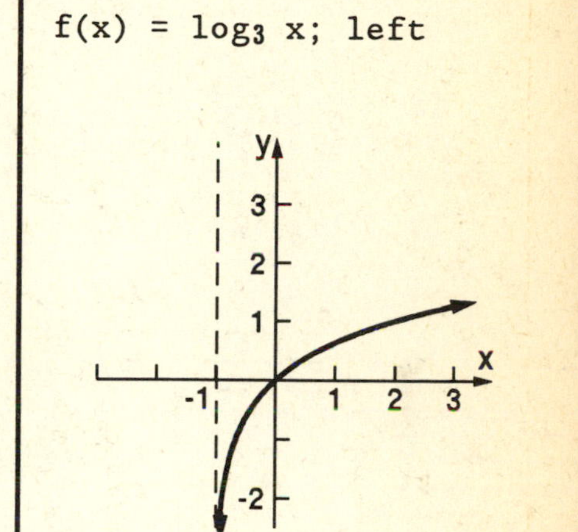

16. Graph $f(x) = \log_2 |x|$.

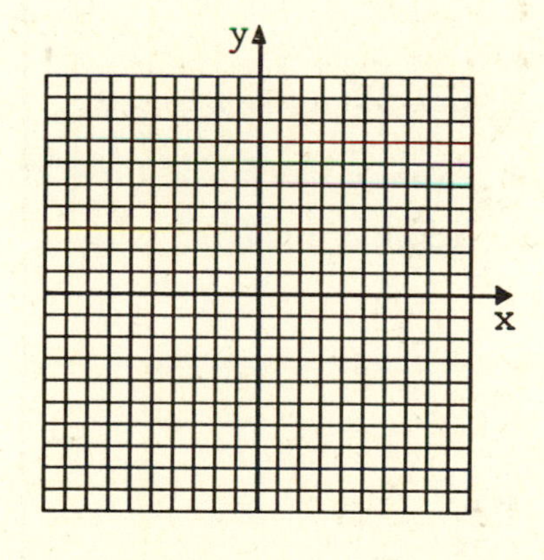

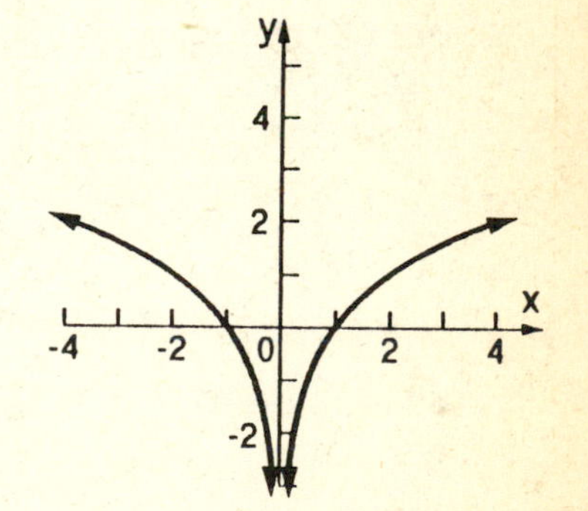

17. Many of the applications of logarithms are based on the following __________ of logarithms. | properties
If x and y are positive, a is any positive real number except ____, and r is any real number, | 1

$\log_a xy =$ ______ + ______	$\log_a x$; $\log_a y$
$\log_a \frac{x}{y} =$ ______ − ______	$\log_a x$; $\log_a y$
$\log_a x^r =$ ______	$r \cdot \log_a x$
$\log_a a =$ ______	1
$\log_a 1 =$ ______.	0

Use the properties of logarithms to simplify the expressions in Frames 18–22.

18. $\log_4 \frac{13}{9} =$ ______	$\log_4 13 - \log_4 9$
19. $\log_7 9 \cdot 4 =$ ______	$\log_7 9 + \log_7 4$
20. $\log_5 \frac{5y}{x} = \log_5 5 + \log_5 y$ ______	$-\log_5 x$
$=$ ______ $+ \log_5 y - \log_5 x$	1
21. $\log_3 \sqrt{5} = \log_3 5^{\_\_\_\_} =$ ______	$\frac{1}{2}$; $\frac{1}{2} \cdot \log_3 5$
22. $\log_9 (8x - 11y) =$ ______	cannot be simplified

Write the expressions of Frames 23 and 24 as a single logarithm.

23. $\log_a 3 + \log_a m - 2 \log_a y =$ ______	$\log_a \frac{3m}{y^2}$
24. $\frac{1}{2} \log_5 z^4x^6 - \frac{3}{4} \log_5 r^8m^{12} =$ ______	$\log_5 \frac{z^2x^3}{r^6m^9}$

Suppose $\log_{10} 7 = .8451$ and $\log_{10} 8 = .9031$. Evaluate Frames 25–28.

25. $\log_{10} 56 = \log_{10}$ (______)	$7 \cdot 8$
$= \log_{10} 7 +$ ______	$\log_{10} 8$
$=$ ______ + ______	.8451; .9031
$=$ ______	1.7482

26. $\log_{10} 49$ = _______	1.6902
27. $\log_{10} \frac{8}{7}$ = _______	.0580
28. $\log_{10} 70 = \log_{10}$ (_______)	7 • 10
= $\log_{10} 7 + \log_{10} 10$ = _______	1.8451
29. To obtain two useful properties in the theorem on inverses, use compositions of ___________ and ___________ functions. This theorem says that for positive a where a ____ 1,	exponential logarithmic ≠
$a^{\log_a x}$ = ____ and $\log_a a^x$ = ____.	x; x

5.3 Evaluating Logarithms; Change of Base

1. Logarithms to base _____ are called ___________ logarithms.	10; common
2. An abbreviation for $\log_{10} x$ is _____________.	log x
3. Since $10^3 = 1000$, we have log 1000 = _____.	3
4. Also, $\log 10^{-4}$ = _____.	-4
5. To find the logarithm of a number which is not a power of _____, use a calculator with a log key. To four decimal places,	10
log 8.41 = _________.	.9248
(By the definition of logarithm, $10^{\_\_\_\_}$ = _____).	.9248; 8.41

Find the logarithms in Frames 6–8 to four decimal places.

6. log 63 __________ — 1.7993

7. log .123 __________ — -.9101

8. log 200 __________ — 2.3010

9. Common logarithms are used in chemistry to find the pH of a solution, defined as pH = ______. A pH value over 7.0 indicates ______ while less than 7.0 indicates ______. $[H_3O^+]$ is the concentration of ______ ____ in the solution.

   $-\log[H_3O^+]$
   alkalinity
   acidity
   hydronium ions

10. Find the pH of a solution with $[H_3O^+] = 2.4 \times 10^{-8}$.

$$\begin{aligned} pH &= -\log(\_\_\_\_\_\_) \\ &= -(\log \_\_\_ + \log \_\_\_) \\ &= -(\_\_\_\_\_\_) \\ &= \_\_\_\_\_\_ \end{aligned}$$

   2.4×10^{-8}
   $2.4;\ 10^{-8}$
   $.3802 - 8$
   7.6198 or 7.6

11. Find the hydronium ion concentration of a solution with pH = 5.4.

$$\begin{aligned} \_\_\_\_\_\_ &= -\log\ [H_3O^+] \\ \_\_\_\_\_\_ &= \log\ [H_3O^+] \end{aligned}$$

   Use the INV and log keys to get ______.

   5.4
   -5.4
   .000004

12. Another important system of logarithms uses the base e, where, to seven decimals,

$$e = \_\_\_\_\_\_.$$

   Logarithms to base e are called ______ logarithms. The natural logarithm of x is written ______.

   2.7182818
   natural
   ln x

13. Many natural logarithms can be found with the ln key of a calculator. For example, to four decimal places,

$$\ln 3.5 = \_\_\_\_\_\_.$$

1.2528

Find the natural logarithms in Frames 14–16 to four decimal places.

14. $\ln 70 = \_\_\_\_\_\_$

4.2485

15. $\ln .9 = \_\_\_\_\_\_$

-.1054

16. $\ln 1.8 = \_\_\_\_\_\_$

.5878

17. For bases of logarithms other than 10 or e, use the ______ of base theorem: If x is any positive number, then

$$\log_a x = \_\_\_\_\_\_.$$

change

$\dfrac{\log_b x}{\log_b a}$

18. Use this theorem to find $\log_7 25$.

With b = ____,

$$\log_7 25 = \frac{\ln 25}{\_\_\_\_}$$

e

ln 7

By calculator,

$$\ln 25 = \_\_\_\_\_\_ \text{ and } \ln 7 = \_\_\_\_\_\_$$

3.2189; 1.9459

Now divide to get

$$\log_7 25 = \_\_\_\_\_\_.$$

1.65

Find the logarithms in Frames 19–22.

19. $\log_8 65 = \dfrac{\ln 65}{\_\_\_\_}$

$= \dfrac{4.1744}{\_\_\_\_}$

$= \_\_\_\_$

ln 8

2.0794

2.01

20. $\log_{15} 3$ = _____	.41
21. $\log_{12} 7$ = _____	.78
22. $\log_{1/2} 9$ = _____	-3.17

5.4 Exponential and Logarithmic Equations

1. To solve equations involving logarithms and exponents, we need two properties. Each has two parts.	
(i) For all $b > 0$ and $b \neq 1$:	
(a) $x = y$ if $b^{—} = b^{—}$.	x; y
(b) $b^x = b^y$ if ___ = ___.	x; y
(ii) For all $b > 0$, $b \neq 1$, and all __________ real numbers x and y:	positive
(a) $x = y$ if $\log_b$ ___ = $\log_b$ ___.	x; y
(b) $\log_b x = \log_b y$ if ___ = ___.	x; y
2. Solve $5^x = 15$.	
Since 5 and 15 are not both powers of the same number, we cannot use Property (i/ii). But we	i
can use Property (i/ii). We take base 10 logarithms of both sides. We have	ii
log _____ = log _____	5^x; 15
or _____ log _____ = log _____.	x; 5; 15
From this $x = \dfrac{\log \_\_\_}{\log \_\_\_}$	15 5
After dividing, x = ________.	1.6826
Solution set: ____________	{1.6826}

3. Solve $3^{x+1} = 20$.

(________) log _____ = log _____ | x + 1; 3; 20

or x + 1 = ________ | $\frac{\log 20}{\log 3}$

$x = \frac{\log \_\_\_}{\log \_\_\_} - 1$ | 20; 3

= ________ - 1 | 2.7268

= ________ | 1.7268

Solution set: ____________ | {1.7268}

4. Solve the equation $2^{y+1} = 15$.

Solution set: ____________ | {-2.9069}

5. To solve log (x + 3) = log (3x + 5), we use part (b) of Property (ii) to write

________ = ________ | x + 3; 3x + 5

or x = _____. | -1

Solution set: _____ | {-1}

6. To solve log (3x - 1) - log x = log 2, use the properties of logarithms to write

log ____________ = log 2 | $\frac{3x - 1}{x}$

This leads to the equation

________ = 2 | $\frac{3x - 1}{x}$

from which x = ____. | 1

Solution set: _____ | {1}

7. To solve $\log_3 x + \log_3 (x - 2) = 1$, first write $1 = \log_3$ ____. Since | 3

$\log_3 x + \log_3 (x - 2) = \log_3$ ________. | x(x - 2)

the original equation can be written as	
$\log_3$ ________ = $\log_3$ ____.	x(x - 2); 3
or ________ = _____.	x(x - 2); 3
Expanding, we get	
________________ = 0.	$x^2 - 2x - 3$
The solutions of this equation are x = _____ or	3
x = _____. Now check these proposed answers in	-1
the original equation. If we let x = 3,	
$\log_3$ ____ + $\log_3$ ____ = 1,	3; 1
which is (*true/false*). If we try x = -1, we	true
see that log x, or log (-1), does not exist.	
Hence the solution set here is _____.	{3}
8. To solve log (x + 2) + log (x - 1) = 1, first	
note that 1 = log _____. Thus	10
log (_______)(_______) = log _____	x + 2; x - 1; 10
or (_______)(_______) = _____.	x + 2; x - 1; 10
Simplifying,	
________________ = 0,	$x^2 + x - 12$
from which x = ____ or x = ____. The only valid	-4; 3
solution for the given equation is x = ____. So	3
the solution set is _____.	{3}
9. To solve log x + log (x + 2) = 1, since	
1 = log 10, write the quadratic equation	
________________ = 0.	$x^2 + 2x - 10$
This equation can be solved by the quadratic	
formula.	
$x = \frac{-(\quad) \pm \sqrt{(\quad)^2 - 4(\quad)(\quad)}}{2(\quad)}$	2; 2; 1; -10 1
x = ________ or x = ________	$-1 + \sqrt{11}$; $-1 - \sqrt{11}$

The only valid solution is ________.	$-1 + \sqrt{11}$
Solution set: ________	$\{-1 + \sqrt{11}\}$

Solve the equations in Frames 10–17.

10. $\log x + \log 2x = 2$	
log ____ = log ____	$2x^2$; 100
x^2 = ____	50
The only valid answer here is x = ____.	$5\sqrt{2}$
Solution set: ______	$\{5\sqrt{2}\}$
11. $\log (x + 2) + \log (x + 5) = 1$	
Solution set: ______	$\{0\}$
12. $\log (x - 1) = \log (x + 2)$	
Solution set: ______	$\emptyset$
13. $\log r + \log (2r + 1) = 1$	
Solution set: ______	$\{2\}$
14. $\log x = 3 - \log (x + 1)$	
Solution set: ______	$\{31.1\}$
15. $\log (x - 2) = 2 - \log x$	
Solution set: ______	$\{11.05\}$
16. $\log_3 p + \log_3 (2p + 1) = 3$	
Use properties of logarithms to get	
$\log_3$ ________ = 3.	$p(2p + 1)$
By the definition of logarithms,	
$p(2p + 1)$ = ____.	3^3
The solution set is ______.	$\{3.43\}$

17. $\log_2 x - \frac{1}{2}\log_2 (x + 2) = 0$

Solution set: ________ | $\{2\}$

5.5 Exponential Growth and Decay

1. At least in theory, interest on a bank deposit could be compounded every instant. The formula for this ____________ compounding is

$$A = \_\_\_\_\_\_\_\_,$$

where P is __________ ________, t is the number of ______, and r is the interest rate per _____.

continuous

Pe^{rt}

present value

years; year

Find the compound interest in Frames 2–4.

2. \$12,000 at 10% compounded continuously for 5 years.

$$A = P \_\_\_\_\_\_\_\_$$
$$= (\_\_\_\_\_\_\_\_)e^{.10(\_\_)}$$
$$= 12{,}000e^{\_\_}$$
$$= 12{,}000(\_\_\_\_\_\_\_\_\_\_)$$
$$= \_\_\_\_\_\_\_\_\_\_$$

e^{rt}

12,000; 5

.5

1.648721271

19,784.66

Now find the interest.

Interest = __________ - \$12,000
= __________

\$19,784.66

\$7784.66

3. \$3150 at 8% compounded continuously for 10 years.

Interest = __________

\$3860.45

4. $19,200 at 6.8% compounded continuously for 4 1/2 years.	
Interest = ________	$6873.26
5. How long would it take for the money in an account that is compounded continuously at 6% interest to triple?	
Here, we want to solve for ____. Since we want to triple the principal, let A = ___ P.	t 3
$\_\_\_\_ = Pe^{\_\_\_\_}$	3P; .06t
$\_\_\_\_ = \_\_\_\_$	3; $e^{.06t}$
Take natural logarithms on both sides.	
$\ln 3 = \ln \_\_\_\_$	$e^{.06t}$
$\ln 3 = \_\_\_\_$	.06t
$\frac{\_\_\_\_}{\_\_\_\_} = t$	ln 3 .06
$\_\_\_\_ \approx t$	18.3102
It will take about ____ years for the amount to triple.	18
6. Suppose the population of a city is given by $$P(t) = 25,000e^{.04t}$$ **where t is time in years. How long will it take for the population of the city to triple?**	
At time t = 0, the population of the city is	
$P(\_\_\_\_) = 25,000e^{.04(\_\_\_\_)}$	0; 0
$= 25,000(\_\_\_\_)$	e^0 or 1
$= \_\_\_\_.$	25,000
When the population triples, it will equal	
____ (25,000) = ________.	3; 75,000

To find the time for the population to triple, we must solve for t in the equation	
$\_\_\_\_\_\_ = 25{,}000e^{.04t}$	75,000
First, divide both sides by ______ to get	25,000
______ = ______	3; $e^{.04t}$
Now take natural logarithms of each side.	
ln 3 = ln ______	$e^{.04t}$
ln 3 = ______	.04t
$\frac{\_\_\_}{\_\_\_} = t$	ln 3 .04
______ ≈ t	27.5
The population will triple in about _____ years.	27.5
7. **A certain radioactive substance is decaying exponentially. If there are 400 grams present initially and 100 grams after 5 days, how much is left after 7 days?**	
Use the equation	
$A(t) = A_0e^{kt}$,	
where A(t) is the amount present after ___ days	t
and A_0 is the amount present ______.	initially
Solve the equation	
100 = ______	$400e^{5k}$
for k.	
Divide both sides by ______.	400
$\_\_\_ = e^{5k}$	$\frac{1}{4}$
Take ______ logarithms of both sides.	natural
$\ln \frac{1}{4} = \_\_\_\_\_\_$	5k
k = ______	-.2773

The equation for decay of this substance is	
$A(t) = 400$ ______.	$e^{-.2773t}$
Substitute ____ for t to find the amount left after 7 days.	7
$A(t) = 400$ ________	$e^{-.2773(7)}$
$A(t) =$ ______	57.4
The amount present after 7 days is ________.	57.4 grams

8. **The amount, y, in grams, of a radioactive substance present at time t in seconds is given by**

$$y = 80e^{-.05t}.$$

Find the half-life of the substance.

The half-life is the time it takes until only ______ the substance is left. There are ____ grams present initially (let t = ____ to see this), so the half-life is found by solving	half; 80 0
_____ $= 80e^{-.05t}$	40
or $\frac{1}{2} =$ ______.	$e^{-.05t}$
Take natural logarithms on each side to get	
$t =$ ______.	13.9
The half-life is about ______ seconds.	13.9

9. **Suppose the cat population in a small area of New York City increases according to the relationship**

$$y = y_0e^{.05t}$$

where t represents time in months and y_0 is the initial population of cats. Find the number of cats present in 6 months if there are 1000 cats now.

Replace t with ____, and y_0 with _______. | 6; 1000

$$y = 1000e^{.05(\_\_\_)} = 1000e^{\_\_\_}$$ | 6; .3

$$y = \_\_\_\_\_\_$$ | 1350

10. In Frame 9, how many cats will be present in one year (12 months)?

_______ | 1822

11. How long will it take for the population of cats in Frame 9 to triple?

We want to know when y = ______. | $3 \cdot y_0$

Start with

$$3 \cdot y_0 = \_\_\_\_\_\_$$ | $y_0e^{.05t}$

or $$3 = \_\_\_\_\_\_.$$ | $e^{.05t}$

Take natural logarithms of both sides.

$$\ln 3 = \ln \_\_\_\_\_$$ | $e^{.05t}$

Now use properties of logarithms to get

$$\ln 3 = .05t \ln e.$$

Since ln e = ___, | 1

$$t = \frac{\ln 3}{.05} = \_\_\_\_\_ \text{ (rounded)}$$ | 22 months

12. Suppose that the rules for a foreign language are remembered according to the formula

$$y = y_0\left(\frac{1 + e}{1 + e^{t+1}}\right)$$

where y is the number of rules remembered at time t, measured in months, and y_0 is the initial number remembered. How long will it take for 1/4 of

the rules to be forgotten? (Hint; we want $y = (3/4)y_0$ because 3/4 of the rules are still remembered.)

____

about .4 months

CHAPTER 5 TEST

Test answers are at the back of this study guide.

Solve each of the following equations.

1. $4^{x+1} = 8^{3x-1}$ 1. ____________

2. $\left(\frac{m}{3}\right)^3 = 5$ 2. ____________

Write each of the following in logarithmic form.

3. $2^6 = w$ 3. ____________

4. $e^t = 3.19$ 4. ____________

Write each of the following in exponential form.

5. $\log_7 \sqrt[3]{2401} = \frac{4}{3}$ 5. ____________

6. $\ln 24 = q$ 6. ____________

7. What is the x-intercept of the graph of $f(x) = \log_a x$? 7. ____________

Graph each of the following functions.

8. $f(x) = 2^x + 2$

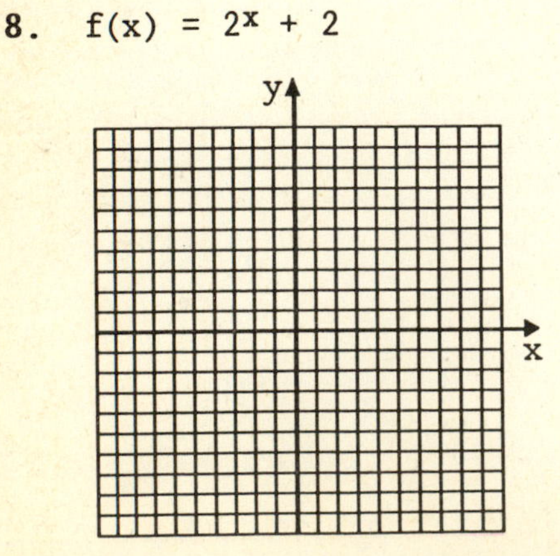

9. $f(x) = \log_2 (1 - x)$

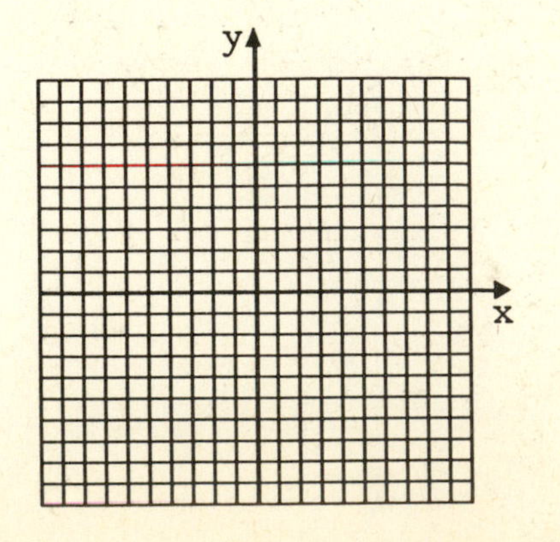

10. Use properties of logarithms to write the following as a sum, difference, or product of logarithms.

$$\log_2 \frac{m^2\sqrt{n^3}}{r^3y}$$

10. ______________________

Find each of the following logarithms. Round to the nearest thousandth.

11. ln 7830

11. ______________________

12. $\log_9 32$

12. ______________________

13. Without solving the equation

$$\log (4 - 3x) = 2,$$

give values of x that cannot be solutions of the equation.

13. ______________________

Solve the following equations. Round answers to the nearest thousandths.

14. $1000 = 5^{3k-1}$

14. ______________________

15. $\log_3 (x + 1) = 2$

15. ______________________

16. $\ln 3x = \ln 4 + \ln (x - 2)$

16. ______________________

17. Suppose the amount, in grams, of a radioactive substance present at time t is

$$A(t) = 200e^{-.2t},$$

where t is measured in hours.

(a) Find the amount present to the nearest tenth, after 5 hr.

17. (a) ______________________

(b) Find the half-life of the substance, to the nearest tenth of an hour.

(b) ____________________

18. How much will $3500 amount to at 5% compounded continuously for 10 yr?

18. ____________________

19. If a population increases at a continuous rate of 14% per year, how many years, to the nearest tenth, will it take the population to triple in size?.

19. ____________________

20. How many years, to the nearest tenth, will be needed for $1000 to increase to $4000 at 8% compounded quarterly?

20. ____________________

CHAPTER 6 PRETEST

Pretest answers are at the back of this study guide.

Graph each polynomial function P as defined.

1. $P(x) = -(x - 1)^3$

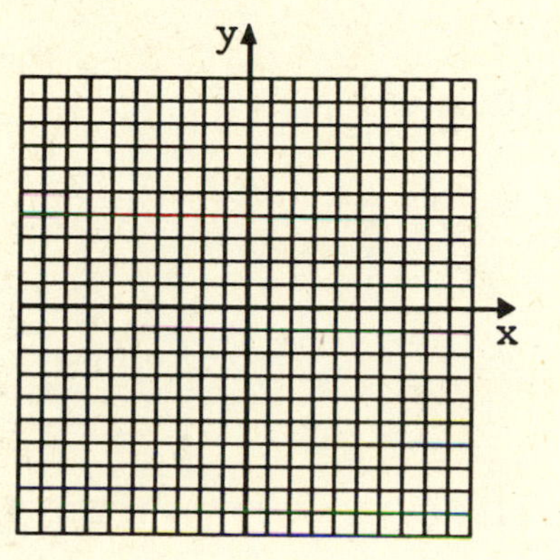

2. $P(x) = x^2(x - 2)(x + 2)$

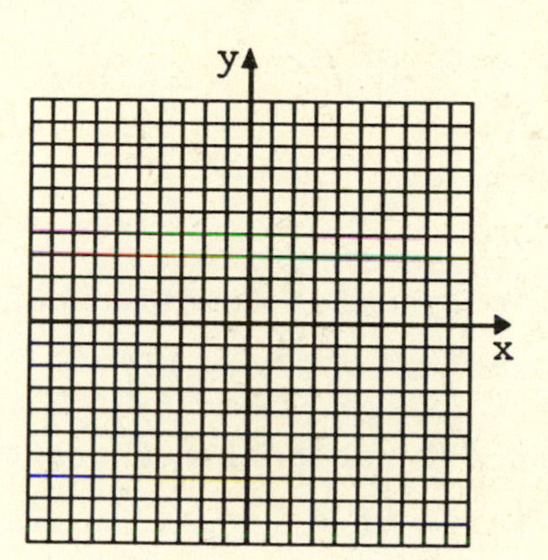

Use synthetic division to perform each division.

3. $(5x^3 + 6x^2 - 3x + 2) \div (x + 2)$ 3. ________

4. $(x^4 - 10x^2 - 8) \div (x - 3)$ 4. ________

5. Use synthetic division to find $P(-3)$ for $P(x) = 3x^4 + 8x^3 + 7x^2 - 1$. 5. ________

6. Find a polynomial function P of lowest degree with 0, 3, −3, and −2 as zeros. 6. ________

7. Is −2 a zero of the function defined by $P(x) = 3x^3 + 4x^2 + 6x + 20$. Why or why not? 7. ________

8. Is $x - 3$ a factor of $P(x) = 4x^2 - 3x + 6$? Why or why not? 8. ________

9. Find a function P defined by a polynomial of degree 3 with $\sqrt{2}$, $-\sqrt{2}$, and 4 as zeros, and $P(1) = -2$.

9. ____________________

10. Factor $P(x) = x^3 + 2x^2 - x - 2$ given that -2 is a zero of P.

10. ____________________

11. For the polynomial function defined by $P(x) = x^3 - x^2 - 17x - 15$,

(a) list all rational numbers that can be zeros of P,

(b) find all rational zeros of P.

11. (a) ____________________

(b) ____________________

12. Explain why the polynomial function defined by $P(x) = x^3 + 2x^2 + x + 7$ cannot have any positive real zeros.

12. ____________________

13. Show that the polynomial function defined by $P(x) = x^2 + 2x^2 - 5x - 6$ has no real zero greater than 3 nor less than -4.

13. ____________________

14. Find the possible numbers of positive and negative real zeros of the polynomial function defined by $P(x) = x^3 - 2x^2 - 3x + 1$.

14. ____________________

Graph the functions defined as follows.

15. $P(x) = 2x^3 + 3x^2 - 3x - 2$

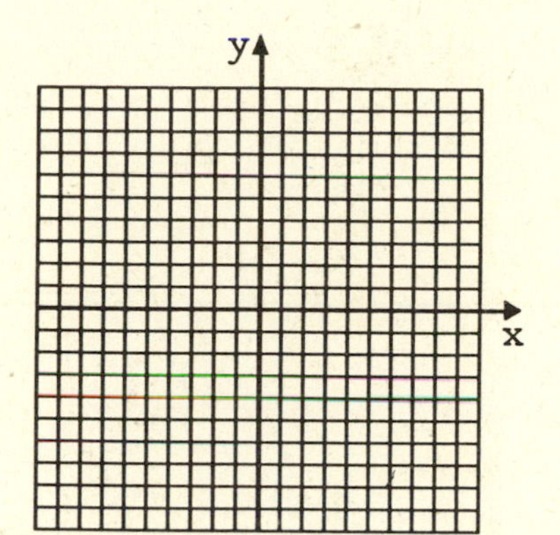

16. $f(x) = \frac{1}{x - 3}$

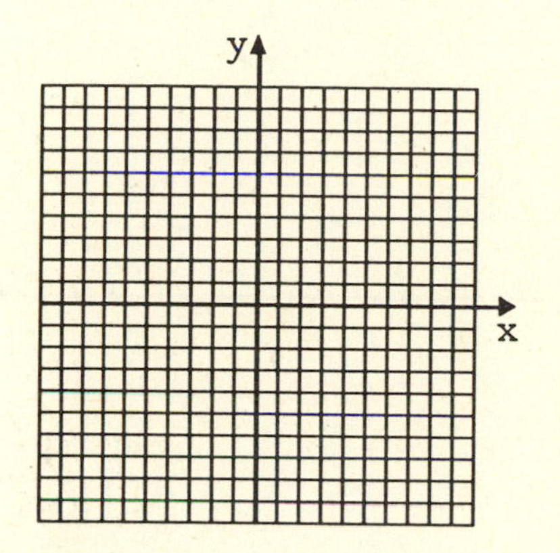

17. $f(x) = \frac{x - 8}{x^2 - 4}$

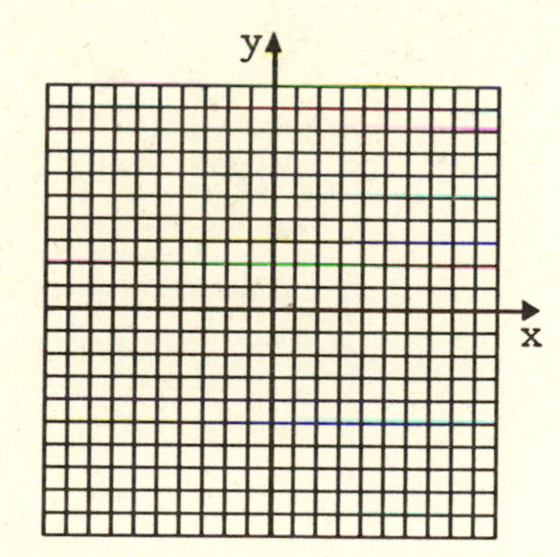

18. $f(x) = \frac{x^2 - 1}{x + 1}$

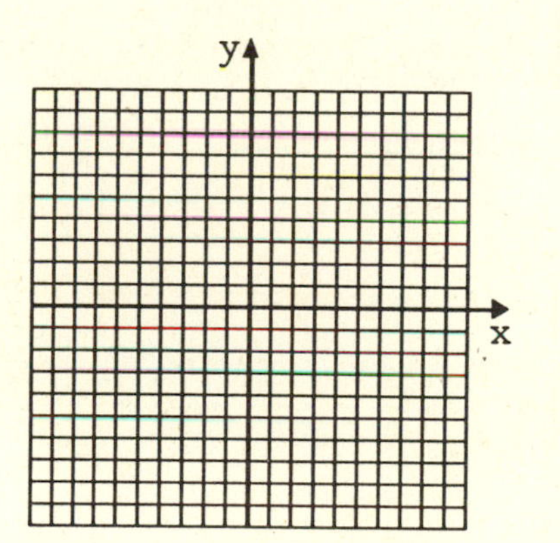

19. What is the equation of the oblique asymptote of the graph of $f(x) = \frac{3x^2 - x + 7}{x - 2}$?

19. ____________________

20. Which one of the functions defined as follows has a graph with no x-intercepts?

(a) $f(x) = x(x + 2)^2$ (b) $f(x) = \frac{4}{x - 3}$

(c) $f(x) = \frac{x - 5}{x - 3}$ (d) $f(x) = x^4 - 3x^2 - 1$

20. ____________________

CHAPTER 6 POLYNOMIAL AND RATIONAL FUNCTIONS

6.1 Graphing Polynomial Functions

1. A ______________ function of degree ______ is a function defined by

$$P(x) = \_\_\_\_\_\_ + a_{n-1}x^{n-1} + \cdots + \_\_\_\_\_\_,$$

for ________ numbers a_n, a_{n-1}, ... , a_1, and a_0, where n is a _______ number and $a_n \neq$ _____.

polynomial; n

a_nx^n; $a_1x + a_0$

complex

whole; 0

2. To graph the _____________ function defined by $P(x) = x^3$, first complete a table of values, as follows.

x	-2	-1	0	1	2
P(x)	____	____	____	____	____

Use these results to complete ordered pairs and then draw the graph.

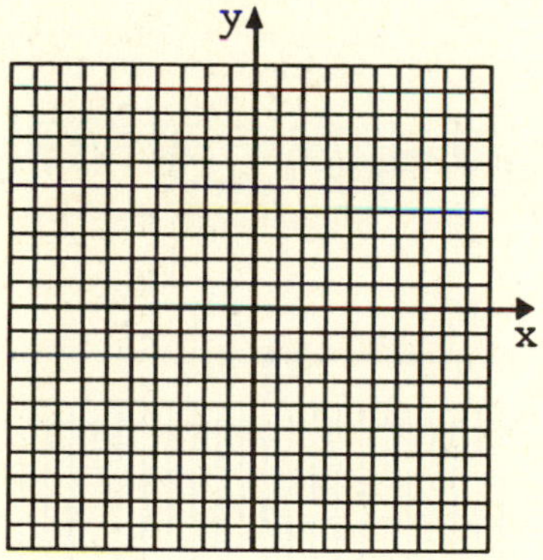

y

x

polynomial

-8; -1; 0; 1; 8

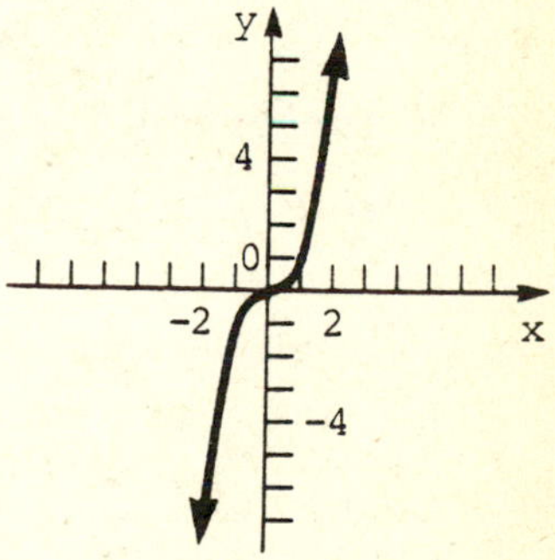

y

4

0

-2

2

-4

x

3. Graph the function defined by $P(x) = x^4$. The graph includes the points (0, ____), (1, ____), (-1, ____), (2, ____), and (-2, ____).

0; 1

1; 16; 16

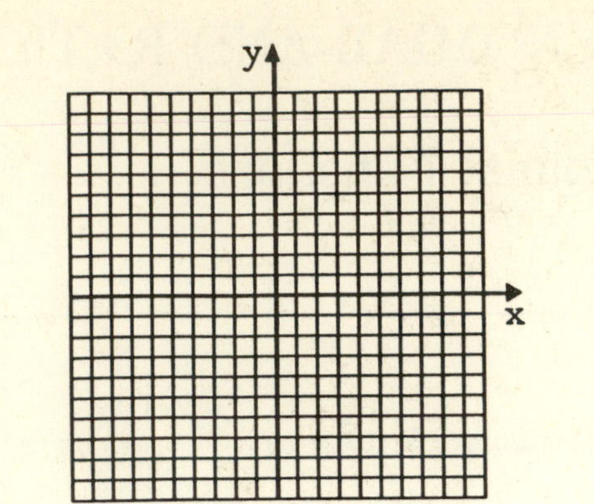

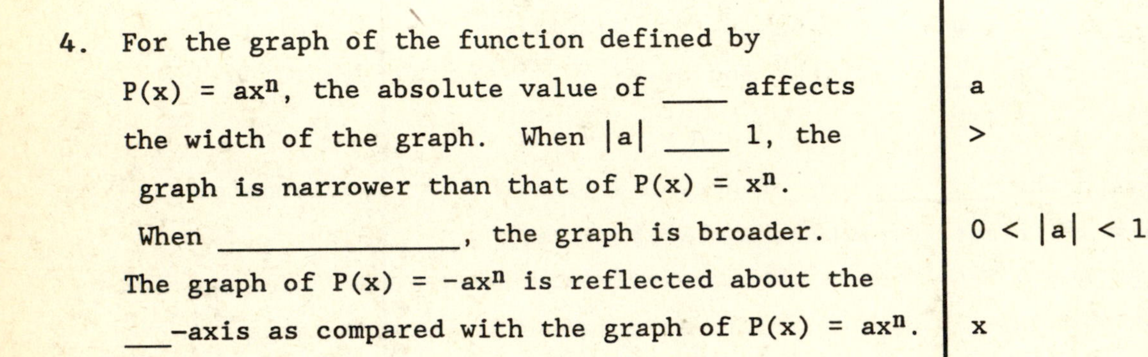

4. For the graph of the function defined by $P(x) = ax^n$, the absolute value of ____ affects the width of the graph. When $|a|$ ____ 1, the graph is narrower than that of $P(x) = x^n$. When ______________, the graph is broader. The graph of $P(x) = -ax^n$ is reflected about the ___-axis as compared with the graph of $P(x) = ax^n$.

a

>

$0 < |a| < 1$

x

In Frames 5 and 6, graph the polynomial functions defined as follows.

5. $P(x) = -\frac{1}{2}x^3$

The graph includes the points (0, ___), (2, ___), and (-2, ___).

0; -4

4

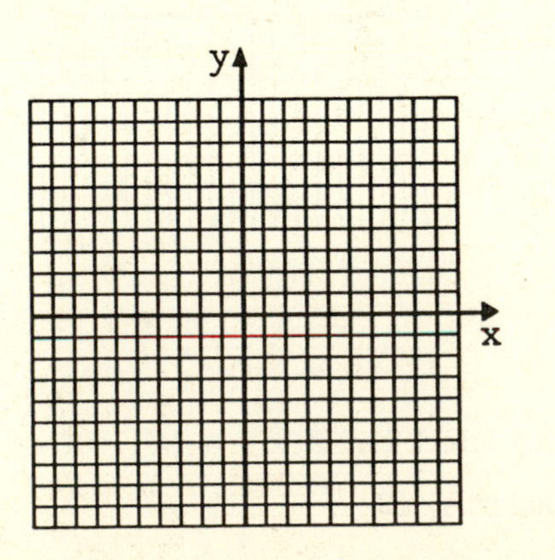

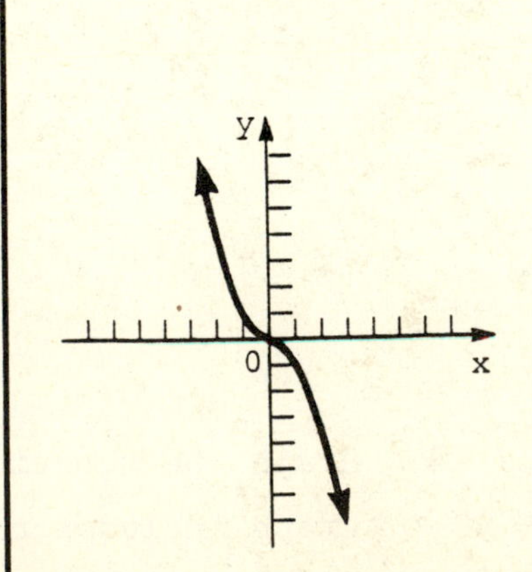

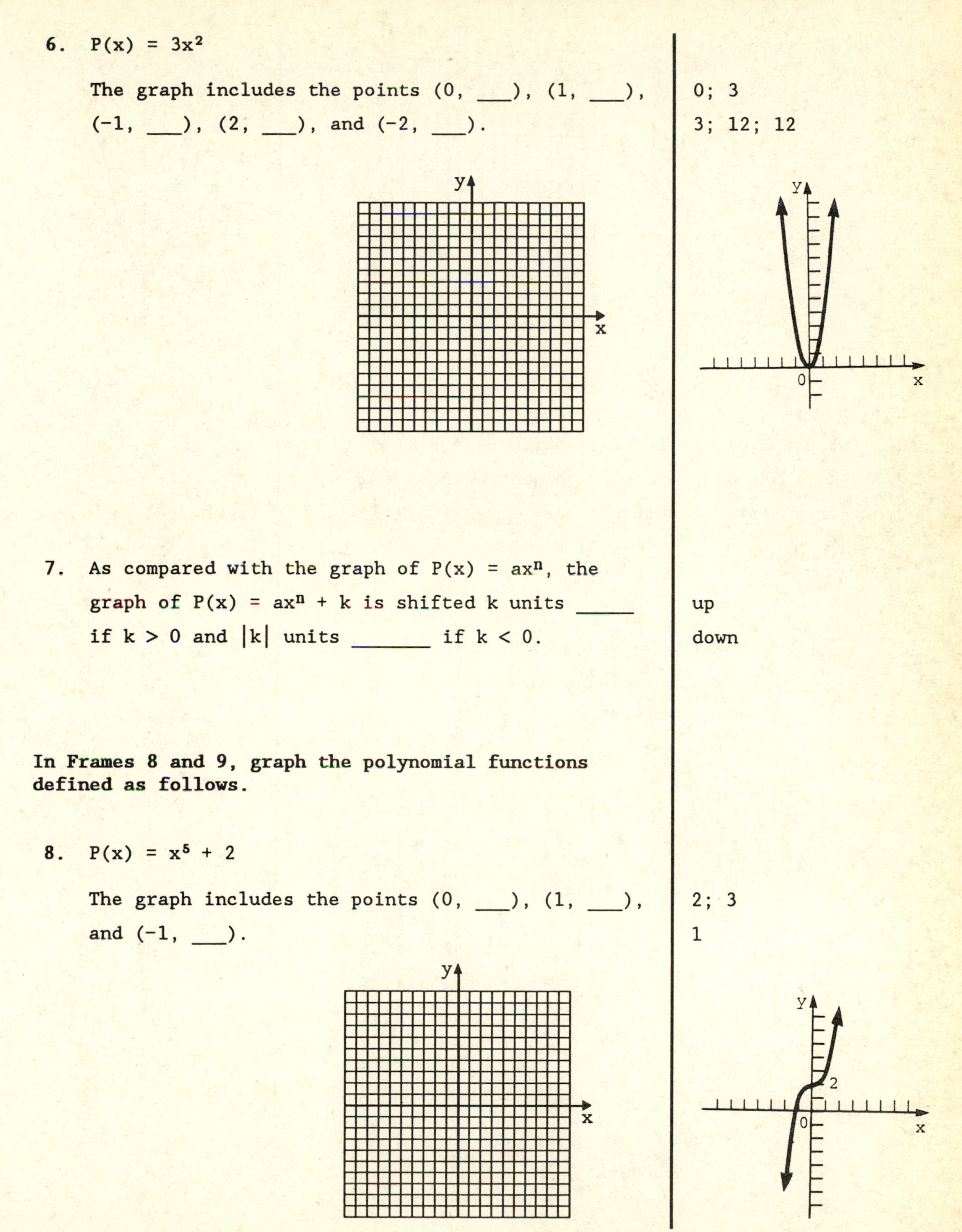

6. $P(x) = 3x^2$

The graph includes the points (0, ___), (1, ___), (-1, ___), (2, ___), and (-2, ___).

0; 3
3; 12; 12

7. As compared with the graph of $P(x) = ax^n$, the graph of $P(x) = ax^n + k$ is shifted k units _____ if $k > 0$ and $|k|$ units _______ if $k < 0$.

up
down

In Frames 8 and 9, graph the polynomial functions defined as follows.

8. $P(x) = x^5 + 2$

The graph includes the points (0, ___), (1, ___), and (-1, ___).

2; 3
1

9. $P(x) = x^2 - 4$

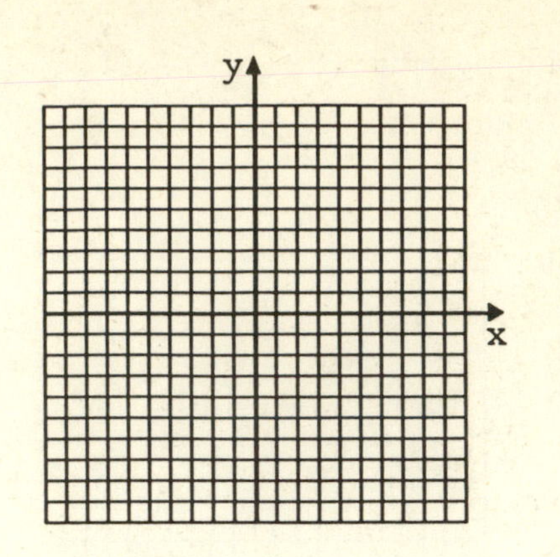

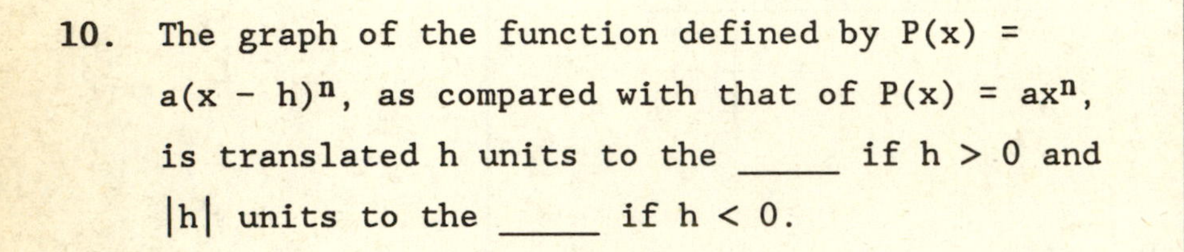

10. The graph of the function defined by $P(x) = a(x - h)^n$, as compared with that of $P(x) = ax^n$, is translated h units to the _____ if $h > 0$ and $|h|$ units to the _____ if $h < 0$.

right
left

11. Graph the function defined by $P(x) = (x - 2)^3$ below. The graph includes the points (0, ___), (1, ___), (2, ___), (3, ___), and (4, ___).

-8
-1; 0; 1; 8

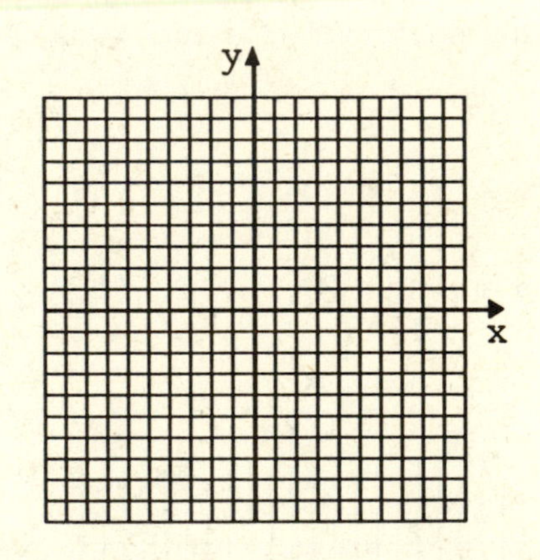

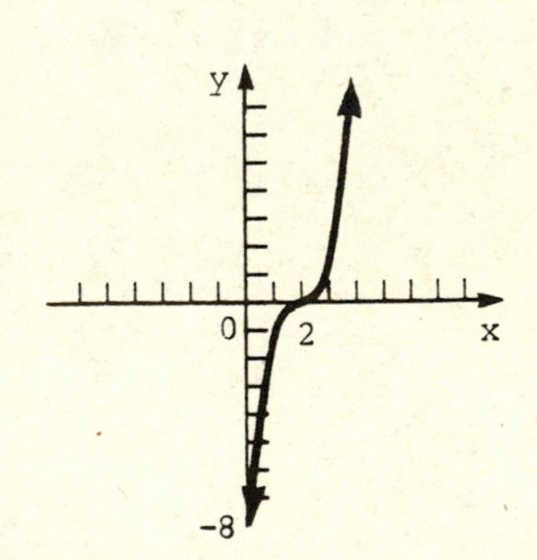

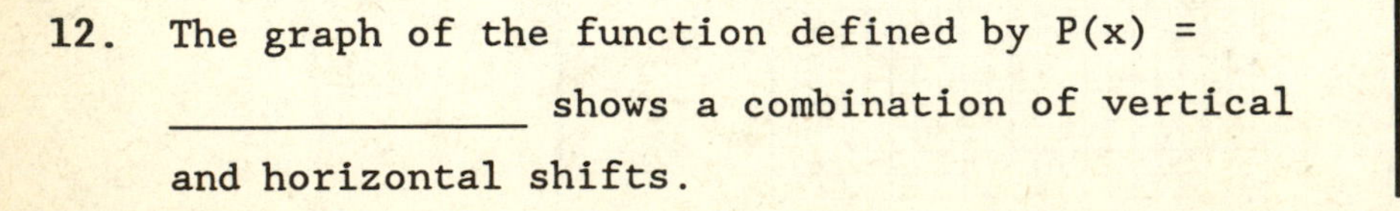

12. The graph of the function defined by $P(x) =$ _______________ shows a combination of vertical and horizontal shifts.

$a(x - h)^n + k$

13. Graph the function defined by $P(x) = 2(x + 1)^5 + 3$ below.

The graph has the same shape as that of $P(x) = 2x^5$, but is shifted 1 unit ________ and 3 units _______. The graph includes the points (0, ___), (-1, ___), and (-2, ___).

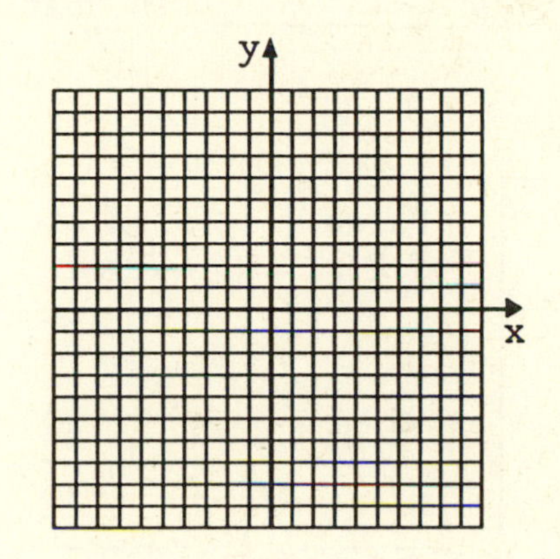

left

up

5; 3; 1

14. The domain of a polynomial function is __________. The range of a polynomial function of odd degree is __________. The range of a polynomial function of even degree will have a range, for some real number k, that takes the form _____________ or _______________.

$(-\infty, \infty)$

$(-\infty, \infty)$

$(-\infty, k]$

$[k, \infty)$

15. A value of x that makes P(x) = _____ is called a _______ of the function P.

0

zero

16. To graph a polynomial function such as the one defined by

$$P(x) = x^2(2x - 3)(x + 2),$$

first find the zeros by setting each ___________ equal to ____. Here the zeros are ____, ____,

factor

0; 0; 3/2

and ____. These zeros divide the x-axis into four intervals:

-2

$(-\infty, -2)$, ________, $(0, \frac{3}{2})$, ________.

$(-2, 0)$; $(\frac{3}{2}, \infty)$

Select a value of x in each interval and determine by ______________ whether the function values are __________ or __________ in that interval.

substitution

positive; negative

Interval	*Test point*	*Value of P(x)*	*Sign of P(x)*
$(-\infty, -2)$	-3	______	+
$(-2, 0)$	-1	______	______
$(0, \frac{3}{2})$	1	______	______
______	2	______	______

81

-5; -

-3; -

$(\frac{3}{2}, \infty)$; 16; +

Sketch the graph of P using the information in the table.

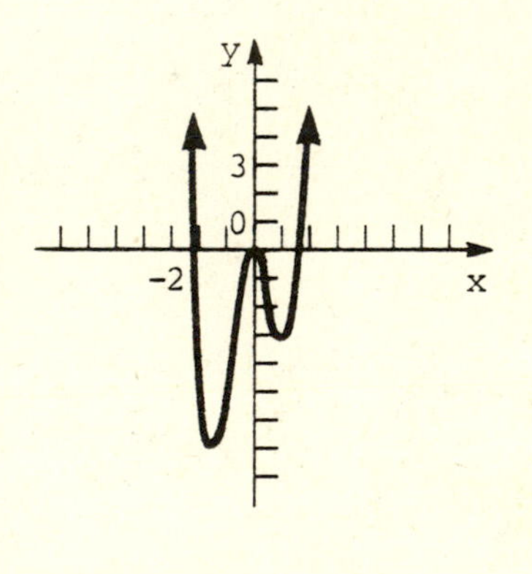

17. A general procedure for graphing polynomial functions follows.

(a) Find __________ by solving $P(x) = 0$.

x-intercepts

(b) Find __________ by evaluating $P(0)$.

y-intercepts

(c) Use test points from the intervals formed by the ____________ to determine whether the graph lies above or below the x-axis in each interval.

x-intercepts

(d) Plot any additional points, as necessary, and join the points with a smooth, unbroken __________.

curve

Graph each polynomial function defined in Frames 18-21.

18. $P(x) = -3x(x - 1)(x + 3)$

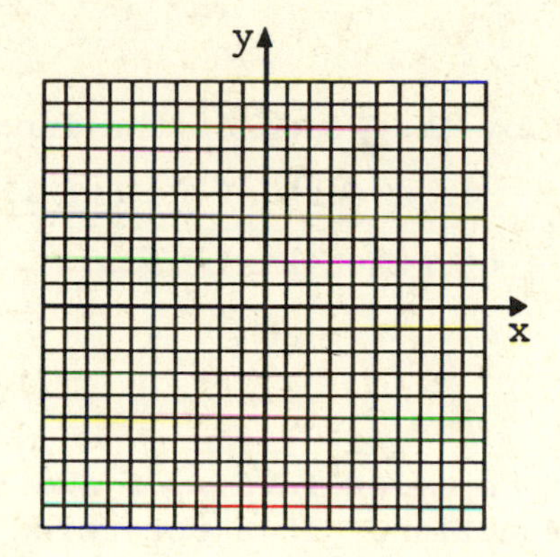

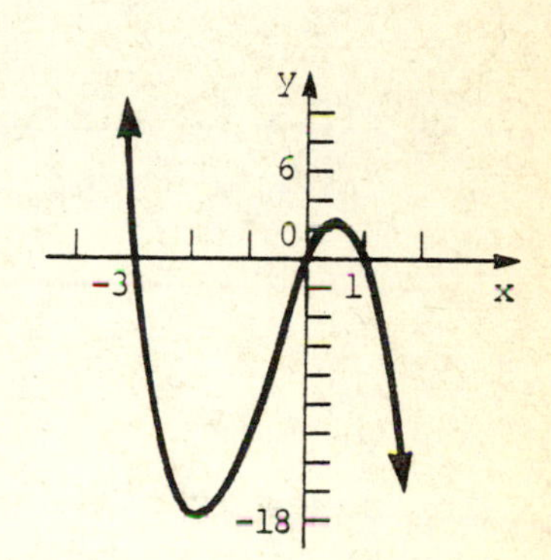

19. $P(x) = x^2(x + 2)(x - 2)^2$

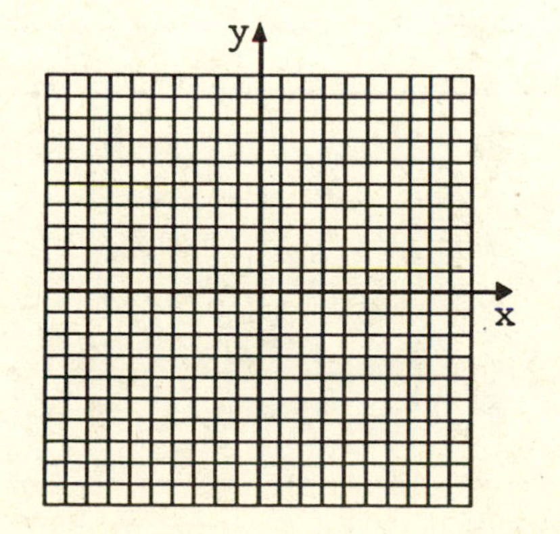

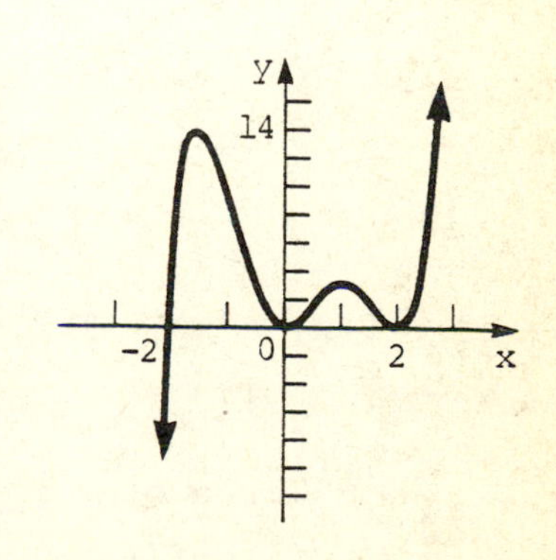

20. $P(x) = x^5 + x^4 - 2x^3$

First factor the polynomial completely.

$P(x) = x^3(\_\_\_\_\_\_\_\_)$

$= x^3(\_\_\_\_\_)(\_\_\_\_\_)$

$x^2 + x - 2$

$x + 2;\ x - 1$

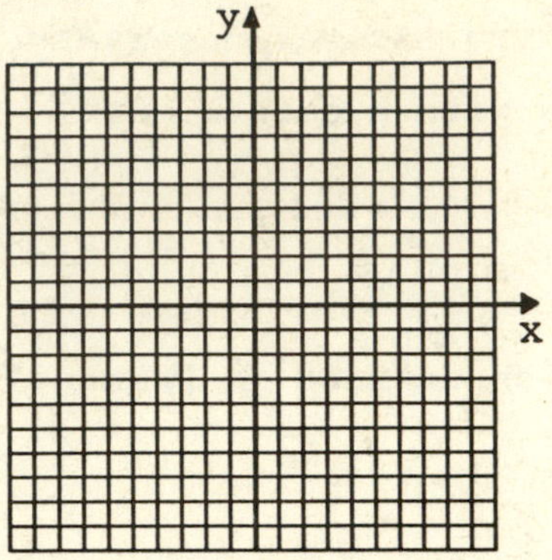

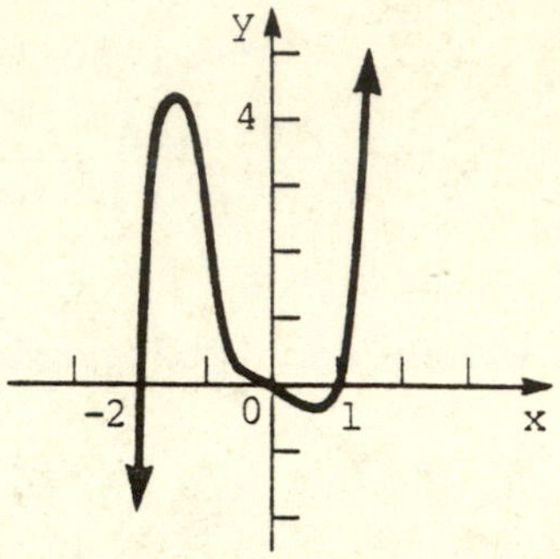

21. If a is an x-intercept of the graph of the function y = P(x), then ____ is a zero of P, and a is a __________ of the equation P(x) = 0.

a

solution

In Frames 22–24, determine the x-intercepts of the graphs of each polynomial function. Then determine the zeros of P and the solutions of P(x) = 0.

22. $P(x) = (x + 3)(x - 2)(x + 1)$

x-intercepts of the graph: ______________

Zeros of P: ______________

Solutions of P(x) = 0: ______________

-3; 2; -1

-3; 2; -1

-3; 2; -1

23. $P(x) = x^4 + 3x^3 - 4x^2$

x-intercepts of the graph: ______________

Zeros of P: ______________

Solutions of P(x) = 0: ______________

0; -4; 1

0; -4; 1

0; -4; 1

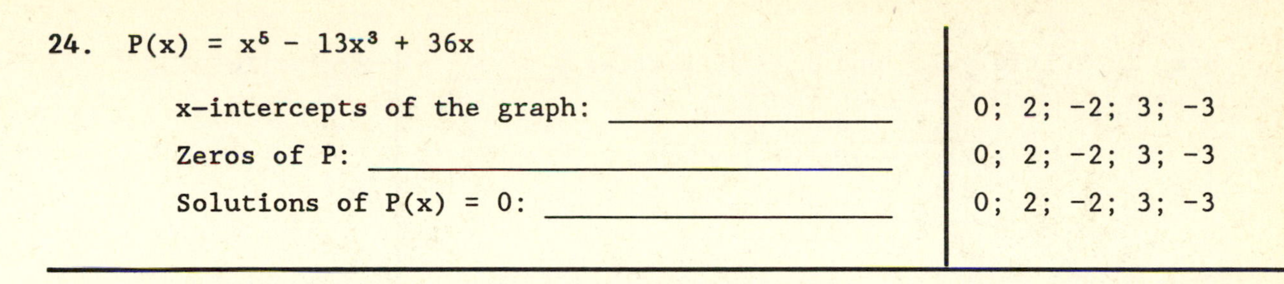

24. $P(x) = x^5 - 13x^3 + 36x$

x-intercepts of the graph: ______________	0; 2; -2; 3; -3
Zeros of P: ______________	0; 2; -2; 3; -3
Solutions of $P(x) = 0$: ______________	0; 2; -2; 3; -3

6.2 Polynomial Division

1. The quotient P/Q is found by multiplying P by the __________ of Q. That is,	reciprocal
$\frac{P}{Q} = P \cdot$ _____ if $Q \neq$ _____.	$\frac{1}{Q}$; 0
2. To divide the polynomial $9x^3y^2 + 3x^2y - 12xy^3$ by the monomial $3xy^2$, simplify each quotient.	
$\frac{9x^3y^2 + 3x^2y - 12xy^3}{3xy^2}$	
$= \frac{9x^3y^2}{3xy^2} + \frac{\_\_\_\_}{3xy^2} - \frac{12xy^3}{\_\_\_\_}$	$3x^2y$; $3xy^2$
= _____ + _____ - _____	$3x^2$; $\frac{x}{y}$; $4y$
3. To divide a polynomial by another polynomial, use an orderly __________ for performing the operation, called a division __________.	procedure algorithm
4. Divide $21x^3 + 16x - 10x^2 + 4$ by $3x - 1$. Begin by writing both polynomials in __________ order.	descending
$3x - 1\overline{)21x^3 - 10x^2 + 16x + 4}$	
The quotient __________ is $7x^2$.	$21x^3/3x$

Step 1 Multiply $7x^2$ and $3x - 1$, getting ________.	$21x^3 - 7x^2$

$$\begin{array}{r} 7x^2 \qquad\qquad\qquad \\ 3x - 1\overline{)21x^3 - 10x^2 + 16x + 4} \\ 21x^3 - \underline{\quad\quad} \qquad\qquad \\ \hline \end{array}$$

	$7x^2$
Step 2 Subtract __________ from $21x^3 - 10x^2$ by	$21x^3 - 7x^2$
changing signs and __________.	adding
Then bring down ______.	$16x$

$$\begin{array}{r} 7x^2 \qquad\qquad\qquad \\ 3x - 1\overline{)21x^3 - 10x^2 + 16x + 4} \\ \underline{21x^3 - 7x^2} \qquad\qquad \\ \underline{\quad\quad} + \underline{\quad\quad} \qquad \end{array}$$

	$-3x^2$; $16x$
Step 3 The quotient of $-3x^2$ and ______ is $-x$.	$3x$
Multiply $-x$ and _______, getting _______.	$3x - 1$; $-3x^2 + x$

$$\begin{array}{r} 7x^2 - x \qquad\qquad \\ 3x - 1\overline{)21x^3 - 10x^2 + 16x + 4} \\ \underline{21x^3 - 7x^2} \qquad\qquad \\ -3x^2 + 16x \qquad \\ \underline{\qquad\qquad\quad} \qquad \\ \hline \end{array}$$

	$-3x^2 + x$
Step 4 Subtract $-3x^2 + x$ from $-3x^2 + 16x$.	
Bring down ____.	4

$$\begin{array}{r} 7x^2 - x \qquad\qquad \\ 3x - 1\overline{)21x^3 - 10x^2 + 16x + 4} \\ \underline{21x^3 - 7x^2} \qquad\qquad \\ -3x^2 + 16x \qquad \\ \underline{-3x^2 + x} \qquad \\ \underline{\quad\quad} + \underline{\quad\quad} \end{array}$$

	$15x$; 4
Step 5 The quotient of _____ and _____ is 5.	$15x$; $3x$
Multiply 5 and _______.	$3x - 1$

$$\begin{array}{r} 7x^2 - x + 5 \qquad \\ 3x - 1\overline{)21x^3 - 10x^2 + 16x + 4} \\ \underline{21x^3 - 7x^2} \qquad\qquad \\ -3x^2 + 16x \qquad \\ \underline{-3x^2 + x} \qquad \\ 15x + 4 \\ \underline{\quad\quad} - \underline{\quad\quad} \\ \hline \end{array}$$

	$15x$; 5

Step 6 Subtract $15x - 5$ from _______, getting _____. | $15x + 4$; 9

$$
\begin{array}{r}
7x^2 - \quad x + 5 \\
3x - 1\overline{)21x^3 - 10x^2 + 16x + 4} \\
\underline{21x^3 - \ \ 7x^2} \qquad\qquad \\
-3x^2 + 16x \qquad \\
\underline{-3x^2 + \quad x} \qquad \\
15x + 4 \\
\underline{15x - 5} \\
\underline{\quad}
\end{array}
$$

9

The quotient is __________, with remainder ______. | $7x^2 - x + 5$; 9

Write this result as

$$\frac{21x^3 - 10x^2 + 16x + 4}{3x - 1} = \underline{\qquad\qquad\qquad}.$$

$7x^2 - x + 5 + \dfrac{9}{3x - 1}$

5. When dividing a polynomial by a polynomial of the form $x - k$, use a shortcut called __________ division. | synthetic

Use synthetic division in Frames 6–10.

6. $\dfrac{x^3 + x^2 - 6x - 18}{x - 3}$

Remember that synthetic division can be used only when the coefficient of x is equal to _____. We change the -3 to a ____ so that we can add during the process instead of __________ as in the long division process of Frame 4. Start with | 1; 3; subtracting

$$\underline{\quad}\overline{)1 \quad 1 \quad -6 \quad -18}$$

3

where 1, 1, -6, and -18 are ______________ of the polynomial to be divided. Bring down the first ____, multiply it by the ____ and add it to the second ____. | coefficients; 1; 3; 1

$$
\begin{array}{l}
3\overline{)\ 1 \quad 1 \quad -6 \quad -18} \\
\quad\ \ \ \underline{\quad\ \ \ \underline{\quad}\qquad\qquad\quad} \\
\quad\ \ \ \underline{\quad}\ \ \underline{\quad}
\end{array}
$$

3

1; 4

Continue this process by multiplying the resulting ____ by ____ and adding it to the ____. | 4; 3; -6

$$\begin{array}{r|rrrr} 3 & 1 & 1 & -6 & -18 \\ & & 3 & 12 & \_\_ \\ \hline & 1 & 4 & 6 & \_\_ \end{array}$$

18

0

Now the coefficients of the ________ are in the last row, and the degree of the quotient is always ________ less than the ________ of the polynomial to be divided. The quotient is x^2 + ____ + ____. The remainder is ____. | quotient; one; degree; 4x; 6; 0

7. $\dfrac{2m^3 + 9m^2 + 11m + 6}{m + 3} =$ ________ | $2m^2 + 3m + 2$

8. $\dfrac{2y^4 + 7y^3 - 5y^2 + 3y + 30}{y + 4} =$ ________ | $2y^3 - y^2 - y + 7 + \dfrac{2}{y + 4}$

9. $\dfrac{3x^5 - x^4 + 6x^3 - 2x^2 + 9x - 3}{x - \frac{1}{3}} =$ ________ | $3x^4 + 6x^2 + 9$

10. $\dfrac{x^5 + 1}{x + 1} =$ ________ | $x^4 - x^3 + x^2 - x + 1$

11. For any polynomial P(x) and any complex number k, there exists a unique polynomial Q(x) and number r such that

____ = (x - k)Q(x) + ____. | P(x); r

12. If we wish to find P(k), we can substitute ____ for ____ in P(x). But using the division equation in Frame 11, P(x) = (x - k)Q(x) + r, we see that if we divide P(x) by x - k, the remainder is ______. This is the ____________ theorem. | k; x; P(k); remainder

In Frames 13–16, use the remainder theorem to find $P(k)$.

13. $k = -4$; $P(x) = 2x^3 + 7x^2 - x + 14$

Use ________ division. — synthetic

$$\begin{array}{r|rrrr} -4 & 2 & 7 & -1 & 14 \\ & & \_\_ & \_\_ & \_\_ \\ \hline & 2 & \_\_ & \_\_ & \_\_ \end{array}$$

−8; 4; −12

−1; 3; 2

As the remainder shows, $P(-4) =$ ____. — 2

14. $k = 3$; $P(x) = 4x^2 - 3x + 2$

$P(3) =$ ______ — 29

15. $k = 1 - i$; $P(x) = 2x^3 + x^2 - 2x + 1$

$$\begin{array}{r|rrrr} 1-i & 2 & 1 & -2 & 1 \\ & & \_\_ & \_\_ & \_\_ \\ \hline & 2 & 3-2i & -1-5i & \end{array}$$

$2 - 2i$; $1 - 5i$; $-6 - 4i$

$P(1 - i) =$ ______ — $-5 - 4i$

16. $k = 2 - i$; $P(x) = 2x^3 - 11x^2 + 22x - 15$

$P(2 - i) =$ ______ — 0

Thus, $2 - i$ is a ______ of $P(x)$. — zero

17. A number k is a zero of $P(x)$ if ________. — $P(k) = 0$

In Frames 18–20, use synthetic division to decide whether or not the given number is a zero of the polynomial function defined by the given polynomial.

18. 4; $P(x) = 2x^2 - 5x - 12$

If $P(4) =$ ____ then ____ is a zero of the polynomial function. Use synthetic division. — 0; 4

$$\begin{array}{r|rrr} 4 & 2 & -5 & -12 \\ & & \_\_ & \_\_ \\ \hline & 2 & 3 & \_\_ \end{array}$$

8; 12

0

The remainder is ____. Therefore $P(4)$ = ____. | 0; 0

Then 4 (*is/is not*) a zero of P. | is

19. 1; $P(x) = x^3 - 3x^2 + 7x - 5$

Is 1 a zero? ______ | yes

20. 2i; $P(x) = 2x^2 + 3x - 4xi - 6i$

First rewrite P(x) as $2x^2 + (3 - 4i)x - 6i$.

Use synthetic division to get

$$\begin{array}{r|rrr} 2i & 2 & 3-4i & -6i \\ & & \_\_\_ & \_\_\_ \\ \hline & 2 & 3 & \_\_\_ \end{array}$$

4i; 6i

0

Then the remainder is ____ and 2i (*is/is not*) a zero of P. | 0; is

6.3 Zeros of Polynomial Functions

1. According to the ________ theorem, $x - k$ is a factor of $P(x)$ if and only if ________. | factor; $P(k) = 0$

2. Is $x + 3$ a factor of $x^4 + 6x^3 + 3x^2 - 26x - 24$? Use ________ division and the ________ theorem to decide. | synthetic; remainder

$$\begin{array}{r|rrrrr} -3 & 1 & 6 & 3 & -26 & -24 \\ & & -3 & -9 & \_\_\_ & \_\_\_ \\ \hline & 1 & 3 & \_\_\_ & \_\_\_ & \_\_\_ \end{array}$$

18; 24

-6; -8; 0

Since the remainder is ______, $P(-3)$ = ______, so $x + 3$ (*is/is not*) a factor of $x^4 + 6x^3 + 3x^2 - 26x - 24$. | 0; 0; is

In Frames 3 and 4 use the factor theorem to find the value of k that makes the second polynomial a factor of the first.

3. $x^3 + 2x + k$; $x + 2$

Use synthetic division until you reach a point where you can determine the necessary k that makes the remainder zero. (Remember that $x - k$ is a factor of $P(x)$ if $P(k) = 0$.)

-2)	1	0	2	k
		-2	4	___
	1	-2	___	___

Since $k - 12 =$ ____, we have $k =$ ____.

-12

6; $k - 12$

0; 12

4. $x^3 - 19x + k$; $x - 3$

3)	___	___	___	___
		___	___	___
	___	___	___	___

$k =$ _____

1; 0; -19; k
3; 9; -30
1; 3; -10; $k - 30$; 30

5. A function defined by a polynomial of degree n has at most ____ distinct zeros.

n

6. The number k is said to be a _______ of _______________ n of the polynomial function P if ________ appears in the factored form of $P(x)$ and there is no power of $(x - k)$ greater than ____ that is also a factor of $P(x)$.

zero
multiplicity
$(x - k)^n$
n

In Frames 7–9, determine the zeros of P and give their multiplicities.

7. $P(x) = (x - 1)^2(x + 3)^4$

Zero: ______; multiplicity: ______
Zero: ______; multiplicity: ______

1; 2
-3; 4

8. $P(x) = x^3(x + 2)^5$

Zero: ______; multiplicity: ______ | 0; 3

Zero: ______; multiplicity: ______ | -2; 5

9. $P(x) = (x - 2i)^5(x + 2i)^5$

Zero: ______; multiplicity: ______ | 2i; 5

Zero: ______; multiplicity: ______ | -2i; 5

In Frames 10–16, find a polynomial of lowest degree with real coefficients having the given zeros and function values.

10. -5 and 4; $P(2) = -7$

The factors of the polynomial must be

$(x - \_\_\_\_)$ and $(x - \_\_\_\_)$, | -5; 4

so that the polynomial has the form $P(x) = a(x + 5)(x - 4)$ for some nonzero real number a.

We want $P(2) =$ ____. Substitute ____ for x. | -7; 2

$P(2) = a(\_\_\_\_\_\_)(\_\_\_\_\_\_) = \_\_\_\_$ | 2 + 5; 2 - 4; -7

____$a =$ ____ | -14; -7

$a =$ ____ | $\frac{1}{2}$

$P(x) =$ ______________ | $\frac{1}{2}(x + 5)(x - 4)$

or ______________ | $\frac{1}{2}x^2 + \frac{1}{2}x - 10$

11. 1, -2, and 3; $P(-1) = 12$

$P(x) =$ ______________ | $\frac{3}{2}x^3 - 3x^2 - \frac{15}{2}x + 9$

12. $2 - i$ and $2 + i$; $P(2) = 2$

We know that $x - (\_\_\_\_)$ and $x - (\_\_\_\_)$ are factors of the polynomial. Then the polynomial of lowest degree is | 2 - i; 2 + i

$P(x) = a(x - 2 + i)(x - 2 - i) = a(\_\_\_\_\_\_\_\_)$. | $x^2 - 4x + 5$

Since $P(2) = 2$, $a =$ ____ and $P(x) =$ __________. | 2; $2x^2 - 8x + 10$

13. $3 + i$ and $3 - i$; $P(1) = 5$ $P(x) =$ ________ | $x^2 - 6x + 10$

14. $2 - i$, $2 + i$, and -2; $P(-1) = 2$
We want $P(x) = a(x - 2 + i)(x - 2 - i)(x + 2)$.
First multiplying $(x - 2 + i)(x - 2 - i)$, we get

$P(x) = a(x^2 - 4x + 5)(x + 2) = a($________$)$. | $x^3 - 2x^2 - 3x + 10$

Since $P(-1) = 2$, $P(x) =$ ________. | $\frac{1}{5}x^3 - \frac{2}{5}x^2 - \frac{3}{5}x + 2$

15. 1, -1, and -3; $P(2) = 15$

$P(x) = a(x - 1)(x + 1)($____$)$ | $x + 3$

$= a($________$)$. | $x^3 + 3x^2 - x - 3$

Since $P(2) = 15$, $P(x) =$ ________. | $x^3 + 3x^2 - x - 3$

16. $1 + i$, $1 - i$, -2, and $\sqrt{2}$; $P(0) = -8\sqrt{2}$

$P(x) = a[x - (1 + i)][x - (1 - i)](x + 2)($____$)$ | $x - \sqrt{2}$

$= a(x^2 - 2x + 2)(x + 2)(x - \sqrt{2})$

$= a($________$)$. | $x^4 - \sqrt{2}x^3 - 2x^2 + (4 + 2\sqrt{2})x - 4\sqrt{2}$

Since $P(0) = -8\sqrt{2}$, $P(x) =$ ________. | $2x^4 - 2\sqrt{2}x^3 - 4x^2 + (8 + 4\sqrt{2})x - 8\sqrt{2}$

17. If $P(x)$ is a polynomial having only real ________ and if $a + bi$ is a zero of P, then the conjugate, ______, is also a zero of P. | coefficients; $a - bi$

In Frames 18–21, one zero is given for the function defined by the given polynomial. Find all others.

18. $P(x) = x^3 + 4x^2 + x - 6;\ -3$

We know ________ is a factor of P(x) since _____ is a zero of P. Use synthetic division to find another polynomial that is a factor of P(x).

−3)	1	4	1	−6
		−3	___	___
	1	1	___	___

The second polynomial is $x^2 + x - 2$ which, factors as $(x + 2)(x - 1)$. The two other zeros are ____ and ____.

x + 3; −3

−3; 6

−2; 0

−2; 1

19. $P(x) = x^3 - x^2 - 4x + 4;\ 1$ ____________

2; −2

20. $P(x) = x^3 - 2x^2 - 14x + 40;\ x = 3 + i$

We know that x − (________) is a factor of P(x) since ________ is a zero of P. Use synthetic division to find another polynomial that is a factor of P(x).

3 + i)	1	−2	−14	40
		3 + i	__________	______
	1	1 + i	__________	______

Now, since 3 + i is a zero, we know that ______ is a zero. Use synthetic division with the quotient $x^2 + (1 + i)x + (-12 + 4i)$ and the zero ______. The final zero is _____.

3 + i

3 + i

2 + 4i; −40

−12 + 4i; 0

3 − i

3 − i; −4

21. $P(x) = x^4 - 6x^3 + 10x^2 + 2x - 15;\ 2 + i$

Use synthetic division to get

2 + i)	1	−6	10	2	−15
		2 + i	−9 − 2i	_____	_____
	1	−4 + i	1 − 2i	_____	_____.

4 − 3i; 15

6 − 3i; 0

Since $2 + i$ is a zero of P, so is ______. Use synthetic division with $2 - i$ on the quotient above. Doing so gives the polynomial ____________ as a result. Now find the zeros of the quadratic polynomial $x^2 - 2x - 3$, which factors as (_______)(_______). The other zeros are _____ and _____.

$2 - i$

$x^2 - 2x - 3$

$x - 3$; $x + 1$; 3

-1

22. Suppose we know that the polynomial function defined by $P(x) = x^3 - 4x^2 - 2x + 20$ has -2 as a zero. We can write this polynomial as a product of linear factors. First use ________________ division to divide $x^3 - 4x^2 - 2x + 20$ by ______.

_____)	1	-4	-2	20
		-2	_____	_____
	1	_____	10	_____

This means that

$$x^3 - 4x^2 - 2x + 20 = (x + 2)(\_\_\_\_\_\_\_\_\_\_\_).$$

Use the quadratic formula to solve $x^2 - 6x + 10 = 0$:

$$x = \_\_\_\_\_\_\_\_ \text{ or } x = \_\_\_\_\_\_\_\_.$$

This means

$$x^3 - 4x^2 - 2x + 20$$
$$= (x + 2)[x - (\_\_\_\_\_\_)][x - (\_\_\_\_\_\_\_)]$$
$$= (x + 2)(\_\_\_\_\_\_\_\_\_\_\_)(x - 3 + i).$$

synthetic

$x + 2$

-2

12; -20

-6; 0

$x^2 - 6x + 10$

$3 + i$; $3 - i$

$3 + i$; $3 - i$

$x - 3 - i$

Write each polynomial of Frames 23–25 as a product of linear factors.

23. $P(x) = x^3 - 3x^2 + 4$, one zero of the function P is 2.

____

$(x - 2)(x + 1)(x - 2)$

24. $P(x) = 4x^3 + 8x^2 - 13x - 3$, one zero of P is -3. Use synthetic division to get

$$4x^3 + 8x^2 - 13x - 3 = (x + 3)(\_\_\_\_\_\_\_\_\_\_)$$

$4x^2 - 4x - 1$

Use the quadratic formula to solve $4x^2 - 4x - 1 = 0$:

$$x = \_\_\_\_\_\_ \text{ or } x = \_\_\_\_\_\_,$$

$\frac{1 + \sqrt{2}}{2}$; $\frac{1 - \sqrt{2}}{2}$

so

$$4x^3 + 8x^2 - 13x - 3$$
$$= a(x + 3)[x - (\_\_\_\_)][x - (\_\_\_\_)]$$

$\frac{1 + \sqrt{2}}{2}$; $\frac{1 - \sqrt{2}}{2}$

for some constant a. To get $4x^3$ in the product on the right, we must choose $a = \_\_\_$. This gives

4

$$= \_\_\_(x + 3)\left[x - \left(\frac{1 + \sqrt{2}}{2}\right)\right]\left[x - \left(\frac{1 - \sqrt{2}}{2}\right)\right]$$

4

$$= (x + 3)(2)\left[x - \left(\frac{1 + \sqrt{2}}{2}\right)\right](2)\left[x - (\_\_\_\_)\right]$$

$\frac{1 - \sqrt{2}}{2}$

$$= (x + 3)[2x - (\_\_\_\_)][2x - (\_\_\_\_)]$$

$1 + \sqrt{2}$; $1 - \sqrt{2}$

$$= (x + 3)(2x - 1 - \sqrt{2})(\_\_\_\_\_\_).$$

$2x - 1 + \sqrt{2}$

25. $P(x) = 18x^3 - 3x^2 - 10x - 2$, one zero of P is $-1/2$.

$$P(x) = \_\_\_\_\_\_\_\_\_\_\_\_\_\_$$

$(2x + 1)(3x - 1 - \sqrt{3}) \cdot (3x - 1 + \sqrt{3})$

6.4 Rational Zeros of Polynomial Functions

1. Let $P(x) = a_n x^n + a_{n-1}x^{n-1} + \cdots + a_1 x + a_0$, where $a_n \neq 0$, define a polynomial function of degree n with ________ coefficients. By the rational zeros theorem, a rational number in lowest terms p/q can be a _____ of P only if p is a factor of ___ and q is a factor of ____.	integer zero; a_0 a_n

Find all rational zeros of the polynomial functions defined in Frames 2–6.

2. $P(x) = x^3 - 5x^2 + 2x + 8$ If p/q is a rational zero of P, then we know by the theorem that p is a factor of a_0 = ______ and q is a factor of a_3 = _____. Since $a_0 = 8$, we know that p must be ______________. Since $a_3 = 1$, we know that q must be _____ and any rational zero p/q of P will come from the list ______________. Use synthetic division to try various values. For example, try 1.	8 1 ±1, ±2, ±4, or ±8 ±1 ±1, ±2, ±4, or ±8
$$\begin{array}{r\|rrrr} 1) & 1 & -5 & 2 & 8 \\ & & 1 & -4 & -2 \\ \hline & 1 & -4 & -2 & 6 \end{array}$$ 1 (*is/is not*) a zero of P.	is not
Now try −1. $$\begin{array}{r\|rrrr} -1) & 1 & -5 & 2 & 8 \\ & & -1 & 6 & -8 \\ \hline & 1 & -6 & 8 & 0 \end{array}$$ −1 (*is/is not*) a zero of P.	is
Try 2. $$\begin{array}{r\|rrrr} 2) & 1 & -5 & 2 & 8 \\ & & 2 & -6 & -8 \\ \hline & 1 & -3 & -4 & 0 \end{array}$$ 2 (*is/is not*) a zero of P.	is

Try −2.

−2)	1	−5	2	8
		−2	14	−32
	1	−7	16	−24

−2 (*is/is not*) a zero of P. — is not

Test ±4 and ±8. The rational zeros are ________. — −1; 2; 4

3. $P(x) = x^3 - x^2 - 4x + 4$

Since $a_0 = 4$, we know p must be ____________. — ±1, ±2, or ±4

Since $a_3 = 1$, we know that q must be ____ and — ±1

p/q will come from the list ______________. — ±1, ±2, or ±4

Use synthetic division. Try −1.

−1)	1	−1	−4	4
		−1	2	2
	1	−2	−2	6

−1 (*is/is not*) a zero of P. — is not

Try 1.

1)	1	−1	−4	4
		1	0	−4
	1	0	−4	0

1 (*is/is not*) a zero of P. — is

Test ±2 and ±4. (Remember if or when you find three zeros, your task is completed. Why?) The zeros are _________. — 1, 2, −2

4. $P(x) = x^4 - 13x^2 + 36$

p will come from the list ±1, ±2, ±3, ±4, ±6, ±9, ±12, ±18, or ±36. To find the possibilities for p with large numbers, it is sometimes helpful to write the number as a product of its prime factors. (In this case $36 = 2 \cdot 2 \cdot 3 \cdot 3$.) Then form all the possible combinations of prime factors. q will again be ±1 so p/q will come from the same list as the one for p. Begin synthetic division, being careful to use zeros for the missing powers of x.

Try -1.

$$\begin{array}{r|rrrrr} -1) & 1 & 0 & -13 & 0 & 36 \\ & & -1 & 1 & 12 & -12 \\ \hline & 1 & -1 & -12 & 12 & 24 \end{array}$$

-1 (*is/is not*) a zero of P. — is not

Try 1.

$$\begin{array}{r|rrrrr} 1) & 1 & 0 & -13 & 0 & 36 \\ & & 1 & 1 & -12 & -12 \\ \hline & 1 & 1 & -12 & -12 & 24 \end{array}$$

1 (*is/is not*) a zero of P. — is not

Try -2.

$$\begin{array}{r|rrrrr} -2) & 1 & 0 & -13 & 0 & 36 \\ & & -2 & 4 & 18 & -36 \\ \hline & 1 & -2 & -9 & 18 & 0 \end{array}$$

-2 (*is/is not*) a zero of P. — is

Continue the trial-and-error process. Now use the reduced polynomial $x^3 - 2x^2 - 9x + 18$ found from the synthetic division for -2. You can stop when you find ______ zeros. The rational zeros are ________________. — four; -2, 2, 3, -3

5. $P(x) = 2x^3 - 7x^2 - 5x + 4$
p will come from the list ±1, ±2, or ±4; q will come from the list ±1, or ±2. Then p/q will come from the list ±1, ±1/2, ±2, or ±4. Test the values from this list to find the rational zeros, if they exist. ____________ — -1, 1/2, 4

6. $P(x) = x^3 - 7x^2 + 13x - 6$
p will come from the list ±1, ±2, ±3, or ±6; q will be ±1, so we use the list for p.

$$\begin{array}{r|rrrr} 1) & 1 & -7 & 13 & -6 \\ & & 1 & -6 & 7 \\ \hline & 1 & -6 & 7 & 1 \end{array}$$

1 (*is/is not*) a zero of P. — is not

-1)	1	-7	13	-6
		-1	8	-21
	1	-8	21	-27

-1 (*is/is not*) a zero of P. — is not

2)	1	-7	13	-6
		2	-10	6
	1	-5	3	0

2 (*is/is not*) a zero of P. — is

Find the rest of the zeros by solving

______________ = 0. — $x^2 - 5x + 3$

They are

x = __________ or x = __________. — $\frac{5 + \sqrt{13}}{2}$, $\frac{5 - \sqrt{13}}{2}$

These additional zeros (*are/are not*) rational. — are not

7. The rational zeros theorem may not be applied if a polynomial function does not have ___________ coefficients. If fractional coefficients appear, then we must first _____________ by a number that will clear all fractions. Then we use the theorem. — integer; multiply

8. Find all rational zeros of the function defined by

$$P(x) = x^3 - \frac{17}{6}x^2 - \frac{13}{3}x - \frac{4}{3}.$$

We must find the values that make P(x) = ____. — 0

Multiply both sides by ____ to eliminate all fractions, obtaining — 6

$$6x^3 - 17x^2 - 26x - 8 = 0.$$

The solutions of this equation are the zeros of P. From the possible rational zeros, we try -1/2.

-1/2)	6	-17	-26	-8
		-3	10	8
	6	-20	-16	0

Therefore, -1/2 (*is/is not*) a zero of P. — is

We may now find the remaining two zeros by factoring ______________.	$6x^2 - 20x - 16$
$6x^2 - 20x - 16 = 2(3x^2 - 10x - 8)$ $= 2(\_\_\_\_\_\_)(\_\_\_\_\_\_)$	$3x + 2$; $x - 4$
Therefore, the three zeros of P are ______, ______, and ______.	$-1/2$ $-2/3$; 4
9. Find the rational zeros of the function defined by $P(x) = x^3 - \frac{9}{2}x^2 - 3x + \frac{5}{2}$. The zeros are _____, _____, and _____.	1/2; −1; 5

6.5 Real Zeros of Polynomial Functions

1. By Descartes' rule of ______, if P(x) defines a polynomial function with ______ coefficients and terms in descending powers of x, then	signs real
(a) the number of __________ real zeros of P either equals the number of _________ in sign occurring in the coefficients of P(x), or is less than the number of variations by a positive _____ integer.	positive variations even
(b) the number of __________ real zeros of P either equals the number of variations in sign of ______, or else is less than the number of variations by a positive _____ integer.	negative P(−x) even

Use Descartes' rule of signs to find the number of positive or negative real zeros for the polynomial functions defined in Frames 2–5.

2. $P(x) = 6x^3 - 7x^2 + 1$

There are ______ variations of sign in $P(x)$. | 2

Then by Descartes' rule of signs, the number of positive real zeros is either ______ or | 2

$2 - 2 =$ _____. Find $P(-x)$. | 0

$P(-x) = 6(-x)^3 - 7(-x)^2 + 1 =$ ______________ | $-6x^3 - 7x^2 + 1$

There is ____ variation of sign in $P(-x)$. Then | 1

by Descartes' rule of signs, the number of negative real zeros is ____. | 1

3. $P(x) = x^3 - 5x^2 - 4x - 1$

$P(x)$ has _____ variation in sign, so that P has | 1

_____ positive real zero. | 1

$P(-x) = (\_\_\_\_)^3 - 5(\_\_\_\_)^2 - 4(\_\_\_\_) - 1$ | $-x$; $-x$; $-x$

$=$ ____________________ | $-x^3 - 5x^2 + 4x - 1$

$P(-x)$ has ____ variations in sign, so that ______ | 2; P

has ____ or ____ negative real zeros. | 2; 0

4. $P(x) = 2x^4 - 3x^3 - 5x^2 + 4x - 1$

There are _____ or _____ positive real zeros. | 3; 1

$P(-x) =$ ______________________ | $2x^4 + 3x^3 - 5x^2 - 4x - 1$

There is 1 ____________ real zero. | negative

5. $P(x) = x^4 - 3x^3 + 11x^2 - 9x - 8$

positive zeros: ___________ | 3 or 1

negative zeros: ___________ | 1

6. By the _________________ value theorem, if $P(x)$ | intermediate

defines a polynomial function with only real coefficients and if $P(a)$ and $P(b)$ have __________ | opposite

signs, then there is at least one _______ zero | real

between _____ and _____. | a; b

In Frames 7–9, show that each polynomial function defined as follows has a real zero between the numbers given.

7. $P(x) = 2x^3 + x^2 - 5x + 2$; 0 and 3/4

   By the intermediate value theorem, if P(0) and P(3/4) are ________ in sign, then there is a zero between them.

   First, $P(0) =$ ____

   and $P(\frac{3}{4}) = 2(\_\_\_) + \_\_\_ - \_\_\_ + 2$

   $= \_\_\_\_.$

   Then there (*is/is not*) a zero between 0 and 3/4.

   opposite

   2

   $\frac{27}{64}$; $\frac{9}{16}$; $\frac{15}{4}$

   $-\frac{11}{32}$

   is

8. $P(x) = 2x^3 + 3x^2 - 5x - 6$; 1 and 2

   $P(1) = -6$ and $P(2) =$ ____

   Then there (*is/is not*) a zero between 1 and 2.

   12

   is

9. $P(x) = 2x^3 + 9x^2 + 13x + 6$; -1.9 and -1.2

   $P(-1.9)$ ___ 0 and $P(-1.2)$ ___ 0

   There (*is/is not*) a zero between -1.9 and -1.2.

   $>$; $<$

   is

10. Suppose that P is a polynomial function with real coefficients and with a positive leading coefficient. Suppose that P(x) is divided synthetically by $x - c$.

    (a) If $c > 0$, and all numbers in the bottom row of the synthetic division are nonnegative, then P has no zero ________________.
    c is called a(n) ________ bound.

    greater than c

    upper

    (b) If $c < 0$, and all the numbers in the bottom row of the synthetic division __________ in sign (with 0 considered positive or negative as needed), then P has no zero ____________________. c is called a(n) _______ bound.

    alternate

    less than c

    lower

11. Show that the function defined by $P(x) = x^3 + 4x^2 + x - 6$ has no zero larger than 2.

To show this, use ____________ division.

synthetic

2)	1	4	1	-6
		2	____	____
	1	6	____	____

12; 26

13; 20

Since 2 is positive and every number in the bottom row is __________, there is no zero greater than _____.

positive

2

12. Show that the function defined by $P(x) = x^4 + 2x^3 - 7x^2 - 8x + 12$ has no zero greater than 3.

3)	1	2	-7	-8	12
		3	15	24	48
	1	5	8	16	60

Every number in the bottom row is positive.

13. Show that the function defined by $P(x) = x^3 + 4x^2 + x - 6$ has no zero less than -5.

-5)	1	4	1	-6
		-5	5	-30
	1	-1	6	____

-36

Since -5 is negative and the numbers in the bottom row ____________ in sign, there is no zero less than _____.

alternate

-5

14. Show that the function defined by $P(x) = x^4 + 2x^3 - 7x^2 - 8x + 12$ has no zero less than -6.

-6)	1	2	-7	-8	12
		-6	24	-102	660
	1	-4	17	-110	672

Signs in the bottom row alternate.

15. Approximate the real zeros of the function defined by $P(x) = 6x^3 - 7x^2 + 1$ to the nearest tenth. We found in Frame 2 that P has 2 or 0 positive real zeros and ____ negative real zero. Let us look for the negative zero first. As a start, use synthetic division with −1 as a divisor.

−1)	6	−7	0	1
		−6	13	_____
	6	−13	13	_____

Since the numbers in the bottom row of the synthetic division alternate in sign, we know that _____ is less than any zero of P. Then, our negative real zero must be between ____ and ____. Is the zero between −.5 and 0? Find P(−.5).

$$P(-.5) = 6(-.125) - 7(.25) + 1 = \_\_\_\_\_\_\_\_$$
$$P(0) = \_\_\_\_\_\_\_\_$$

Then, we know, since the signs are ____________, that our negative real zero lies between ______ and 0. Now test −.3.

$$P(-.3) = 6(-.3)^3 - 7(-.3)^2 + 1 = \_\_\_\_\_$$

Since P(−.3) ___ 0 and P(−.5) ___ 0, the negative zero lies between _____ and _____. Try −.4.

$$P(-.4) = \_\_\_\_\_\_\_\_$$

P(_____) is closer to 0 than P(_____), so to the nearest tenth our zero is _____. Continue in this way to find the two positive real zeros, ________.

1
−13
−12
−1
−1; 0
−1.5
1
opposite
−.5
.208
$>$; $<$
−.3; −.5
−.504
−.3; −.4
−.3
1 and .5

16. Earlier in this chapter, we graphed functions of the polynomials that could be factored. However, we cannot factor

$$P(x) = 6x^3 - 23x^2 + 12x + 20,$$

so we must graph the function using theorems of this section.

First note that we have _____ variations of sign in P(x) and since

P(-x) = ______________________,

we have _____ variation in sign for P(-x). We have two or no positive real zeros and one negative real zero. Let's evaluate some points, using the shortened version of synthetic division.

2

$-6x^3 - 23x^2 - 12x + 20$

1

x				P(x)	Ordered pair
	6	-23	12	20	
-1	6	-29	41	-21	________
0	6	-23	12	20	________
1	6	-17	-5	15	________
2	6	-11	-10	0	________
3	6	-5	-3	11	________
4	6	1	16	84	________

(-1, -21)
(0, 20)
(1, 15)
(2, 0)
(3, 11)
(4, 84)

Now we analyze our chart. The first thing we notice is that ____ is a zero of P. Notice that the row of the synthetic division of -1 has alternating signs, which means that _____ is less than any zero of P. Also, P(-1) and P(0) vary in sign so our negative zero is between _____ and _____.

2

-1

-1

0

The last row of the synthetic division of 4 has all nonnegative numbers so _____ is larger than any zero of P. Let's do one further test between 2 and 3. (We choose this area because 0 can be negative or positive as needed and P(2) = 0.) Testing 5/2 (which is 2.5), we get

4

```
5/2) 6   -23    12    20
          15   -20   ____
     ----------------------
     6    -8    -8   ____.
```

-20

0

This is the third zero. Sketch the graph and check your answer. Notice that for $x < -1$ the values will become increasingly more negative and for $x > 3$ the values will become increasingly more __________.

positive

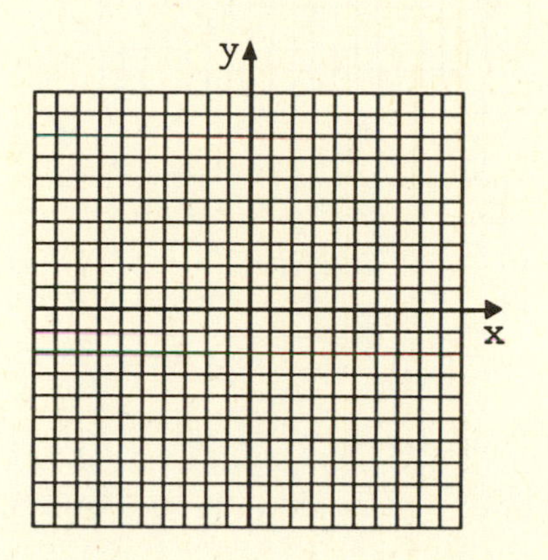

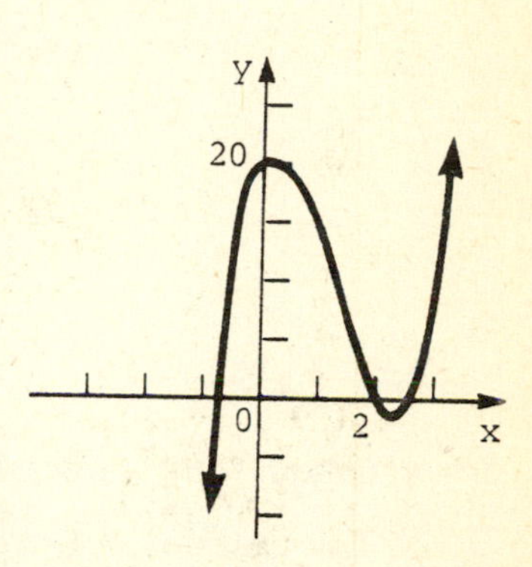

17. Graph the function defined by $P(x) = x^3 + x^2 - 2x$. $P(x)$ has ____ variation in sign, so we expect to find one real positive zero.

1

$P(-x) =$ ______________

$-x^3 + x^2 + 2x$

which also has one variation in sign, so we expect to find _____ real negative zero(s). We may have a polynomial of third degree with only two real zeros, but we must work carefully. Let's find a few points.

one

x				P(x)	Ordered pair
	1	1	-2	0	
-2	1	-1	0	0	__________
-1	1	0	-2	2	__________
0	1	1	-2	0	__________
1	1	2	0	0	__________
2	1	3	4	8	__________

(-2, 0)
(-1, 2)
(0, 0)
(1, 0)
(2, 8)

Sketch the graph, testing the area between $x = 0$ and $x = 1$.

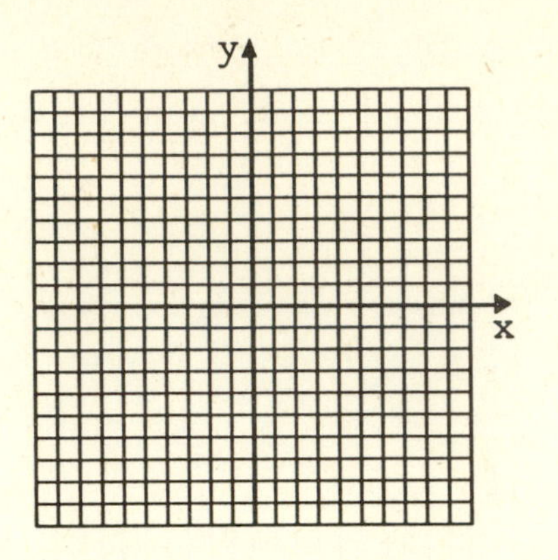

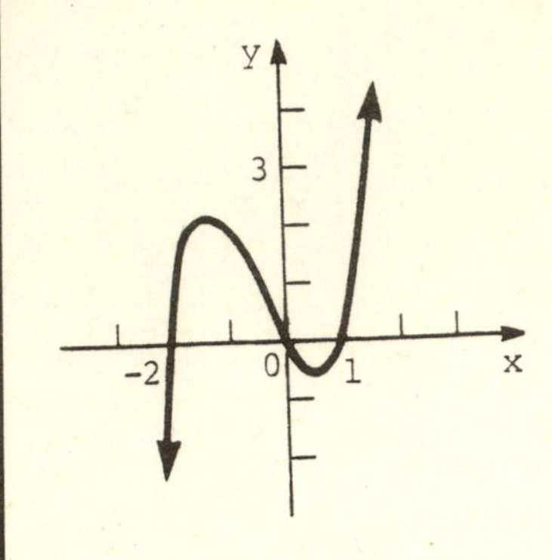

18. Graph the function defined by $P(x) = x^3 + 1$. The number -1 is a zero of this function, but how do we find any other zeros? We can use synthetic division with ____ to find the quadratic polynomial that is the quotient. | -1

-1)	1	0	0	1
		-1	1	-1
	1	-1	1	0

Then $P(x) = (x + 1)($________$)$. (We could have used our knowledge of factoring the sum of two ________.) Using the quadratic formula on ______________ = 0, we find complex solutions. Then _____ is the only real zero of P. Plot a few points and sketch the graph.

$x^2 - x + 1$

cubes

$x^2 - x + 1$

-1

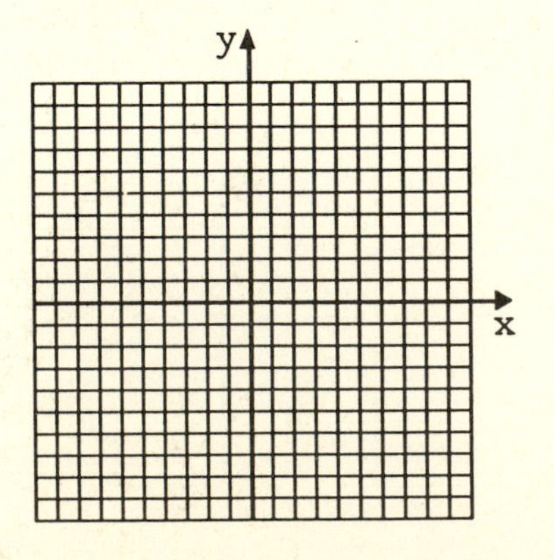

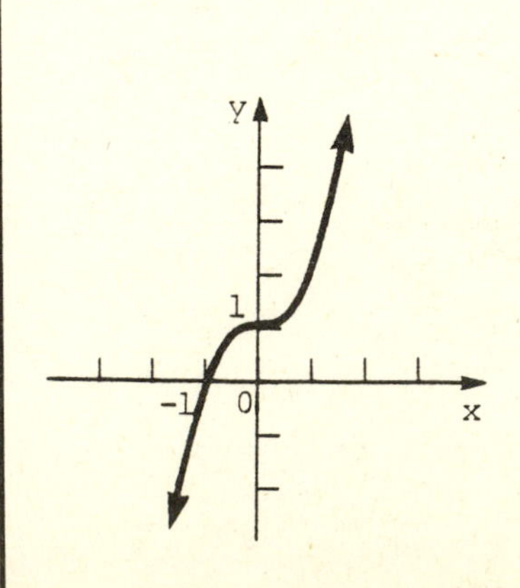

19. Graph the function defined by $P(x) = 2x^3 + x^2 - 5x + 2$.

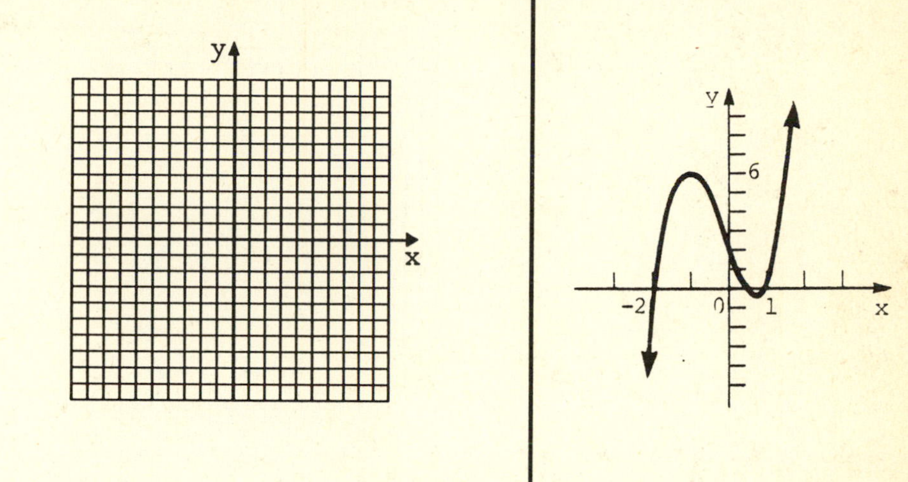

6.6 Graphing Rational Functions

1. A function f of the form p/q defined by

$$f(x) = \frac{p(x)}{q(x)},$$

where p(x) and q(x) are ____________, is called a ____________ function.

polynomial

rational

In Frames 2–4, use reflections and translations to graph each rational function.

2. $f(x) = \frac{3}{x} + 1$

Compared with the graph of $f(x) = 1/x$, the graph will be points _____, times as far away from the x-axis translated 1 unit _____. The graph includes the points

3

up

(1, ____), (3, ____), (-1, ____) and (-3, ____).

y
x

4; 2; -2; 0

y
5
y = 1
-3
0
x

3. $f(x) = \frac{1}{x - 2}$

Compared with the graph of $f(x) = 1/x$, the graph of this function will be translated ____ units ________.

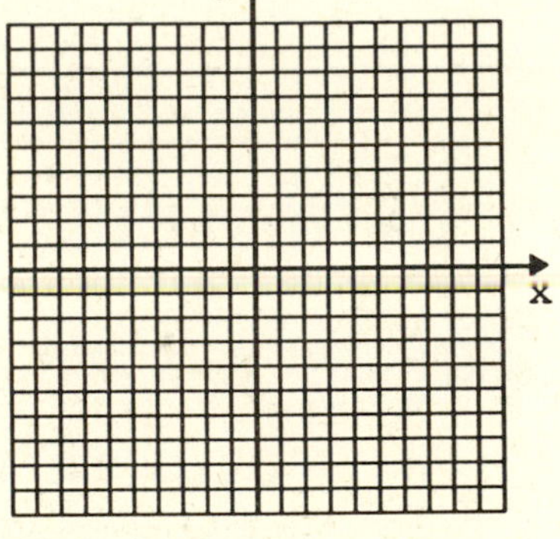

2

right

y
5
0
5
x
x = 2

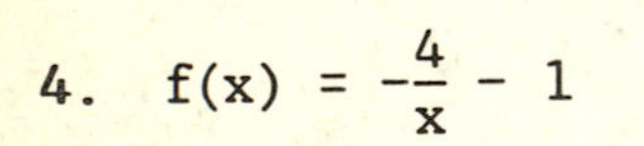

4. $f(x) = -\frac{4}{x} - 1$

y
x

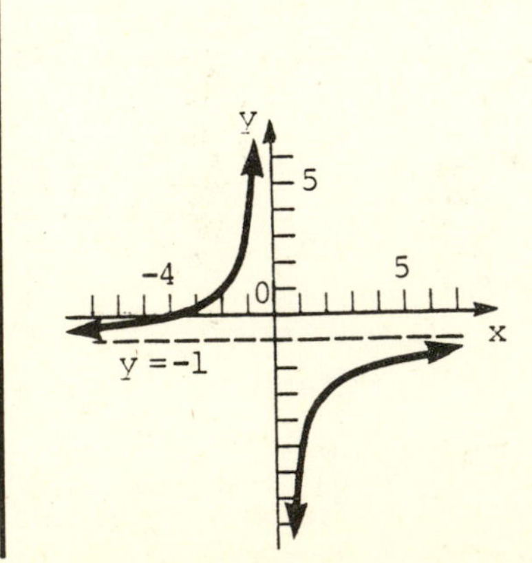

5. Vertical asymptotes are found by determining the values of x which make the __________ equal to 0 but do not make the numerator equal to ____. __________, or in some cases __________, asymptotes are found by considering what happens to f(x) as ____ gets larger and larger, written __________.

denominator

0; Horizontal

oblique

$|x|$

$|x| \to \infty$

Identify all vertical and horizontal asymptotes in Frames 6–8.

6. $f(x) = \frac{6x - 5}{2x + 7}$

To find the vertical asymptote, set ______ = 0, and solve for ____. To find the horizontal asymptote, divide each term in the numerator and denominator by ____, the largest power of x in the expression. As $|x|$ gets larger, the value of f(x) approaches ____.

Vertical: __________

Horizontal: __________

$2x + 7$

x

x

3

$x = -7/2$

$y = 3$

7. $f(x) = \frac{9}{x - 8}$

Vertical: __________

Horizontal: __________

$x = 8$

$y = 0$

8. $f(x) = \frac{3x - 5}{9x + 11}$

Vertical: __________

Horizontal: __________

$x = -11/9$

$y = 1/3$

9. Find all asymptotes of $f(x) = \frac{2x^2 - 1}{x + 3}$.

The vertical asymptote is ______, found by setting the denominator equal to zero. Dividing each term by ____, the largest power of x in the expression, indicates that there is no

$x = -3$

x^2

______________ asymptote. (Note that the degree of the numerator is __________ than the degree of the denominator.)
If we divide the numerator by the denominator, $f(x) =$ _______________. Then, as $|x|$ gets larger and larger, the graph of $f(x)$ gets closer to the oblique asymptote, the line ______________.

horizontal

greater

$2x - 6 + \frac{17}{x + 3}$

$y = 2x - 6$

10. The graph of a rational function (*may/may not*) intersect a vertical asymptote; it (*may/may not*) intersect a nonvertical asymptote.

may not

may

11. Graph $f(x) = \frac{3}{2x + 1}$.

The vertical asymptote is __________.
The horizontal asymptote is __________.
$f(0) =$ _____, so the y-intercept is __________.
The x-intercept is found by solving $f(x) = 0$, which gives $3 \neq 0$, so there (*is/is not*) an x-intercept. The graph (*does/does not*) intersect its horizontal asymptote, $y = 0$. Plot a point using an x-value in each interval determined by the x-intercept(s) and vertical asymptotes.
The intervals for this function are _________ and _________.
Complete the sketch.

y

x

$x = -\frac{1}{2}$

$y = 0$

3; 3

is not

does not

$(-\infty, -1/2)$

$(-1/2, \infty)$

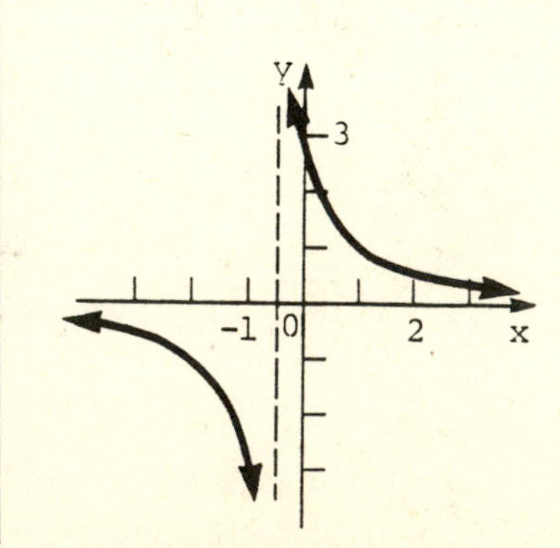

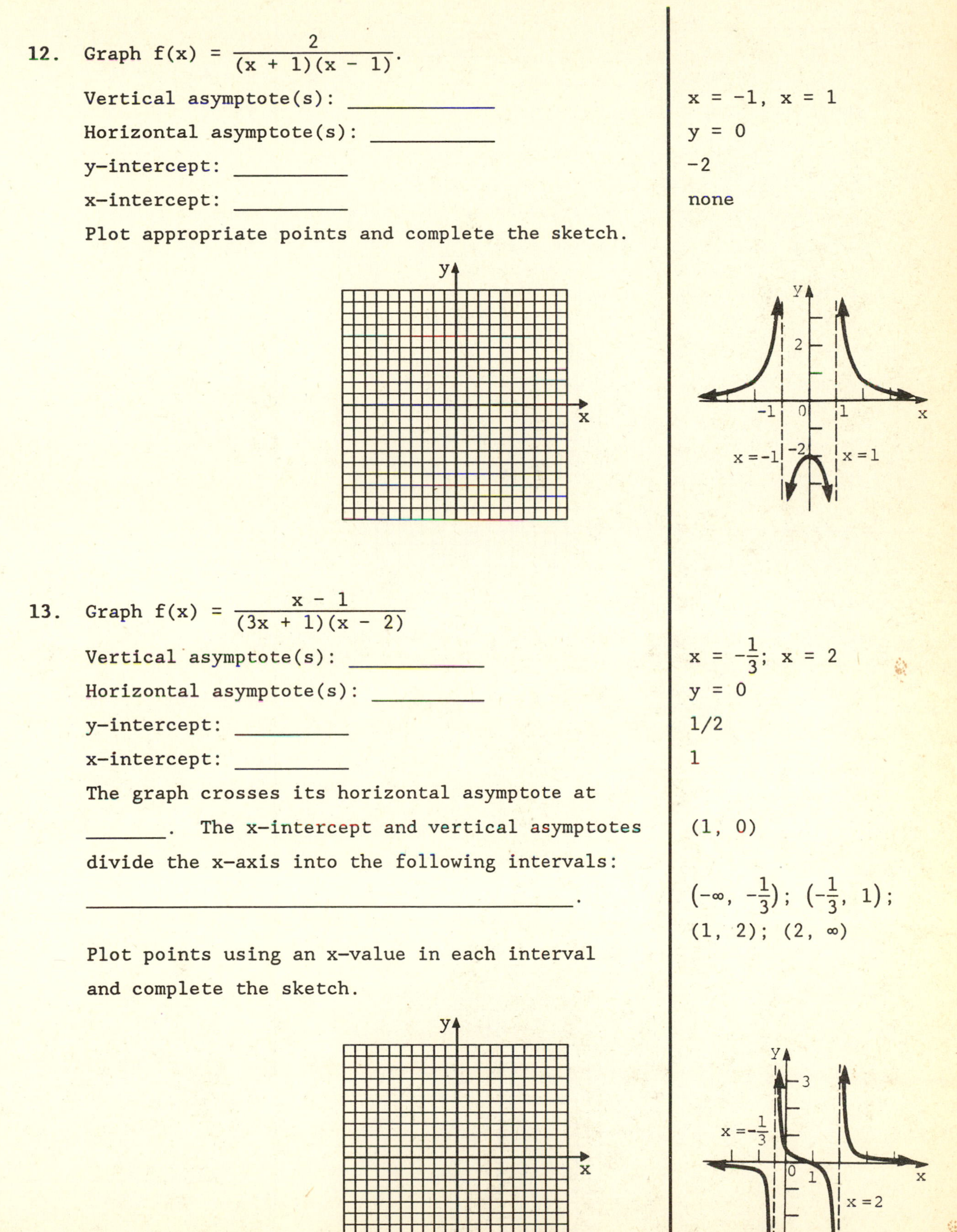

12. Graph $f(x) = \frac{2}{(x + 1)(x - 1)}$.

Vertical asymptote(s): ____________

Horizontal asymptote(s): __________

y-intercept: __________

x-intercept: __________

Plot appropriate points and complete the sketch.

$x = -1$, $x = 1$

$y = 0$

-2

none

13. Graph $f(x) = \frac{x - 1}{(3x + 1)(x - 2)}$

Vertical asymptote(s): ____________

Horizontal asymptote(s): __________

y-intercept: __________

x-intercept: __________

The graph crosses its horizontal asymptote at _______. The x-intercept and vertical asymptotes divide the x-axis into the following intervals: ____.

Plot points using an x-value in each interval and complete the sketch.

$x = -\frac{1}{3}$; $x = 2$

$y = 0$

1/2

1

(1, 0)

$(-\infty, -\frac{1}{3})$; $(-\frac{1}{3}, 1)$; $(1, 2)$; $(2, \infty)$

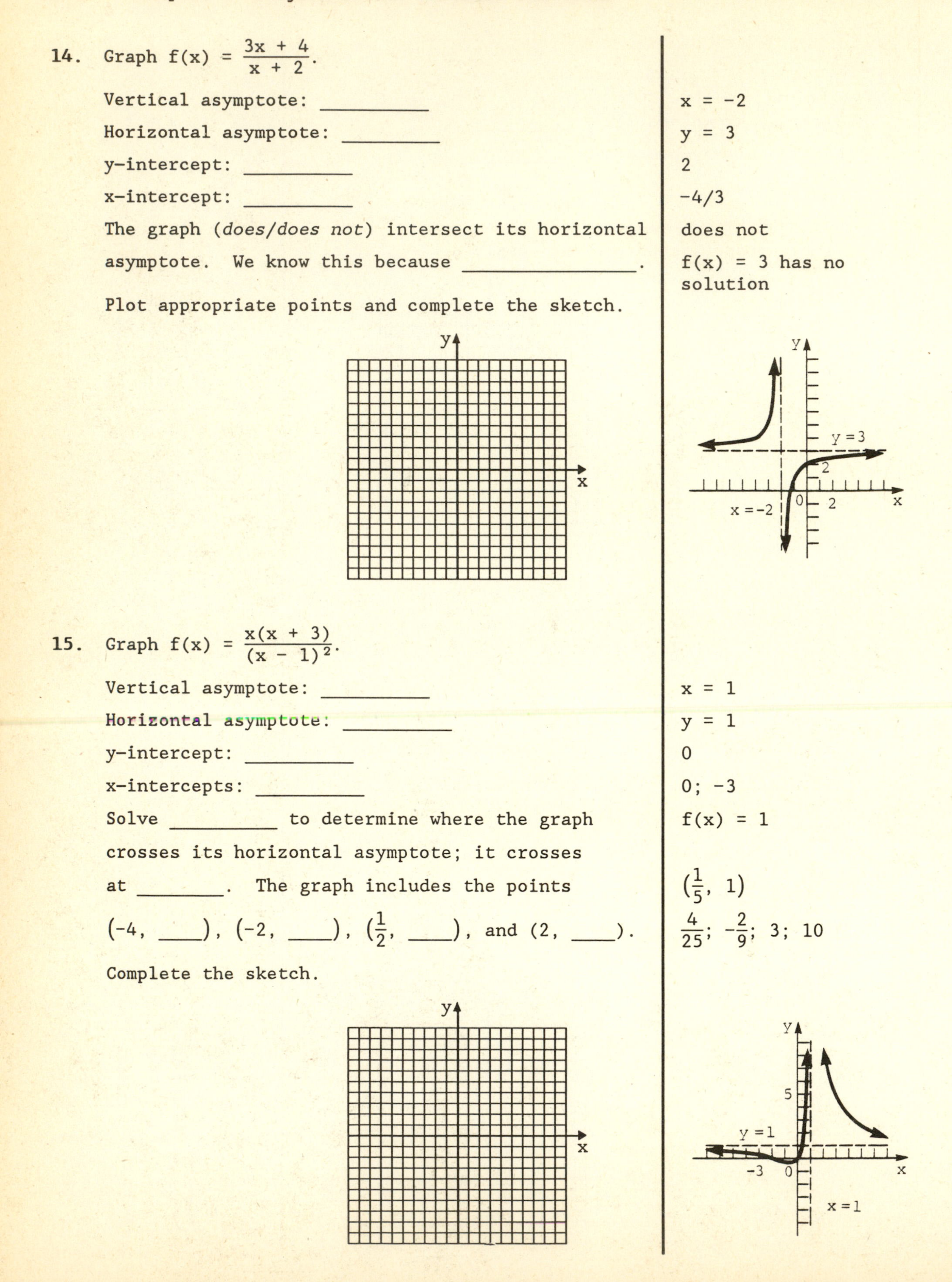

14\. Graph $f(x) = \frac{3x + 4}{x + 2}$.

Vertical asymptote: ______

Horizontal asymptote: ______

y-intercept: ______

x-intercept: ______

The graph (*does/does not*) intersect its horizontal asymptote. We know this because ______.

Plot appropriate points and complete the sketch.

Answers: $x = -2$; $y = 3$; 2; $-4/3$; does not; $f(x) = 3$ has no solution

15\. Graph $f(x) = \frac{x(x + 3)}{(x - 1)^2}$.

Vertical asymptote: ______

Horizontal asymptote: ______

y-intercept: ______

x-intercepts: ______

Solve ______ to determine where the graph crosses its horizontal asymptote; it crosses at ______. The graph includes the points $(-4,\ \_\_\_\_)$, $(-2,\ \_\_\_\_)$, $(\frac{1}{2},\ \_\_\_\_)$, and $(2,\ \_\_\_\_)$.

Complete the sketch.

Answers: $x = 1$; $y = 1$; 0; 0; -3; $f(x) = 1$; $(\frac{1}{5}, 1)$; $\frac{4}{25}$; $-\frac{2}{9}$; 3; 10

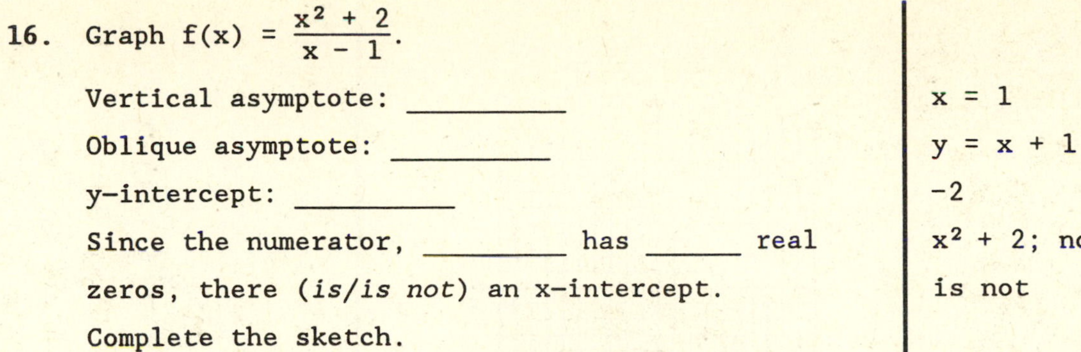

16. Graph $f(x) = \dfrac{x^2 + 2}{x - 1}$.

Vertical asymptote: __________

Oblique asymptote: __________

y-intercept: __________

Since the numerator, _________ has ______ real zeros, there (*is/is not*) an x-intercept.

Complete the sketch.

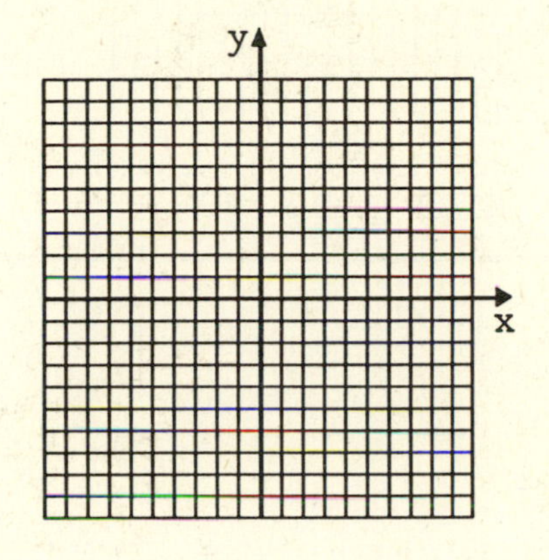

$x = 1$

$y = x + 1$

-2

$x^2 + 2$; no

is not

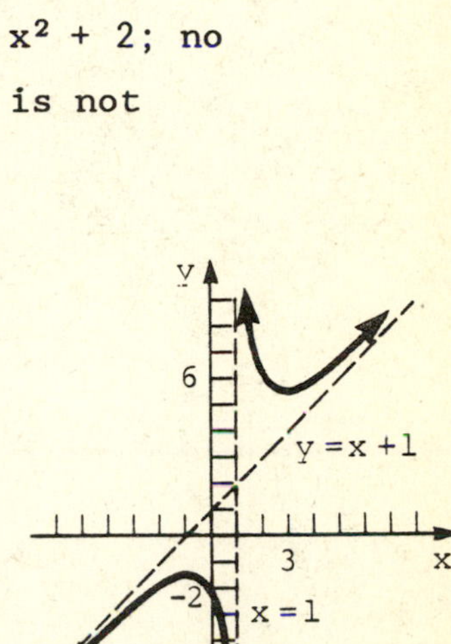

17. Graph $f(x) = \dfrac{x^2 - 9}{x + 3}$.

First factor the numerator and write f(x) in lowest terms.

$f(x) =$ ________, $x \neq$ ______.

A "_______" will appear in the graph at x = ____.

Sketch the graph.

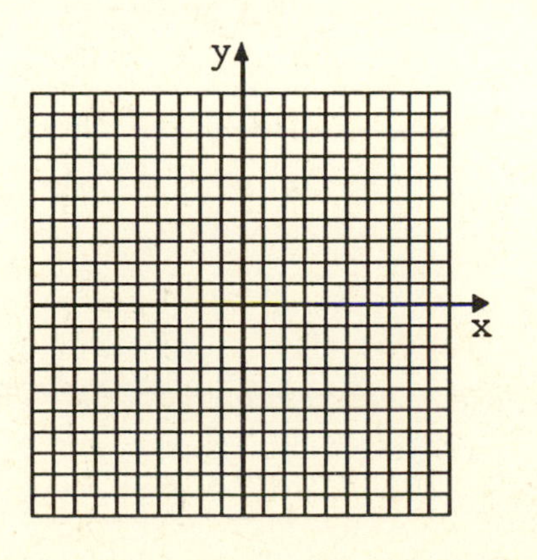

$x - 3$; -3

hole; -3

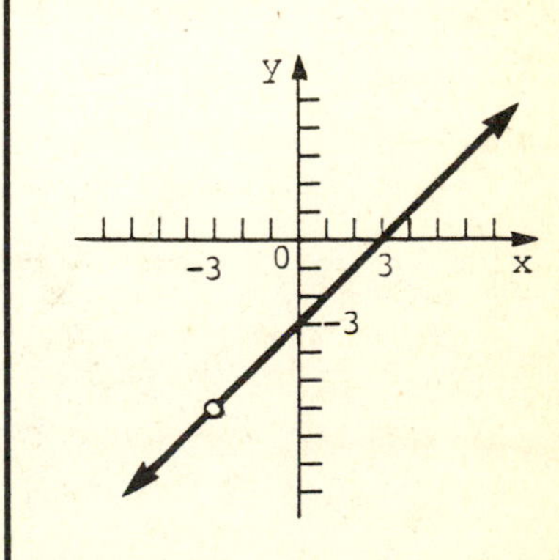

18. Graph $f(x) = \dfrac{25 - x^2}{5 - x}$.

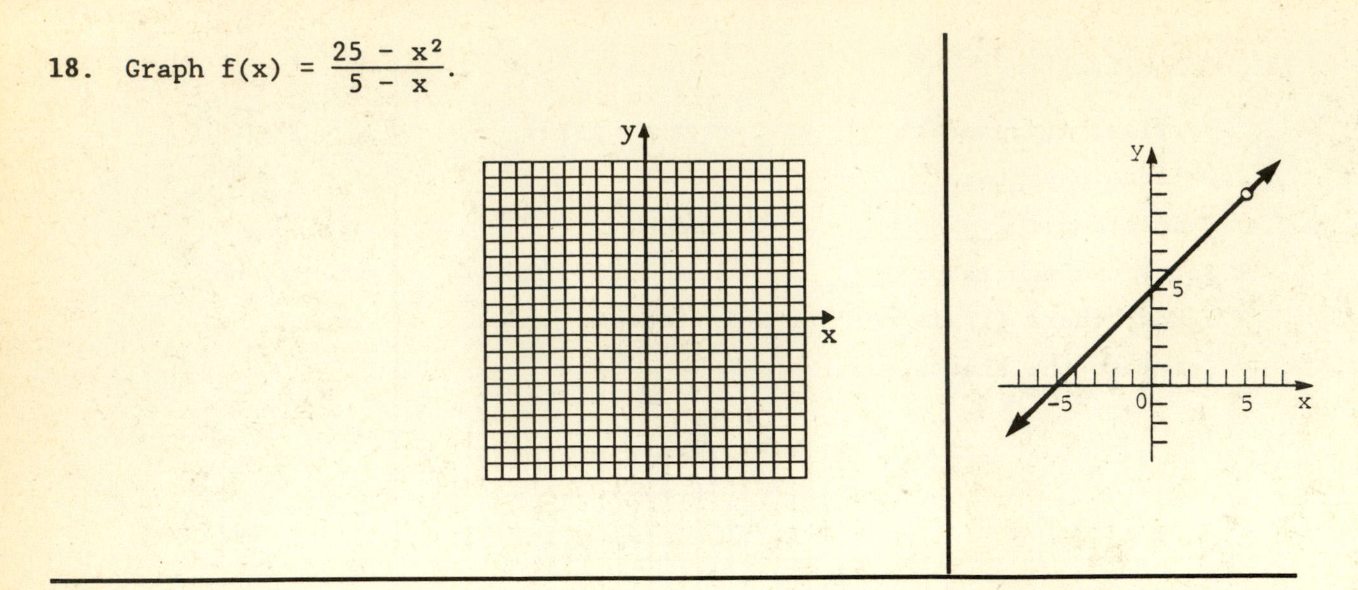

Chapter 6 Test

Test answers are at the back of this study guide.

Graph each polynomial function P as defined.

1. $P(x) = (2 + x)^2$

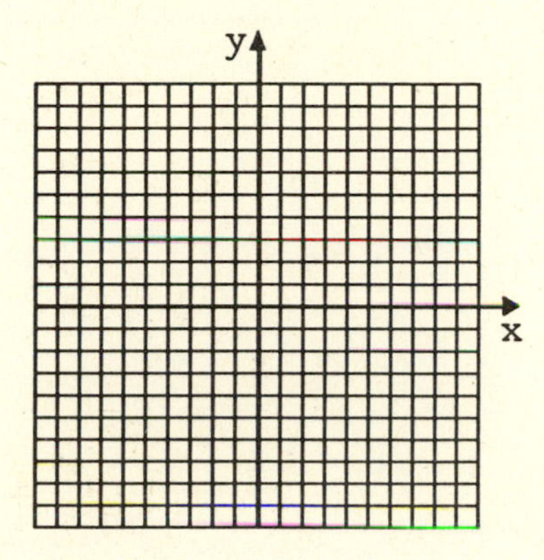

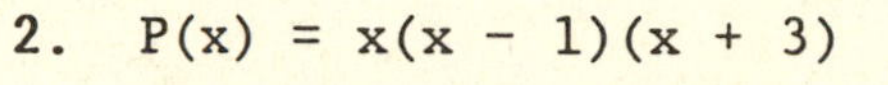

2. $P(x) = x(x - 1)(x + 3)$

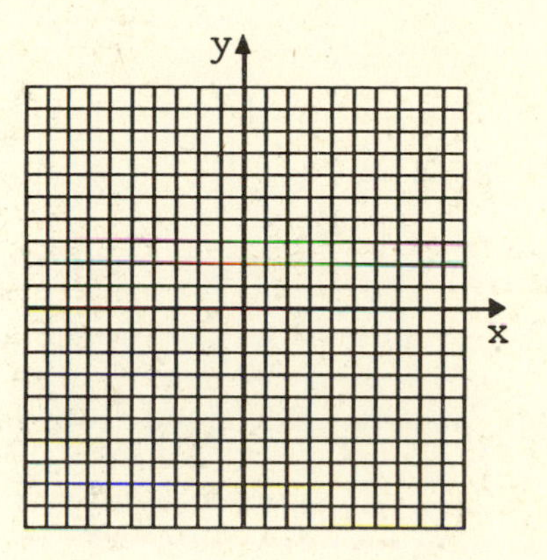

Use synthetic division to perform each division.

3. $(4x^3 - 3x^2 + 6x - 1) \div (x + 2)$ — 3. ____________

4. $(x^5 - 20x^2 - 62) \div (x - 3)$ — 4. ____________

5. Use synthetic division to find $P(-4)$ for

$$P(x) = x^5 + 4x^4 + x^2 - 10.$$

5. ____________

6. Find a polynomial function of lowest degree with 0, -1, -2, -3, and 2 as zeros. — 6. ____________

7. Is 4 a zero of the function P defined by

$$P(x) = x^4 - 3x^3 + 5x^2 - x + 12?$$

Why or why not? — 7. ____________

8. Is $x + 3$ a factor of

$$P(x) = 2x^3 + 5x^2 - 4x - 3?$$

Why or why not? 8. ____________

9. Find a function P defined by a polynomial of degree 4 with -1, 4, and i as zeros and with $P(-2) = 64$. 9. ____________

10. Factor $P(x) = x^3 - 4x^2 + x + 6$ given that 2 is a zero of P. 10. ____________

11. For the polynomial function defined by

$$P(x) = 2x^3 + 3x^2 - 3x - 2,$$

(a) list all rational numbers that can possibly be zeros of P, 11. (a) ____________

(b) find all rational zeros of P. (b) ____________

12. Explain why the polynomial function defined by $P(x) = x^4 + 3x^3 - 2x^2 + x - 1$ can have at most one negative real zero.

12. ____________

13. Show that the polynomial function defined by $P(x) = x^3 - 2x^2 - 4x + 4$ has no real zero greater than 4 nor less than -2.

13. ____________

14. Find the possible numbers of positive and negative real zeros of the polynomial function defined by $P(x) = 4x^4 - 8x^3 + 17x^2 - 2x - 14$.

14. ____________

Graph each function defined as follows.

15. $P(x) = 8x^3 - 6x^2 - 11x + 3$

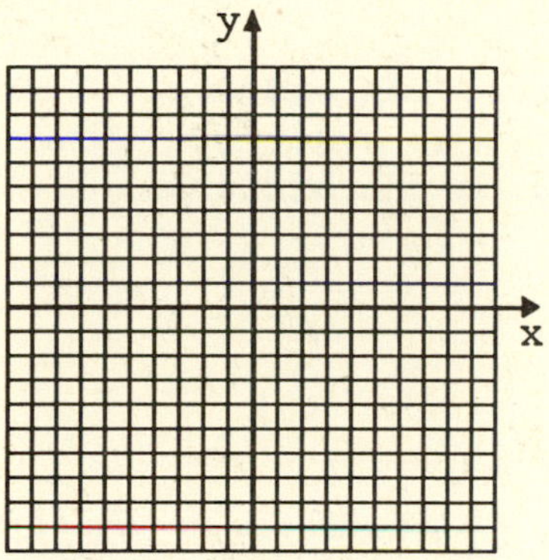

16. $f(x) = \frac{1}{2 - x}$

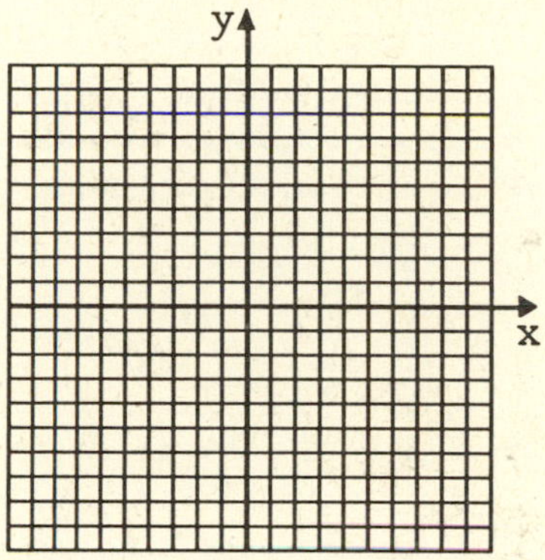

17. $f(x) = \frac{x^2 - 4}{x - 1}$

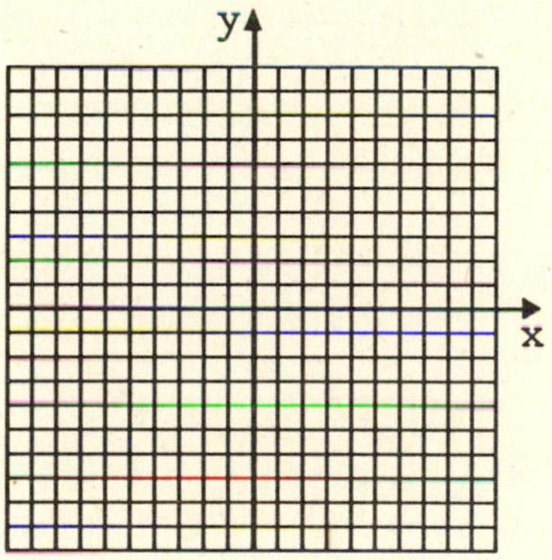

18. $f(x) = \frac{3x}{x^2 - 1}$

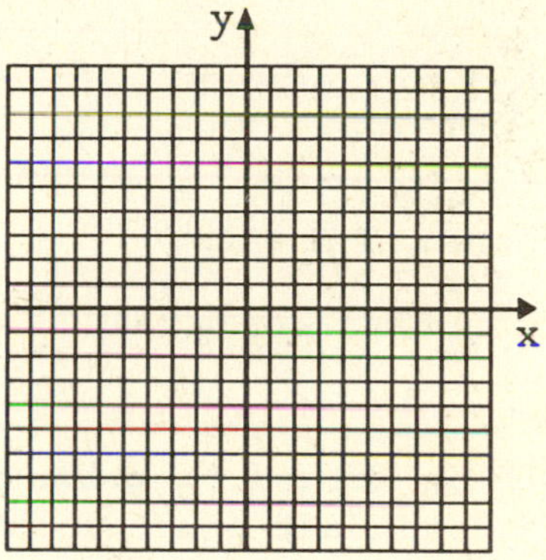

19. What is the equation of the oblique asymptote of the graph of

$$f(x) = \frac{3x^2 - 2x + 1}{x - 1}?$$

19. ____________________

20. Which one of the functions defined as follows has a graph with no x-intercepts?

(a) $f(x) = \frac{x + 1}{x - 2}$ (b) $f(x) = x(x - 3)^3$

(c) $f(x) = \frac{4}{x^3 - 1}$ (d) $f(x) = \frac{3x}{x^2 - 1}$

20. ____________________

CHAPTER 7 PRETEST

Pretest answers are at the back of this study guide.

Use the substitution method to solve each of the following systems. Identify any systems with dependent equations or any inconsistent systems. If a system has dependent equations, express the solution set with y arbitrary.

1. $$\begin{aligned} 2x + 3y &= 5 \\ -2x + y &= 7 \end{aligned}$$

1. ______________

2. $$\begin{aligned} 2x - 3y &= -5 \\ x + 2y &= -6 \end{aligned}$$

2. ______________

3. $$\begin{aligned} -x + 4y &= 7 \\ 2x - 8y &= 1 \end{aligned}$$

3. ______________

Use the addition method to solve each of the following systems. Identify any systems with dependent equations or any inconsistent systems. If a system has dependent equations, express the solution set with y arbitrary.

4. $$\begin{aligned} 2x + 6y &= -8 \\ -x - 3y &= 4 \end{aligned}$$

4. ______________

5. $$\begin{aligned} 4x + 5y &= -5 \\ 2x - 2y &= 2 \end{aligned}$$

5. ______________

6. $$\begin{aligned} 6x + 3y &= 12 \\ 3x + 2y &= 7 \end{aligned}$$

6. ______________

7. A student says that $\{(1, 1)\}$ is the solution set of the system

$$\begin{aligned} 2x - 3y &= -1 \\ -4x + 6y &= 2 \end{aligned}$$

because (1, 1) satisfies both equations at the same time. Is he correct? Explain.

7. ____________________

8. On Saturday Drew bought 3 plants and 2 hanging baskets at the Garden Center for a total of \$22.50. On Sunday he went back and bought 2 more plants and 2 more hanging baskets for a total of \$19. If each plant cost the same price and each hanging basket cost the same price, how much did each cost?

8. ____________________

Use the addition method to solve the following linear systems.

9.
$$\begin{aligned} 2x + y - 2z &= -2 \\ x + 2y + 3z &= 0 \\ -2x - y + 3z &= 3 \end{aligned}$$

9. ____________________

10.
$$\begin{aligned} 4p - 3q + r &= -11 \\ p + 2q - 2r &= 10 \\ 3p - q - r &= -1 \end{aligned}$$

10. ____________________

11. Find a solution for the following system in terms of the arbitrary variable z.

$$\begin{aligned} 3x - 2y - 2z &= 1 \\ x + y - z &= 2 \end{aligned}$$

11. ____________________

12. The sum of three numbers is 2. The first number is the sum of the other two. The third number is three times the first number. Find the three numbers by using a system of equations.

12. ______________________

Use the Gauss-Jordan method to solve each of the following systems.

13. $4x + 2y = 20$
$x + y = 4$

13. ______________________

14. $2x + 3y + 2z = -1$
$-2x + 2y + 3z = 6$
$x - 2y - 2z = -4$

14. ______________________

Solve each of the following nonlinear systems of equations.

15. $y = -x^2$
$y - x = -2$

15. ______________________

16. $x^2 + y^2 = 16$
$(x - 2)^2 + y^2 = 4$

16. ______________________

17. If a system of two nonlinear equations contains an equation whose graph is a parabola and another equation whose graph is a line, can the system have exactly one solution? If so, describe the graph of the situation.

17. ______________________

18. Use a system of equations to find two numbers such that their sum is 0 and the sum of their squares is 242.

18. ______________________

19. Graph the solution set of the following system of inequalities.

$$\frac{x^2}{16} + \frac{y^2}{9} \geq 1$$

$$\frac{x^2}{36} + y^2 \leq 1$$

19.

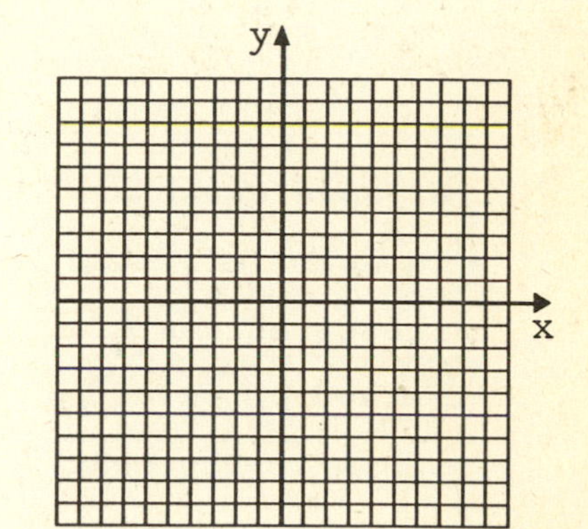

20. A company makes two models of widgets, A and B, which must be assembled and packed. The time it takes to assemble model A is 8 minutes and model B takes 2 minutes. The time it takes to pack model A is 2 minutes and model B takes 3 minutes. Each hour there are available 32 minutes of assembly time and 18 minutes of packing time. If both models of widgets sell for $12 each, how many of each model should be made per hour to obtain maximum income? What is the maximum hourly income?

20. ______________________

____

CHAPTER 7 SYSTEMS OF EQUATIONS AND INEQUALITIES

7.1 Linear Systems with Two Variables

1. A __________ of equations is made up of _______ or more equations that have the same __________. To solve a system, find all values that make all the equations true at the same _________. One method of solution, the substitution method, is shown in the next few frames.

system; two
variables
time

2. Solve the system

$$3x + 4y = 17$$
$$y = -5x.$$

To solve this system by the substitution method, use the top equation and replace y with _______. This gives

$$3x + 4(\_\_\_\_\_) = 17$$
$$\_\_\_\_\_\_\_\_ = 17$$
$$x = \_\_\_\_\_.$$

To find y, use the equation ________ and replace x with ____.

$$y = -5x$$
$$y = -5(\_\_\_\_)$$
$$y = \_\_\_\_$$

The solution set is _____________.
The graphs of the two equations in the system are ________. They intersect in a single ________ whose coordinates give the _________ of the system.

$-5x$
$-5x$
$-17x$
-1
$y = -5x$
-1
-1
5
$\{(-1, 5)\}$
lines
point; solution

Solve the systems of Frames 3 and 4 by the substitution method.

3. $8x - 3y = -1$

$x = 4 - y$

Replace x with _______ in the first equation. | $4 - y$

$8(\_\_\_\_\_\_\_) - 3y = -1$ | $4 - y$

$32 \_\_\_\_\_\_\_ - 3y = -1$ | $- 8y$

$32 - \_\_\_\_\_ = -1$ | $11y$

$-11y = \_\_\_\_\_$ | -33

$y = \_\_\_\_\_$ | 3

Use x = _______ to find that x = ___. The solution set is _________. | $4 - y$; 1 | $\{(1, 3)\}$

4. $2x - 5y = -43$

$x + 2y = 10$

First, solve the second equation for x.

$x = \_\_\_\_\_\_\_\_$ | $10 - 2y$

Now replace x with ________ in the first equation and complete the solution. Solution set: ______ | $10 - 2y$ | $\{(-4, 7)\}$

5. As an alternative to the substitution method, systems can also be solved by the ___________ method. For example, to solve the system | addition

$x - 2y = 7$

$2x + 4y = 6,$

we can eliminate the variable x by multiplying both sides of the first equation by _____ and _______ the two equations. | -2 | adding

$\_\_\_\_\_\_\_\_\_\_\_\_\_\_\_$ | $-2x + 4y = -14$

$2x + 4y = 6$

$\_\_\_\_\_\_\_\_\_\_\_\_\_\_\_$ | $8y = -8$

Then $y =$ ____. Substituting ____ for y in the first equation we get	-1; -1
$$x - 2(-1) = 7$$ $$x = \_\_\_\_.$$	5
Check the answer by substituting the value for x and y in both equations.	
$(\_\_\_) - 2(\_\_\_\_\_) = \_\_\_\_$	5; -1; 7
$2(\_\_\_\_) + 4(\_\_\_\_\_) = \_\_\_\_$	5; -1; 6
The solution set is ________.	$\{(5, -1)\}$

Solve the systems of Frames 6–9.

6. $4x + 2y = 20$ $3x + y = 14$	
We choose to eliminate the variable y this time because it is easier. Multiply the second equation by _____ and add.	-2
$4x + 2y = 20$ ______________ $-2x \quad = -8$	$-6x - 2y = -28$
$x =$ ____	4
Substitute ____ for x in either of the original equations to find that $y =$ ____. The solution set for the system is ________.	4 2 $\{(4, 2)\}$
7. $-3x + 2y = 23$ $2x + 3y = 2$	
(Hint: Multiply each equation by a different number in order to eliminate one of the variables.)	
Solution set __________	$\{(-5, 4)\}$

8. $4x - 3y = 6$
$-8x + 6y = 5$

Eliminate the variable x by multiplying both sides of the first equation by 2 and adding.

____
$-8x + 6y = 5$
____

Here we get a false statement. This means that the system has ____ solution, and the graphs of the equations are distinct ____________ lines. Such a system is ______________. The solution set is ____.

$8x - 6y = 12$

$0 = 17$

no

parallel

inconsistent

Ø

9. $15x - 12y = 30$
$5x - 4y = 10$

Eliminate x by multiplying both sides of the second equation by -3 and adding.

$15x - 12y = 30$
____
____

A true result means that a solution of one equation is a solution of the _______. The solution set is _______________ many ordered pairs. The equations of this system are _________________. The graphs are the _______ line.

$-15x + 12y = -30$
$0 = 0$

other

infinitely

dependent

same

Solve the following applications using a system of two equation and two unknowns.

10. Kay went to Organic Garden Supply to buy fertilizer for her garden. She bought 3 one-pound bags of blood meal and 2 ten-pound bags of cottonseed meal, all for $12.90. These were the last bags the store had. She needed 5 lb of

blood meal and 50 lb of cottonseed meal so she went to a second store and bought 2 one-pound bags of blood meal and 3 ten-pound bags of cottonseed meal, all for $16.10. The prices were the same at both stores. What is the price of blood meal? Cottonseed meal?

Let b = the cost of a one-pound bag of blood meal and c = the cost of a ten-pound bag of cottonseed meal. Then we have two equations.

________________ (from Organic Garden Supply)	$3b + 2c = 12.90$
________________ (from the other store)	$2b + 3c = 16.10$
Solving this system we get b = ____ and c = ____.	1.30; 4.50
Then blood meal is _____ per pound and cottonseed	$1.30
meal is _____ per ten pounds.	$4.50

11. **During the summer, Carol and Dave worked for a veterinarian and between them earned $4560. Carol worked 10 days less than Dave but made $12 more each day because of her experience. The total number of days worked by both was 70. Find the number of days worked and the daily wage of Carol, then find the number of days worked and the daily wage of Dave.**

First write two equations to find the number of days worked. Let c = the number of days that Carol worked and d = the number of days that Dave worked. The two equations are

________________	$c + d = 70$
________________.	$d = c + 10$
Carol worked ____ days;	30
Dave worked ____ days.	40

Use these answers to create equations to find the daily wage.

If x = Carol's wage and y = Dave's wage, the two equations are

______________ | $30x + 40y = 4560$

______________. | $x = y + 12$

Carol's wage was ____ per day; | \$72

Dave's wage was ____ per day. | \$60

12. Anna Wilson has \$37,000 to invest. She invests part at 11% and the rest at 14%. Her total annual income from interest is \$4670. Find the amount invested at each rate.

Let x = the amount invested at 11% and y = the amount invested at 14%. The two equations are

______________ | $.11x + .14y = 4670$

______________. | $x + y = 37{,}000$

Amount at 11% _______ | \$17,000

Amount at 14% _______ | \$20,000

13. A system of equations can be used to find the _______________ price when supply and demand are equal. | equilibrium

14. Suppose the supply and demand equations for a product are

supply: $p = 1.5q + 4$

demand: $p = 50 - .5q$

where ____ is the price and ____ is the number of units produced. Find the equilibrium supply or demand and price. | p; q

Solve the system using substitution.

______________ | $1.5q + 4 = 50 - .5q$

q = ______ | 23

The equilibrium supply or demand is _____ units.	23
Substituting 23 for q in either equation gives ______ as the equilibrium price.	38.5

7.2 Linear Systems with Three Variables

1. This section discusses systems of three equations and three __________. Most of these systems can be solved by the elimination method.	variables
2. Solve the system $\begin{aligned} 2x + 2y + z &= -3 \\ x - 2y - 2z &= -5 \\ x + 2y - 3z &= 4. \end{aligned}$	
This system can be solved by the _______________ method. One way is to add the first and second equations; then add the second and third equations.	elimination
First: $\begin{aligned} 2x + 2y + z &= -3 \\ x - 2y - 2z &= -5 \end{aligned}$ ________________	$3x - z = -8$
Then: $\begin{aligned} x - 2y - 2z &= -5 \\ x + 2y - 3z &= 4 \end{aligned}$ ________________	$2x - 5z = -1$
We now have two equations and two unknowns. $\begin{aligned} 3x - z &= -8 \quad (*) \\ 2x - 5z &= -1 \end{aligned}$	

Multiply the first equation by ___ and the second equation by ____; then add.	2 -3
____________	$6x - 2z = -16$
$-6x + 15z = 3$	
____________	$13z = -13$
$z =$ ____	-1
Substituting in equation (*), we get	
$3x - (\_\_\_\_) = -8$	-1
$3x =$ ____	-9
$x =$ ____.	-3
Now use one of the original equations and substitute the values for both x and z to find y.	
$y =$ ____.	2
The solution set is ______________. Check your answer by substituting the values in the three equations.	$\{(-3, 2, -1)\}$
3. Solve the system $4x + 3y + 2z = 20$, $2x + y + 3z = 11$, $-2x + 2y + z = -1$. Solution set: __________	$\{(3, 2, 1)\}$
4. Solve the system $x - 3y + 4z = -11$, $5x + y - 7z = 9$, $3x - 2y + 5z = -5$. Solution set: __________	$\{(1, 4, 0)\}$
5. Solve the system $5x - 2y + z = 9$, $3x + y - 4z = 1$, $-6x - 2y + 8z = -5$. This system is ______________.	inconsistent
Solution set: __________	∅

6. Sometimes, a system has more variables than ________. Such a system can never have a ________ solution. For example, let us solve the system | equations; single

$$\begin{aligned} 2x - y - z &= 2 \\ x + 2y - 3z &= 6. \end{aligned}$$

Let us eliminate x by multiplying the second equation by ____ and adding. | -2

$$2x - y - z = 2$$

____

____

____

| $-2x - 4y + 6z = -12$

| $-5y + 5z = 1-$

Solve this result for y. (We could have used __.) | z

$y =$ ______ | $z + 2$

Substitute $z + 2$ for y in the first equation, and solve for x.

$2x - (\_\_\_\_\_) - z = 2$ | $z + 2$

$x =$ ______ | $z + 2$

With z arbitrary, the solution set is written $\{(\_\_\_\_\_\_, \_\_\_\_\_\_, z)\}$. | $z + 2$; $z + 2$

7. Solve the system

$$\begin{aligned} x + 2y - 3z &= -10 \\ 2x - y - z &= -5. \end{aligned}$$

Write the solution with y arbitrary.

$(\_\_\_\_\_\_, y, \_\_\_\_\_\_)$ | $y - 1$; $y + 3$

8. Find values of a, b, and c so that the graph of the equations $y = ax^2 + bx + c$ goes through the points $(-3, 1)$, $(-2, 0)$, and $(0, 4)$.

Form these equations.

$1 = a(-3)^2 + b(-3) + c$ or ________ | $1 = 9a - 3b + c$

$0 = a(-2)^2 + b(-2) + c$ or ________ | $0 = 4a - 2b + c$

$4 = a(0)^2 + b(0) + c$ or ______ | $4 = c$

Solve this system of equations for the values of a, b, and c.

a = ____, b = ____, c = ____ | 1; 4; 4

9. **The sum of three numbers is 30. The middle number is 3 less than the largest number and 1 more than twice the smallest number. Find the numbers.**

Let x = the smallest number,

y = the ________ number, | middle

z = the ________ number. | largest

Now write three equations.

"The sum of three numbers is ____" becomes | 30

____________ = 30. | $x + y + z$

"The middle number is ____ less than the largest" is | 3

y = ________. | $z - 3$

"The middle number is ___ more than ________ the smallest" is | 1; twice

y = ________. | $1 + 2x$

To find the three numbers, solve the system

____

____

__________________. | $x + y + z = 30$
 $y = z - 3$
 $y = 1 + 2x$

The numbers are ____, ____, and ____. | 5; 11; 14

10. **A pension fund invests $110,000. Part of the money is put in bonds. $10,000 more than this amount goes in stocks. Finally, $10,000 more than the total of the other two investments goes into a real estate deal. Find the amount in each investment.**

Let x = the amount invested in bonds,

y = the amount invested in stocks,

z = the amount invested in real estate.

Write a system of equations.

______________ | $x + y + z = 110{,}000$

______________ | $y = x + 10{,}000$

______________ | $z = x + y + 10{,}000$

Solve the system.

Bonds: __________ | \$20,000

Stocks: __________ | \$30,000

Real estate: __________ | \$60,000

7.3 Solution of Linear Systems by Matrices

1. To solve systems of equations with ______ methods, start by writing the __________ matrix for the system. | matrix; augmented

For example, the system

$$\begin{aligned} 2x + y - 3z &= 1 \\ 3x - 2y + z &= 5 \\ 5x - 3y - z &= 8 \end{aligned}$$

has the augmented matrix

$$\left[\begin{array}{ccc|c} 2 & \_\_ & -3 & \_\_ \\ 3 & \_\_ & \_\_ & \_\_ \\ \_\_ & -3 & \_\_ & \_\_ \end{array}\right]$$

1; 1
−2; 1; 5
5; −1; 8

2. Since the rows of an augmented matrix can be treated just like the ______________ of the system, the following matrix ________ transformations may be used.

   (1) Any two rows may be ______________.

   (2) The elements of any row may be ____________ by a nonzero __________.

   (3) Any row may be changed by __________ to its elements a multiple of the corresponding elements of another row.

equations

row

interchanged

multiplied

number

adding

Use the row transformations to change each of the following augmented matrices in Frames 3–5.

3. $\left[\begin{array}{cc|c} 6 & 3 & 0 \\ 12 & 4 & 2 \end{array}\right]$

   Multiply the first row by −2 and add to the second row.

$$\left[\begin{array}{cc|c} 6 & 3 & 0 \\ \_\_ & \_\_ & \_\_ \end{array}\right]$$

0; −2; 2

4. $\left[\begin{array}{ccc|c} 2 & 1 & 3 & 2 \\ 1 & 2 & 4 & -1 \\ -2 & 1 & 2 & 3 \end{array}\right]$

   Multiply the third row by 5 and add to the first row.

$$\left[\begin{array}{ccc|c} \_\_ & \_\_ & \_\_ & \_\_ \\ 1 & 2 & 4 & -1 \\ -2 & 1 & 2 & 3 \end{array}\right]$$

−8; 6; 13; 17

5. $\left[\begin{array}{rrr|r} -2 & 6 & 4 & 8 \\ 2 & 1 & -2 & 1 \\ 1 & 4 & 1 & -2 \end{array}\right]$

Multiply the first row by -1/2 .

$\left[\begin{array}{rrr|r} \_\_ & \_\_ & \_\_ & \_\_ \\ 2 & 1 & -2 & 1 \\ 1 & 4 & 1 & -2 \end{array}\right]$

1; -3; -2; -4

Write the augmented matrix for the systems of Frames 6 and 7.

6. $4x + 3y = 10$
$2x + 3y = 8$

$\left[\begin{array}{rr|r} \_\_ & \_\_ & \_\_ \\ \_\_ & \_\_ & \_\_ \end{array}\right]$

4; 3; 10
2; 3; 8

7. $x + 2y = 5$
$2x + y = 7$

$\left[\begin{array}{rr|r} \_\_ & \_\_ & \_\_ \\ \_\_ & \_\_ & \_\_ \end{array}\right]$

1; 2; 5
2; 1; 7

In Frames 8 and 9, write the system of equations associated with the augmented matrix.

8. $\left[\begin{array}{rr|r} 2 & 1 & 6 \\ 1 & 2 & 6 \end{array}\right]$

Here we reverse the procedure from above and get

$2x + y =$ ____
$x + 2y =$ ____.

6
6

9. $\left[\begin{array}{rr|r} 1 & 4 & 7 \\ 0 & 1 & 4 \end{array}\right]$ ____________

$x + 4y = 7$
$y = 4$

Use the Gauss-Jordan method to solve each of the systems of equations of Frames 10-13.

10. $3x + 2y = 4$
$2x + y = 3$

Begin by writing the augmented matrix.

$$\left[\begin{array}{cc|c} 3 & \underline{\quad} & \underline{\quad} \\ \underline{\quad} & 1 & 3 \end{array}\right]$$

2; 4
2

Our goal is to transform the augmented matrix into

$$\left[\begin{array}{cc|c} \underline{\quad} & 0 & k \\ 0 & \underline{\quad} & j \end{array}\right]$$

1
1

so that x = ___ and y = ___, and the system is solved. Given the augmented matrix

k; j

$$\left[\begin{array}{cc|c} 3 & 2 & 4 \\ 2 & 1 & 3 \end{array}\right],$$

multiply by the first row by _____ and get

1/3

$$\left[\begin{array}{cc|c} 1 & \underline{\quad} & \underline{\quad} \\ 2 & 1 & 3 \end{array}\right].$$

2/3; 4/3

We now have the necessary 1 in the first row, first column position. Now get ______ in the second row, ______ column position. Multiply the first row by ____ and add to the second row.

0
first
-2

$$\left[\begin{array}{cc|c} 1 & 2/3 & 4/3 \\ 0 & \underline{\quad} & \underline{\quad} \end{array}\right]$$

-1/3; 1/3

Multiply the second row by ____.

-3

$$\left[\begin{array}{cc|c} 1 & 2/3 & 4/3 \\ 0 & \underline{\quad} & \underline{\quad} \end{array}\right]$$

1; -1

The last step is to multiply the second row by $-2/3$, add it to the first row, and read the result.	
$x =$ ____	2
$y =$ ____	-1
The solution set is ________.	$\{(2, -1)\}$

11. $2x - 5y = 16$
$3x + y = 7$

Solution set: _______	$\{(3, -2)\}$

12. $2x + y + 3z = -7$
$x + 2y + 4z = -1$
$-2x + y + 2z = 12$

Solution set: _______	$\{(-5, 0, 1)\}$

13. $y + z = 3$
$x + 2z = 5$
$y - 2z = 0$

Form the augmented matrix.

$$\left[\begin{array}{ccc|c} 0 & 1 & \_\_\_ & 3 \\ 1 & 0 & \_\_\_ & 5 \\ 0 & 1 & \_\_\_ & \_\_\_ \end{array}\right]$$

1
2
-2; 0

Because of the ____ in row one, column one, interchange the first and second rows.	0

$$\left[\begin{array}{ccc|c} 1 & 0 & 2 & 5 \\ \_\_\_ & \_\_\_ & \_\_\_ & \_\_\_ \\ 0 & 1 & -2 & 0 \end{array}\right]$$

0; 1; 1; 3

Multiply the second row by ____ and add to the	-1
_______ row.	third

$$\left[\begin{array}{ccc|c} 1 & 0 & 2 & 5 \\ 0 & 1 & 1 & 3 \\ 0 & \_\_\_ & \_\_\_ & \_\_\_ \end{array}\right]$$

0; -3; -3

Finish the solution and check your answer by substituting into all three equations. The solution set is ____________.	$\{(3, 2, 1)\}$

14. Use the Gauss–Jordan method to solve

$$\begin{aligned} 2x - y &= 4 \\ -6x + 3y &= 1. \end{aligned}$$

Write the augmented matrix.

$$\left[\begin{array}{cc|c} \_\_ & \_\_ & \_\_ \\ \_\_ & \_\_ & \_\_ \end{array}\right]$$

2; -1; 4
-6; 3; 1

Multiply the elements of the first row by 1/2.

$$\left[\begin{array}{cc|c} \_\_ & \_\_ & \_\_ \\ -6 & 3 & 1 \end{array}\right]$$

1; -1/2; 2

Multiply the elements of the first row by 6 and add the result to the corresponding elements in the __________ row. This transformation is abbreviated ____________.

second
6R1 + R2

$$\left[\begin{array}{cc|c} 1 & -1/2 & 2 \\ \_\_ & \_\_ & \_\_ \end{array}\right]$$

0; 0; 13

The next step would be to get ___ in the second row, second column. Because of the ________ in the second row, it is impossible to do this. The second row corresponds to the equation

1
zeros

____ + ____ = ____,

0x; 0y; 13

which has ___ solution.
The system is ____________, and the solution set is ___.

no
inconsistent
Ø

15. The Gauss–Jordan method can be used to indicate a solution of two equations in three variables with an ____________ variable.
Solve this system.

arbitrary

$$\begin{aligned} 2x - 3y + z &= 10 \\ x - 2y + 4z &= 6 \end{aligned}$$

Write the augmented matrix.

$$\left[\begin{array}{ccc|c} \_\_ & \_\_ & \_\_ & \_\_ \\ \_\_ & \_\_ & \_\_ & \_\_ \end{array}\right]$$

2; -3; 1; 10
1; -2; 4; 6

Interchange the rows.

$$\left[\begin{array}{ccc|c} \_\_ & \_\_ & \_\_ & \_\_ \\ \_\_ & \_\_ & \_\_ & \_\_ \end{array}\right]$$

1; -2; 4; 6
2; -3; 1; 10

$$\left[\begin{array}{ccc|c} 1 & -2 & 4 & 6 \\ \_\_ & \_\_ & \_\_ & \_\_ \end{array}\right] \quad -2R1 + R2$$

0; 1; -7; -2

$$\left[\begin{array}{ccc|c} \_\_ & \_\_ & \_\_ & \_\_ \\ 0 & 1 & -7 & -2 \end{array}\right] \quad 2R2 + R1$$

1; 0; -10; 2

It is impossible to go further with this method.
The two equations are:

____

____

$x - 10z = 2$
$y - 7z = -2$

Solve for x and y so that the solution is written with z arbitrary.

Solution set: ____________

$\{(2+10z, -2+7z, z)\}$

7.4 Nonlinear Systems of Equations

1. A system of __________ where one equation is not a first-degree equation is called a __________ system of equations.

equations
nonlinear

Solve each of the nonlinear system of equations in Frames 2–8.

2. $y = -x + 4$
   $y = (x - 2)^2$

   It is helpful to visualize the graphs of these equations. The graph of $y = -x + 4$ is a ______. — line

   The graph of $y = (x - 2)^2$ is a ______________. — parabola

   These graphs may have ___, ___, or ___ points of intersection. — 0; 1; 2

   The easiest solution method to use here is substitution of two equal values of ___. — y

   $(x - 2)^2 =$ _______ — $-x + 4$

   Square on the left.

   ____________ $= -x + 4$ — $x^2 - 4x + 4$

   ____________ $= 0$ — $x^2 - 3x$

   ____________ $= 0$ — $x(x - 3)$

   $x =$ _____ or $x =$ _____ — 0; 3

   Now substitute and solve for y in both equations. Check.

   $y = (0 - 2)^2 = 4$ or $y = (3 - 2)^2 =$ _____ — 1

   $y = 0 + 4 =$ ____ or $y = -3 + 4 = 1$ — 4

   The solution set is ______________. — $\{(0, 4), (3, 1)\}$

3. $y = (x + 1)^2$
   $y = -x + 1$

   Solution set: ____________ — $\{(0, 1), (-3, 4)\}$

4. $x^2 + y^2 = 9$
   $2x^2 - y^2 = 3$

   The graph of $x^2 + y^2 = 9$ is a ________ and the graph of $2x^2 - y^2 = 3$ is a ____________. These graphs may have ____, ____, ____, ____, or ____ points of intersection. — circle; hyperbola; 0; 1; 2; 3; 4

Use the addition method.

$$\begin{aligned} x^2 + y^2 &= 9 \\ \underline{\qquad\qquad} &= 3 \\ \hline 3x^2 &= \underline{\quad} \\ x^2 &= \underline{\quad} \\ x = 2 \quad \text{or} \quad x &= \underline{\qquad} \end{aligned}$$

$2x^2 - y^2$

12

4

-2

To find y, substitute the x–values into either equation.

$$\begin{aligned} 2^2 + y^2 &= 9 \\ y^2 &= \underline{\qquad} \\ y = \sqrt{5} \quad \text{or} \quad y &= -\sqrt{5} \end{aligned}$$

5

and

$$\begin{aligned} (-2)^2 + y^2 &= 9 \\ y &= \underline{\qquad\quad} \quad \text{or} \quad \underline{\qquad\quad} \end{aligned}$$

$\sqrt{5}$; $-\sqrt{5}$

The solution set is $\underline{\qquad\qquad\qquad}$.

$\{(2, \sqrt{5}), (2, -\sqrt{5})$ $(-2, \sqrt{5}), (-2, -\sqrt{5})\}$

5. $x^2 + 2xy + y^2 = 20$
   $x^2 - xy + y^2 = 5$

Eliminate the x^2 and y^2 variables by addition.

$$\begin{aligned} x^2 + 2xy + y^2 &= 20 \\ \underline{-x^2 + xy - y^2} &\underline{= -5} \\ \underline{\qquad\qquad\qquad\qquad} \end{aligned}$$

$3xy = 15$

Solve for y.

$$y = \underline{\qquad\quad} \text{ with } x \neq 0$$

$5/x$

Substitute in the first equation.

$$\begin{aligned} x^2 + 2x(\underline{\qquad}) + (\underline{\qquad})^2 &= 20 \\ x^2 + 10 + \underline{\qquad\qquad} &= 20 \\ x^4 + 10x^2 + 25 &= 20x^2 \\ \underline{\qquad\qquad\qquad} &= 0 \\ (x^2 - 5)(x^2 - 5) &= 0 \\ x = \underline{\quad} \text{ or } \underline{\quad} \end{aligned}$$

$5/x$; $5/x$

$25/x^2$

$x^4 - 10x^2 = 25$

$\sqrt{5}$; $-\sqrt{5}$

Substitute these values for x into $y = 5/x$ to find y.

Solution set: ______________ | $\{(\sqrt{5}, \sqrt{5}), (-\sqrt{5}, -\sqrt{5})\}$

6. $x^2 + y^2 = 9$
$y - |x| = -3$

The substitution method is required here. The equation $y - |x| = -3$ can be rewritten as $y + 3 = |x|$. Then by the definition of absolute value,

$x = y + 3$ or $x = -(\_\_\_\_) = \_\_\_\_$. | $y + 3$; $-y - 3$

Substitute into $x^2 + y^2 = 9$.

$(y + 3)^2 + y^2 = 9$ or $(\_\_\_\_)^2 + y^2 = 9$ | $-y - 3$

$y^2 + 6y + 9 + y^2 = 9$ or $\_\_\_\_ + y^2 = 9$ | $y^2 + 6y + 9$

Since these two equations are the same, we need to solve only one:

$\_\_\_\_ = 0$ | $2y^2 + 6y$

$\_\_\_\_ = 0$ | $2y(y + 3)$

$y = 0$ or $y = \_\_\_$. | -3

Thus the solution set is ________________. | $\{(3, 0), (-3, 0), (0, -3)\}$

7. $x^2 - 4x + y^2 = 1$
$x^2 \quad - y^2 = 5$

The graph of the first equation is a ________ and the graph of the second is a __________. These graphs will intersect in ___, ___, ___, ___, or ___ points. | circle; hyperbola; 0; 1; 2; 3; 4

Solution set: __________ | $\{(3, 2), (3, -2)$ $(-1, 2i), (-1, -2i)\}$

The graphs of the equations intersect in only ____ points since only ____ solutions are real numbers.	two two
8. $y = -x^2$ $y = 2x + 5$ Using substitution, we have	
$-x^2 =$ ____________	$2x + 5$
$0 =$ ____________.	$x^2 + 2x + 5$
Using the quadratic formula, we have	
$x =$ ____________	$\frac{-2 \pm \sqrt{-16}}{2}$
which is a(n) ____________ number.	imaginary
Simplify:	
$x = \frac{-2 \text{ \_\_\_\_\_\_}}{2} = \frac{2(\text{\_\_\_\_\_\_})}{2} =$ ________.	$\pm 4i$; $-1 \pm 2i$; $-1 \pm 2i$
Since $y = 2x + 5$,	
$y = 2($______$) + 5 =$ ________.	$-1 \pm 2i$; $3 \pm 4i$
The solution set is ______________.	$\{(-1 + 2i, 3 + 4i), (-1 - 2i, 3 - 4i)\}$
Since imaginary solutions occur in this system, the graphs do not __________.	intersect

7.5 Systems of Inequalities; Linear Programming

1. A system containing at least one inequality is called a system of ______________. To graph such a system, find the ______________ of the graphs of the inequalities.	inequalities intersection

Graph the solution sets of the systems of inequalities in Frames 2–6.

2. $y - \frac{1}{2}x > 1$

   $y + \frac{1}{2}x \le 1$

   First, note that this is a system of linear inequalities and we graph the boundaries, being careful to use dashed lines when appropriate. Solve each inequality for ___ and shade the proper area. The point where the dashed line intersects the solid line is shown as a(n) ______ circle, since it (*does/does not*) belong to the solution set.

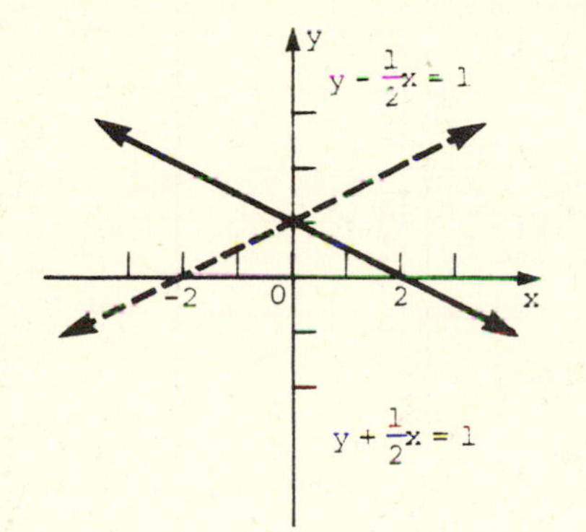

y

open

does not

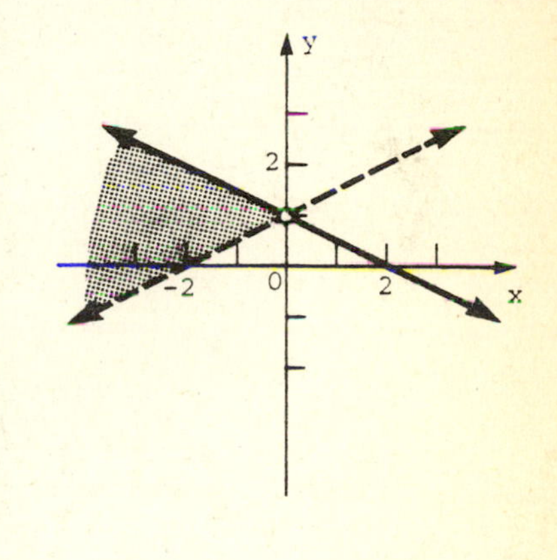

3. $x^2 + y^2 \le 9$

   $y \ge 2^x$

   First sketch the graphs, starting with a ________ with center at _______ and radius 3 as the boundary of one inequality of the system and the graph of an ______________ function as the boundary of the second inequality of the system. Now use (0, 0) as a test point, and shade the appropriate region.

circle

(0, 0)

exponential

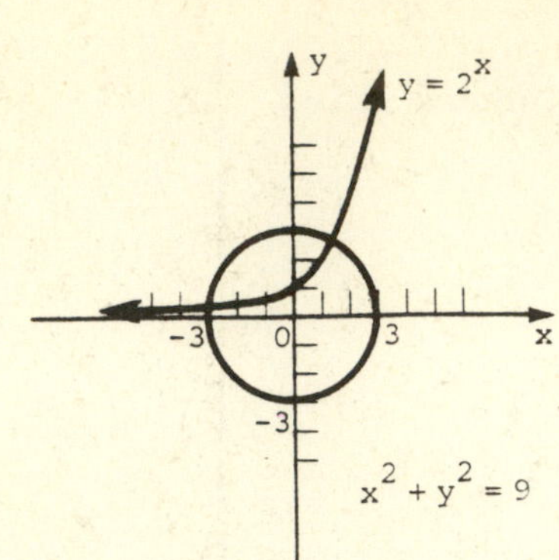

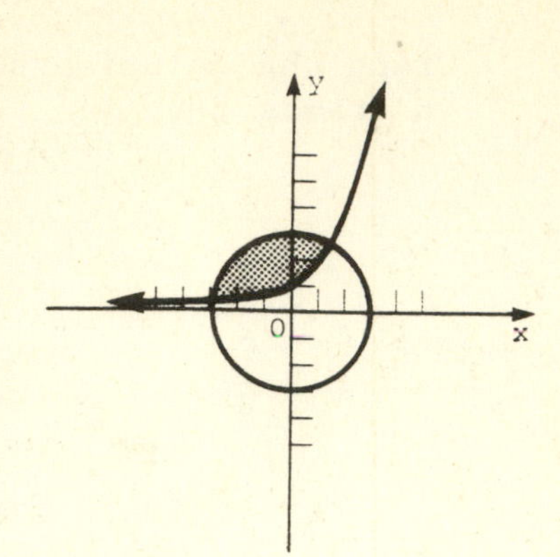

4. $x^2 - y^2 \leq 1$

   $y < 1$

   The two boundaries are a ____________ with x inter- cepts (1, 0) and (-1, 0) and the ____________ line $y = 1$.

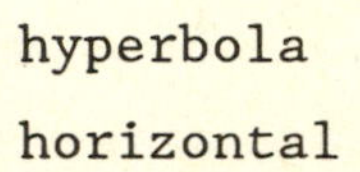
hyperbola
horizontal

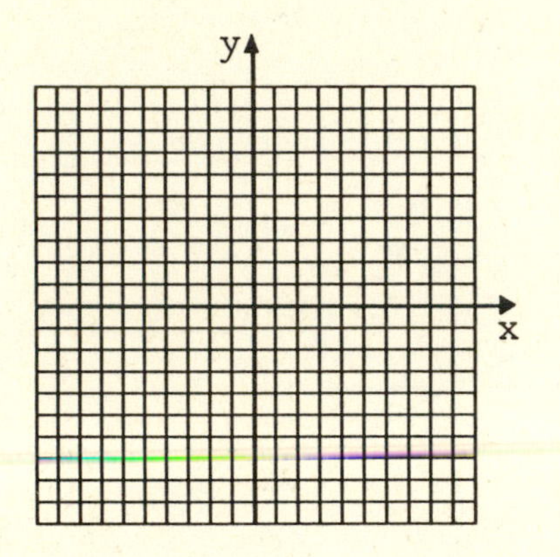

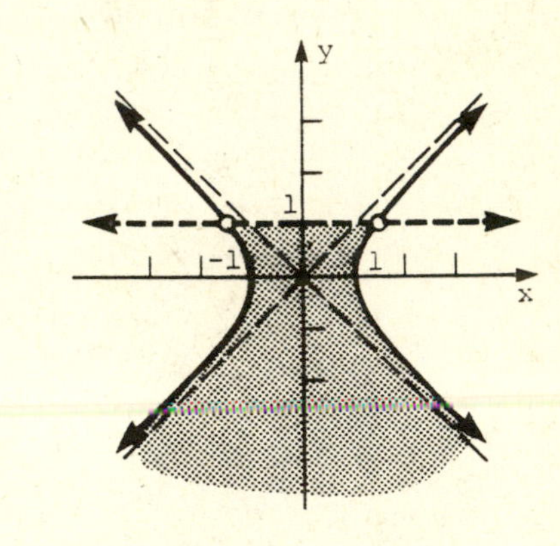

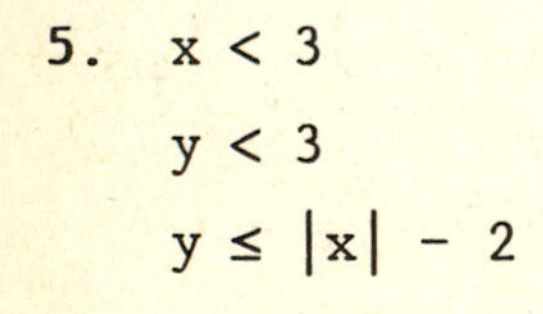

5. $x < 3$

   $y < 3$

   $y \leq |x| - 2$

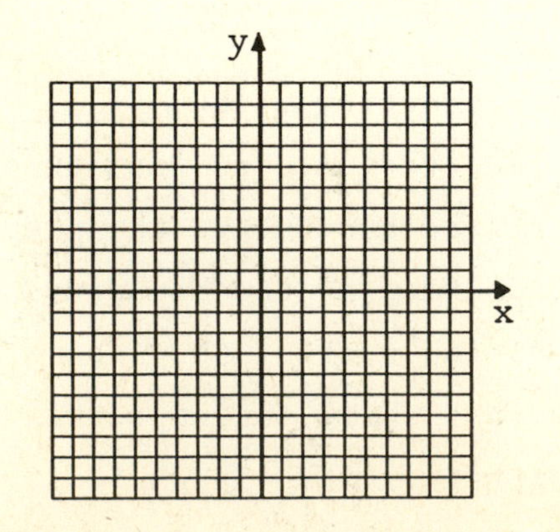

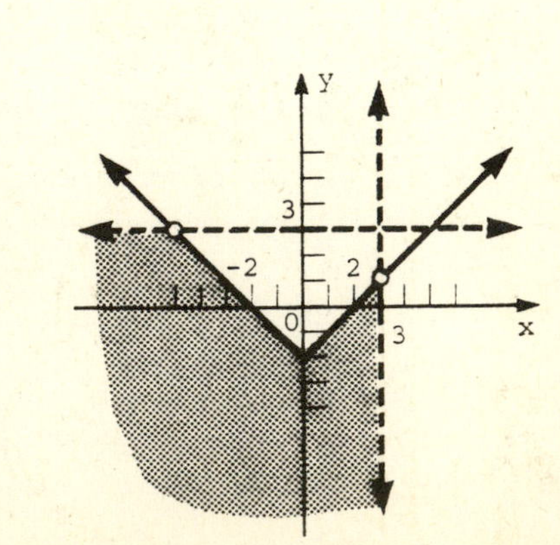

6. 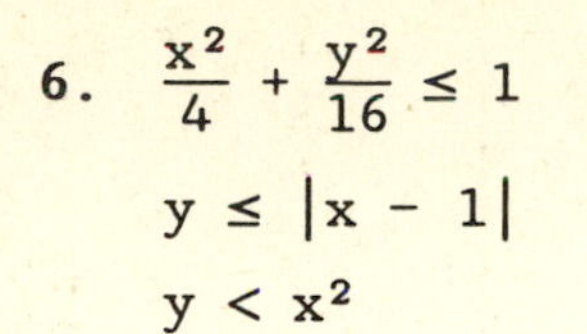

$$\frac{x^2}{4} + \frac{y^2}{16} \le 1$$

$$y \le |x - 1|$$

$$y < x^2$$

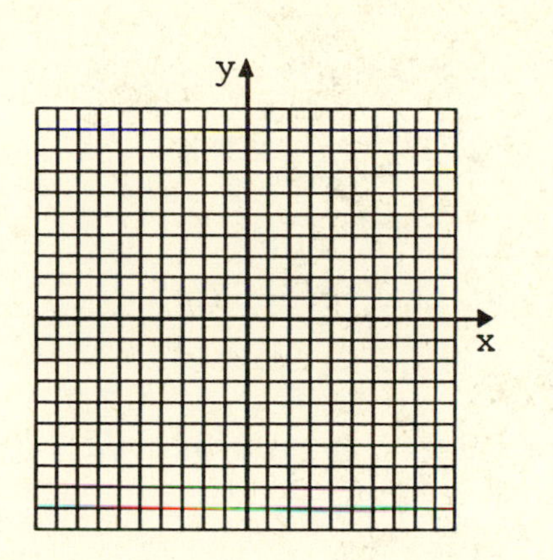

7. The method of linear programming requires that we find the maximum or ________________ values of a linear expression, subject to restrictions, or ________________. For example, let us find the maximum value of the expression $5x + 2y$ subject to

minimum

constraints

$$2x + 3y \le 6$$
$$4x + y \le 6$$
$$x \ge 0$$
$$y \ge 0.$$

First, graph the intersection of these inequalities. This intersection is the region of _______ solutions. Also, locate every _______ (or _______ point). To maximize $5x + 2y$, try every ________.

feasible

vertex; corner

vertex

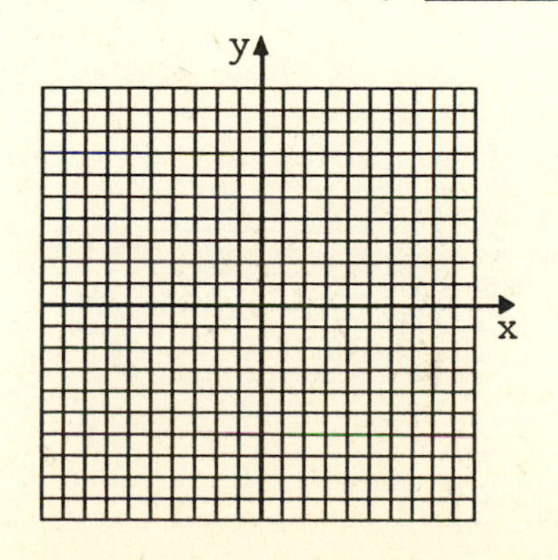

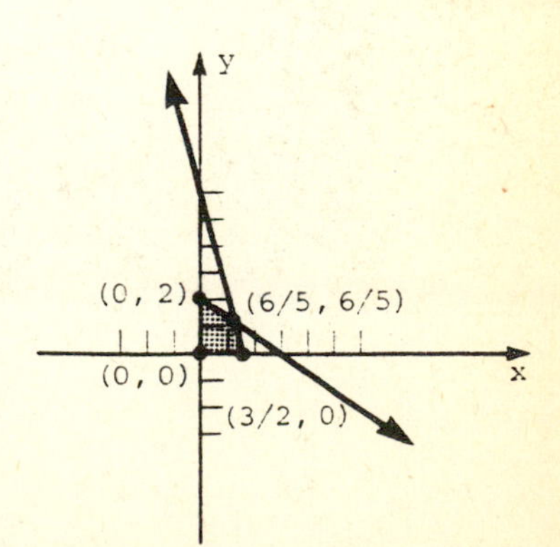

Point	*Value of 5x + 2y*	
(0, 0)	$5 \cdot 0 + 2 \cdot 0 =$ ____	0
(0, 2)	$5 \cdot 0 + 2 \cdot 2 =$ ____	4
(6/5, 6/5)	$5 \cdot \frac{6}{5} + 2 \cdot \frac{6}{5} = \frac{42}{5} =$ ____	8.4
(3/2, 0)	$5 \cdot \frac{3}{2} + 2 \cdot 0 = \frac{15}{2} =$ ____	7.5

The maximum is ______ at the vertex __________. 8.4; (6/5, 6/5)

8. Find the minimum value of 2x + y subject to

$$3x - y \geq 12$$
$$x + y \leq 15$$
$$x \geq 2$$
$$y \geq 5.$$

The minimum value is ________ at the vertex ____________. 49/3 (17/13, 5)

9. **Vince and Muriel produce coffee tables and end tables in their basement. Vince spends 6 hr building a coffee table and 2 hr building an end table. Muriel spends 2 hr finishing either kind of table. Vince can spend 24 hr per week building tables and Muriel can spend 16 hr per week finishing tables. If the profit on a coffee table is $50 and on an end table is $20, how many of each kind of table should they produce each week for maximum profit? How much is that profit?**

Let x = the number of coffee tables and y = the number of end tables they should produce in a week. Then the amount of profit is given by

_______ + _______. 50x; 20y

Since Vince spends ____ hr building a coffee table and ____ hr building an end table and can spend 24 hr per week, the inequality representing the hours Vince spends is

6
2

______ + ______ ≤ ______.

6x; 2y; 24

Since Muriel spends ____ hr finishing each table and can spend _____ hr per week, the inequality representing the hours Muriel spends is

2
16

______ + ______ ≤ ______.

2x; 2y; 16

Also, $x \geq 0$ and $y \geq 0$ since a negative number of tables cannot be build. Sketch the solution of the system.

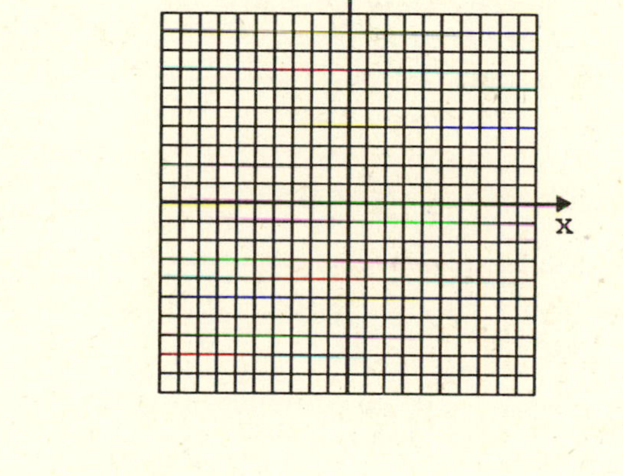

y
(0, 8)
(2, 6)
(4, 0)
x

The number of each kind of table Vince and Muriel should build for maximum profit is found by checking the coordinates of the ________ in the profit function ____________ to find the maximum profit.

vertices
50x + 20y

Point	Profit
(____, 0)	______
(0, ____)	______
(___,___)	______

4; $200
8; $160
2; 6; $220

Vince and Muriel should produce ____ coffee tables	2
and ____ end tables for ______ profit.	6; $220

CHAPTER 7 TEST

Test answers are the back of this study guide.

Use the substitution method to solve each of the following systems. Identify any systems with dependent equations or any inconsistent systems. If a system has dependent equations, express the solution set with y artitrary.

1. $3x - y = -7$
   $-2x + 3y = 14$

   1. ______________

2. $2x + 3y = 0$
   $2x - y = 16$

   2. ______________

3. $-9x + 3y = -12$
   $3x - y = 4$

   3. ______________

Use the addition method to solve each of the following systems. Identify any systems with dependent equations or any inconsistent systems. If a system has dependent equations, express the solution set with y arbitrary.

4. $3x + 3y = 0$
   $-6x + 5y = -22$

   4. ______________

5. $5x - 3y = 2$
   $-10x + 6y = 8$

   5. ______________

6. $-x + 4y = 16$
   $-5x + 3y = 12$

   6. ______________

7. A student says that (1, 1) is a possible solution of the system

$$3x - 4y = 1$$
$$6x - 8y = 2$$

since the system has infinitely many solutions. Is she correct? Explain.

   7. ______________

8. Joyce bought 2 small cans of fruit and 3 large cans, paying $3.47. Later she bought 7 small and 4 large cans, paying $6.75. Find the cost of a small can and the cost of a large can.

8. ______________________

Use the addition method to solve the following systems.

9.
$$\begin{aligned} 2x - 3z &= -4 \\ 2y + z &= 8 \\ x + y &= 4 \end{aligned}$$

9. ______________________

10.
$$\begin{aligned} 3k + 2m - n &= 7 \\ k - 4m + 2n &= 0 \\ 2k - 3m + n &= 3 \end{aligned}$$

10. ______________________

11. Find a solution for the following system in terms of the arbitrary variable z.

$$\begin{aligned} 4x - 2y + z &= 0 \\ x + 2y - 3z &= 2 \end{aligned}$$

11. ______________________

12. The sum of three numbers is 2. The first two numbers are equal. The difference between the first and third number is −2. Use a system of equations to find the three numbers.

12. ______________________

Use the Gauss-Jordan method to solve each of the following systems.

13. $x + y = 4$
$2x + 3y = 5$

13. ____________________

14. $2x + 3y - 4z = -5$
$x + 2y + z = -2$
$3x - y + z = -13$

14. ____________________

Solve each of the following nonliner systems of equations.

15. $y = x^2 - 2$
$y - x = 0$

15. ____________________

16. $x^2 + y^2 = 9$
$y + x = 3$

16. ____________________

17. If a system of two nonlinear equations consists of two equations whose graphs are both circles, can the system have exactly one solution? If so describe the graph of the solution.

17. ____________________

18. Use a system of equations to find two numbers such that their sum is 2 and the second number subtracted from the square of the first number is 28.

18. ____________________

19. Graph the solution set of the following system of inequalities.

$$y^2 \le 4 - x^2$$
$$x^2 - y^2 \ge 1$$

19.

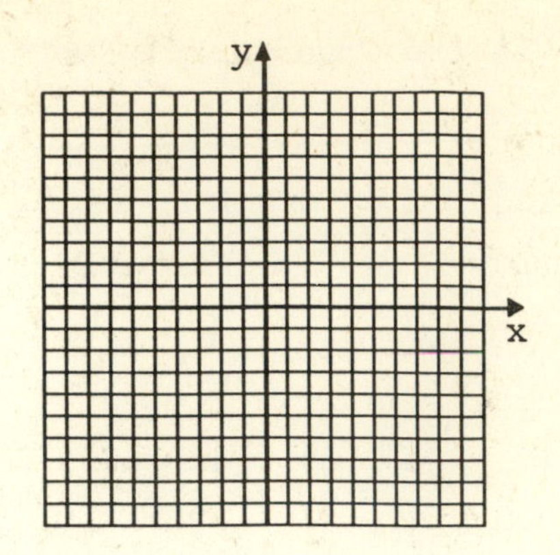

20. A company produces two models of lamps, Model X and Model Y. There are two stations on the assembly line. The assembly of one Model X requires 40 minutes at station 1 and 2 hours at station 2. The assembly of one Model Y requires 30 minutes at station 1 and 1 hour at station 2. Station 1 can be operated for no more than 4 hours per day and station 2 can be operated for no more than 10 hours per day. If the profit is $36 on each Model X and $24 on each Model Y, how many of each model should be assembled each day for maximum profit? What is the maximum daily profit?

20. ______________________

____

CHAPTER 8 PRETEST

Pretest answers are at the back of this study guide.

1. Find the values of the variables in the following equation.

$$\begin{bmatrix} x+3 & y+2 \\ 4 & 2 \end{bmatrix} = \begin{bmatrix} -3 & 5 \\ q-1 & z+1 \end{bmatrix}$$

1. ______________________

Perform each of the following operations, whenever possible.

2. $\begin{bmatrix} 3 & 4 & 5 \\ 1 & 0 & -6 \end{bmatrix} + \begin{bmatrix} 0 & -7 & 1 \\ -1 & -14 & 7 \end{bmatrix}$

2. ______________________

3. $\begin{bmatrix} 2 & 8 & 1 \\ 3 & 7 & 0 \\ -5 & 4 & -9 \end{bmatrix} - \begin{bmatrix} 1 & 7 & 0 \\ -2 & 2 & -5 \\ 1 & 10 & -3 \end{bmatrix}$

3. ______________________

4. At the Craft Fair Loretta bought 2 baskets, 4 pots, and 1 picture. Levi bought 3 baskets, 2 pots, and 2 pictures. Write this information first as a 3×2 matrix and then a 2×3 matrix.

4. ______________________

Find each of the following matrix products, whenever possible.

5. $\begin{bmatrix} 1 & 3 \\ 2 & 4 \end{bmatrix} \begin{bmatrix} -1 \\ 8 \end{bmatrix}$

5. ______________________

6. $\begin{bmatrix} 3 & 2 & -1 \\ 2 & 0 & -3 \end{bmatrix} \begin{bmatrix} 0 & 2 \\ -1 & 2 \\ 0 & 4 \end{bmatrix}$

6. ______________________

7. Is multiplication of matrices associative?

7. ______________

8. Decide whether or not the following pair of matrices are inverses.

$$\begin{bmatrix} 1 & 1 & 0 \\ 0 & 0 & 1 \\ -1 & 0 & 1 \end{bmatrix} \quad \begin{bmatrix} 0 & 1 & -1 \\ 1 & -1 & 1 \\ 0 & 1 & 0 \end{bmatrix}$$

8. ______________

Find the inverse, if it exists, for each of the following.

9. $\begin{bmatrix} 6 & 3 \\ -3 & 2 \end{bmatrix}$

9. ______________

10. $\begin{bmatrix} 5 & 2 \\ 25 & 10 \end{bmatrix}$

10. ______________

11. $\begin{bmatrix} 6 & 1 & 1 \\ 1 & 0 & 0 \\ 3 & 2 & 1 \end{bmatrix}$

11. ______________

Use matrix inverses to solve each of the following.

12. $$\begin{aligned} 2x - y &= 3 \\ x + y &= 0 \end{aligned}$$

12. ______________

13. $$\begin{aligned} 2x + y - 2z &= 2 \\ x - 2y + z &= 0 \\ -x - y - z &= 6 \end{aligned}$$

13. ______________

Evaluate each of the following determinants.

14. $\begin{vmatrix} 7 & 1 \\ -2 & 3 \end{vmatrix}$

14. ______________

15. $\begin{vmatrix} 2 & 1 & 0 \\ -1 & 0 & -2 \\ 0 & 1 & 1 \end{vmatrix}$

15. ______________

16. Solve the following determinant equation for x.

$$\begin{vmatrix} 1 & 0 & -2x \\ 3 & 2 & -1 \\ 1 & 6 & x \end{vmatrix} = 66$$

16. ______________

Give the property or properties that justify each of the following statements.

17. $\begin{vmatrix} 2 & 3 & 1 \\ 1 & 0 & 4 \\ 2 & 3 & 1 \end{vmatrix} = 0$

17. ______________

18. $\begin{vmatrix} 2 & 7 & 1 \\ -1 & 0 & 3 \\ 2 & 1 & -6 \end{vmatrix} = 2\begin{vmatrix} 2 & 7 & 1 \\ -1 & 0 & 3 \\ 1 & 1/2 & -3 \end{vmatrix}$

18. ______________

Solve each of the following systems by Cramer's rule. Identify any systems with dependent equations or any inconsistent systems.

19. $\begin{aligned} 8x + 3y &= -2 \\ -2x + y &= 4 \end{aligned}$

19. ______________

20. $$\begin{aligned} 2x + 3y + z &= 3 \\ -2x + 5y - 2z &= -2 \\ x + 2y - 3z &= 5 \end{aligned}$$

20. ______________________

CHAPTER 8 MATRICES

8.1 Basic Properties of Matrices

Frame	Answer
1. A matrix is classified by its ________, which tells the number of ________ and ________ in the matrix. A matrix with m rows and n columns is called a(n) ________ matrix. The number of ________ is always given first.	size rows; columns $m \times n$ rows
Classify the matrices in Frames 2–4 by size.	
2. $\begin{bmatrix} 9 & 7 & -10 & 14 \\ 5 & -3 & 4 & 8 \end{bmatrix}$ is a ________ matrix.	2×4
3. $\begin{bmatrix} 3 & -2 \\ 2 & -3 \end{bmatrix}$ is a ________ matrix. Since the number of rows equals the number of columns, this matrix is a ________ matrix.	2×2 square
4. $\begin{bmatrix} 1 \\ 9 \\ -7 \end{bmatrix}$ is a ________ matrix. This matrix has only one ________, so it is called a ________ matrix.	3×1 column column
5. Two matrices are equal if their corresponding ________ are equal.	elements

Find the value of the variables in Frames 6–8.

6. $\begin{bmatrix} 3 & y \\ z & -1 \end{bmatrix} = \begin{bmatrix} q & 2 \\ 4 & t \end{bmatrix}$

Since the two matrices are ______, we know that their corresponding ____________ must be equal. Therefore, q = ____; y = ____; z = ____; t = ____.

equal
elements
3; 2; 4; -1

7. $\begin{bmatrix} 9 & 0 & -6 \\ 7 & 5 & -3 \end{bmatrix} = \begin{bmatrix} x & 0 & z \\ q & 5 & s \end{bmatrix}$

x = ____, z = ____
q = ____, s = ____

9; -6
7; -3

8. $\begin{bmatrix} x - 2 & 10 \\ 8 & y + 3 \end{bmatrix} = \begin{bmatrix} 1 & z \\ w + 5 & 0 \end{bmatrix}$

x = ____, z = ____
w = ____, y = ____

3; 10
3; -3

9. Remember that two matrices that are not the same ________ cannot be equal.

size

10. To add two m × n matrices, ______ the corresponding ________ to get a(n) ______ matrix.

add
elements; m × n

11. To subtract an m × n matrix B from another m × n matrix A, add the ____________ inverse of _____ to _____. That is, subtract the corresponding ____________ of ____ from those of ____.

additive; B
A
elements; B; A

Add or subtract in Frames 12–16.

12. $\begin{bmatrix} 3 & -4 \\ -8 & 7 \\ 5 & 2 \end{bmatrix} + \begin{bmatrix} 1 & -1 \\ 6 & -9 \\ -1 & 4 \end{bmatrix} = \begin{bmatrix} \_\_\_\_ & -4 + (-1) \\ \_\_\_\_ & 7 + (-9) \\ \_\_\_\_ & 2 + 4 \end{bmatrix}$

$= \begin{bmatrix} \_\_\_\_ & \_\_\_\_ \\ \_\_\_\_ & \_\_\_\_ \\ \_\_\_\_ & \_\_\_\_ \end{bmatrix}$

3 + 1
-8 + 6
5 + (-1)

4; -5
-2; -2
4; 6

13. $\begin{bmatrix} 11 & 7 \\ -2 & 3 \end{bmatrix} - \begin{bmatrix} 9 & -1 \\ 4 & -2 \end{bmatrix} = \begin{bmatrix} \_\_\_\_ & \_\_\_\_ \\ \_\_\_\_ & \_\_\_\_ \end{bmatrix}$

2; 8
-6; 5

14. $\begin{bmatrix} 4 & 1 & 2 \\ 5 & 0 & -4 \\ 3 & -1 & 5 \end{bmatrix} + \begin{bmatrix} -5 & -7 & 1 \\ 2 & 0 & 9 \\ -1 & 8 & 8 \end{bmatrix} - \begin{bmatrix} -7 & -2 & 1 \\ 0 & 2 & 9 \\ 1 & -4 & 3 \end{bmatrix}$

= __________

$\begin{bmatrix} 6 & -4 & 2 \\ 7 & -2 & -4 \\ 1 & 11 & 10 \end{bmatrix}$

15. $[1 \quad 0 \quad 1] + \begin{bmatrix} 2 \\ 3 \\ 5 \end{bmatrix}$ __________

The sum cannot be found since the sizes are not the same.

16. $\begin{bmatrix} -2x & 3y & 4z \\ q & -5r & 9t \end{bmatrix} - \begin{bmatrix} 3x & 2y & 3z \\ -9q & -5r & 6t \end{bmatrix} =$ __________

$\begin{bmatrix} -5x & y & z \\ 10q & 0 & 3t \end{bmatrix}$

17. **Mark and Marion are each doing repairs and refurbishing their houses. At the builder's supply Mark bought 3 gal of flat white paint, 2 qt of yellow enamel, and 3 door hinges. Marion bought 5 gal of beige flat paint, 1 qt of white enamel, and 2 door hinges. Write this information first as a 2 × 3 matrix and then as a 3 × 2 matrix.**

The 2 × 3 matrix will have what Mark bought as one row and what Marion bought as the other row. The columns will be flat paint, enamel, and hinges. We get

$$\begin{bmatrix} 3 & \_\_\_ & 3 \\ 5 & 1 & \_\_\_ \end{bmatrix}.$$

2

2

Find the 3 × 2 matrix. ______________

$$\begin{bmatrix} 3 & 5 \\ 2 & 1 \\ 3 & 2 \end{bmatrix}$$

18. Given $A = \begin{bmatrix} 4 & -2 \\ 1 & 5 \\ 0 & 3 \end{bmatrix}$,

find the matrix $-A$ such that $A + (-A) = 0$. ______________

$$\begin{bmatrix} -4 & 2 \\ -1 & -5 \\ 0 & -3 \end{bmatrix}$$

19. In work with matrices, a real number is called a __________.

scalar

20. To find the product of a scalar and a __________, find the product of the __________ and each __________ of the matrix.

matrix

scalar

element

Let $M = \begin{bmatrix} -1 & 2 \\ 4 & 3 \end{bmatrix}$ and $N = \begin{bmatrix} 0 & -2 \\ 1 & 5 \end{bmatrix}$. Find each of the following matrices.

21. $-2M$. ______________

$$\begin{bmatrix} 2 & -4 \\ -8 & -6 \end{bmatrix}$$

22. $-N$. ______________

$$\begin{bmatrix} 0 & 2 \\ -1 & -5 \end{bmatrix}$$

23. 3M - 2N. ________ | $\begin{bmatrix} -3 & 10 \\ 10 & -1 \end{bmatrix}$

8.2 Matrix Products

1. To multiply matrices A and B, the number of_______ of A must match the number of _______ B. The product has as many rows as ___ and as many columns as ___.

   columns
   rows
   A
   B

2. The element for the first row, first column of a product matrix is found by _____________ the elements of the first ______ of A and the first __________ of B. The _______ of these products gives the necessary element.

   multiplying
   row
   column; sum

3. To find the product $\begin{bmatrix} 2 & 1 & -4 \\ 3 & -1 & 2 \end{bmatrix}\begin{bmatrix} 5 & 7 \\ 8 & 0 \\ -1 & 2 \end{bmatrix}$, use the first row of the first matrix and the first column of the second and multiply as follows:

   2(____) + 1(____) + (-4)(____) = ____. — 5; 8; -1; 22

   The number ____ goes in the first row, first column of the product matrix. Now find the entry for the first row, second column. — 22

   2(____) + 1(____) + (-4)(____) = ____. — 7; 0; 2; 6

   Find the remaining entries:

   3(____) + (-1)(____) + 2(____) = ____ — 5; 8; -1; 5

   3(____) + (-1)(____) + 2(____) = ____. — 7; 0; 2; 25

Write the product:

$$\begin{bmatrix} 2 & 1 & -4 \\ 3 & -1 & 2 \end{bmatrix} \begin{bmatrix} 5 & 7 \\ 8 & 0 \\ -1 & 2 \end{bmatrix} = \_\_\_\_\_\_\_\_\_\_.$$

$\begin{bmatrix} 22 & 6 \\ 5 & 25 \end{bmatrix}$

Find each product in Frames 4–8.

4. $\begin{bmatrix} 1 & 3 \\ 2 & 5 \end{bmatrix} \begin{bmatrix} -1 & 2 \\ -2 & 0 \end{bmatrix}$

Multiply the first element of the first ______ of A with the first element of the first __________ of B.

row
column

$$1(-1) = \_\_\_\_$$

-1

Add this to the product of the _________ element of the first row of A and the second element of the first __________ of B.

second
column

$$(1)(-1) + (3)(-2) = \_\_\_\_$$

-7

Then ____ is the first row, first column entry of the product matrix. Complete the product matrix.

-7

$$\begin{bmatrix} 1 & 3 \\ 2 & 5 \end{bmatrix} \begin{bmatrix} -1 & 2 \\ -2 & 0 \end{bmatrix}$$

$$= \begin{bmatrix} -7 & (\_)(\_) + (\_)(\_) \\ (2)(-1) + (5)(-2) & (\_)(\_) + (\_)(\_) \end{bmatrix},$$

1; 2; 3; 0
2; 2; 5; 0

which is $\begin{bmatrix} \_\_\_\_ & \_\_\_\_ \\ \_\_\_\_ & \_\_\_\_ \end{bmatrix}$.

-7; 2
-12; 4

Notice that A is a _______ matrix, B is a _______ matrix, and AB is a ______ matrix.

2×2; 2×2
2×2

5. $AB = \begin{bmatrix} 1 & -2 & 0 \\ 4 & 3 & 1 \end{bmatrix} \begin{bmatrix} 1 & 3 & 1 \\ 0 & 2 & -1 \\ 5 & 0 & 3 \end{bmatrix}$

A is a ______ matrix and B is a ______ matrix. AB exists and will be a ______ matrix. It will have the same number of rows as ___ and the same number of columns as ___. In order for the product AB to exist the number of __________ of A must be the same as the number of ______ of B. Find AB.

____

Answers: 2×3; 3×3 / 2×3 / A / B / columns / rows / $\begin{bmatrix} 1 & -1 & 3 \\ 9 & 18 & 4 \end{bmatrix}$

6. $\begin{bmatrix} 1 & 4 \\ 0 & 2 \\ 2 & 3 \end{bmatrix} \begin{bmatrix} 1 & -1 & -2 & 3 \\ 2 & 1 & 4 & -1 \end{bmatrix} =$ ______________

Answer: $\begin{bmatrix} 9 & 3 & 14 & -1 \\ 4 & 2 & 8 & -2 \\ 8 & 1 & 8 & 3 \end{bmatrix}$

7. $\begin{bmatrix} -5 & 0 & 3 & 2 \end{bmatrix} \begin{bmatrix} -1 \\ 23 \\ 2 \\ 4 \end{bmatrix} =$ ______

Answer: $[19]$

8. $\begin{bmatrix} 4 & 3 \\ 2 & -1 \end{bmatrix} \begin{bmatrix} -1 & 0 & -2 & 1 \\ -2 & 0 & 1 & 1 \end{bmatrix} =$ ______________

Answer: $\begin{bmatrix} -10 & 0 & -5 & 7 \\ 0 & 0 & -5 & 1 \end{bmatrix}$

Let $A = \begin{bmatrix} -5 & 2 \\ 3 & 0 \end{bmatrix}$ and let $B = \begin{bmatrix} -2 & 3 \\ 2 & 1 \end{bmatrix}$.

9. Find AB. ______________

Answer: $\begin{bmatrix} 14 & -13 \\ -6 & 9 \end{bmatrix}$

10. Find BA. ______________

Answer: $\begin{bmatrix} 19 & -4 \\ -7 & 4 \end{bmatrix}$

11. In Frames 9 and 10, does AB = BA? ______
Therefore, multiplication of matrices (*is/is not*) ______________.

Answer: No / is not; commutative

Let $A = \begin{bmatrix} 1 & 3 \\ 2 & 1 \end{bmatrix}$ and $B = \begin{bmatrix} 3 & 2 \\ 2 & 3 \end{bmatrix}$.

12. Find (A - B)(A - B) ____________

$\begin{bmatrix} 4 & -4 \\ 0 & 4 \end{bmatrix}$

13. Find $A^2 - 2AB + B^2$. ($A^2 = AA$) ____________

$\begin{bmatrix} 2 & -4 \\ 0 & 6 \end{bmatrix}$

14. Did you get the same answer in Frames 12 and 13? ____. The reason the answers are different can be found by multiplying out (A - B)(A - B) as

No

$(A - B)(A - B) = A^2 =$ ____ - ____ $+ B^2$

AB; BA

Since, in general, $AB \neq$ ____,

BA

$AB - BA \neq$ ______.

-2AB

15. **Beads For You sells assorted beads to local craftsmen. The cost of a half dozen beads is given by matrix A.**

	per half dozen
glass	$ .30
plastic	$ .15
wood	$ .60
silver	$2.20

= A

Jason makes necklaces which he has decided to sell for $10 apiece plus the cost of materials. He has three styles. The amount in half dozens of each bead required for these styles is shown in matrix B.

	glass	plastic	wood	silver
Style a	2	2	1	0
Style b	0	0	3	2
Style c	0	3	0	1

= B

Find the price Jason charges for each style.

a = ______ $11.50

b = ______ $16.20

c = ______ $12.65

8.3 Multiplicative Inverses of Matrices

1. If A is an $n \times n$ matrix, then the ______ matrix, written ___, is the matrix such that

$$AI = \_\_\_\_ \text{ and } IA = \_\_\_\_.$$

identity

I

A; A

2. Write the 3×3 identity matrix I.

I = ______

$$\begin{bmatrix} 1 & 0 & 0 \\ 0 & 1 & 0 \\ 0 & 0 & 1 \end{bmatrix}$$

3. Let $A = \begin{bmatrix} -4 & 7 \\ 2 & 9 \end{bmatrix}$ and $I = \begin{bmatrix} 1 & 0 \\ 0 & 1 \end{bmatrix}$

Find IA. ______

$$\begin{bmatrix} -4 & 7 \\ 2 & 9 \end{bmatrix}$$

4. Find AI. ______

$$\begin{bmatrix} -4 & 7 \\ 2 & 9 \end{bmatrix}$$

5. The multiplicative ______ of a matrix A, written ______, is a matrix such that

$$AA^{-1} = \_\_\_\_ \text{ and } \_\_\_\_ = I.$$

inverse

A^{-1}

I; $A^{-1}A$

6. To find A^{-1} for a given matrix A, first form the ________ matrix, [A|____].	augmented; I
7. Then use row transformations to change [A|I] into [___|B].	I
8. Then B = _____. Verify this statement by showing _____ = _____ = I.	A^{-1} BA; AB
9. The following matrix _____ transformations may be used.	row
(1) Any two rows may be __________.	interchanged
(2) The elements of any row may be ________ by a nonzero __________.	multiplied scalar
(3) A multiple of the elements of one row may be _______ to the elements of another row.	added
10. To find the inverse of $\begin{bmatrix} 4 & -5 \\ -3 & 4 \end{bmatrix}$, start with the __________ matrix	augmented
$\left[\begin{array}{cc\|cc} 4 & -5 & \_\_ & \_\_ \\ -3 & 4 & \_\_ & \_\_ \end{array}\right].$	1; 0 0; 1
We need ____ in the first row, first column. To get ____, multiply each element in the first row by _____.	1 1 1/4
$\left[\begin{array}{cc\|cc} \_\_ & \_\_ & \_\_ & \_\_ \\ -3 & 4 & 0 & 1 \end{array}\right]$	1; -5/4; 1/4; 0
Now we need ___ in the ________ row, first column. To get 0, multiply each element of the _____ row by ___, and add the result to the second row.	0; second first 3
$\left[\begin{array}{cc\|cc} 1 & -5/4 & 1/4 & 0 \\ \_\_ & \_\_ & \_\_ & \_\_ \end{array}\right]$	0; 1/4; 3/4; 1

Get ____ in the second row, second column, by multiplying each element of the second row by ____.	1 4
$\begin{bmatrix} 1 & -5/4 & \vert & 1/4 & 0 \\ \_\_ & \_\_ & \vert & \_\_ & \_\_ \end{bmatrix}$	0; 1; 3; 4
Finally, get ____ in the ______ row, second column, by multiplying each element of the second row by _____ and adding the result to the first row.	0; first 5/4
$\begin{bmatrix} \_\_ & \_\_ & \vert & \_\_ & \_\_ \\ 0 & 1 & \vert & 3 & 4 \end{bmatrix}$	1; 0; 4; 5
The inverse is given to the _______ of the bar:	right
$A^{-1} = \begin{bmatrix} \_\_ & \_\_ \\ \_\_ & \_\_ \end{bmatrix}$	4; 5 3; 4
To check, show that _______ = I and _______ = I.	AA^{-1}; $A^{-1}A$

Find each inverse in Frames 11–17.

11. $A = \begin{bmatrix} 3 & 1 \\ 0 & 2 \end{bmatrix}$	
Form the augmented matrix [A|____].	I
$\begin{bmatrix} 3 & 1 & \vert & 1 & 0 \\ \_\_ & \_\_ & \vert & \_\_ & \_\_ \end{bmatrix}$	0; 2; 0; 1
Use row operations to form an augmented matrix [I|B] where $B = A^{-1}$. To do this, first multiply each element of the first row by ______ to get ___ in the first row, first column position.	1/3 1
$\begin{bmatrix} \_\_ & \_\_ & \vert & \_\_ & \_\_ \\ 0 & 2 & \vert & 0 & 1 \end{bmatrix}$	1; 1/3; 1/3; 0
Since 0 is the first element of the second row we now have the first column of I completed.	

Now operate on the matrix to obtain the necessary 1 as the second row, second ________ element. (Note that we always obtain the 1 in its proper position in the row before obtaining the zeros.) Multiply each element of the second row by _____.

column

1/2

$$\left[\begin{array}{cc|cc} 1 & 1/3 & 1/3 & 0 \\ \_\_ & \_\_ & \_\_ & \_\_ \end{array}\right]$$

0; 1; 0; 1/2

Multiply each element of the second row by ______ and adding the result to the first row gives

-1/3

$$\left[\begin{array}{cc|cc} 1 & 0 & \_\_ & \_\_ \\ 0 & 1 & 0 & 1/2 \end{array}\right].$$

1/3; -1/6

So A^{-1} = ________________. Check your answer.

$\begin{bmatrix} 1/3 & -1/6 \\ 0 & 1/2 \end{bmatrix}$

12. $A = \begin{bmatrix} 1 & 2 \\ 1 & 3 \end{bmatrix}$ A^{-1} = ______________

$\begin{bmatrix} 3 & -2 \\ -1 & 1 \end{bmatrix}$

13. $A = \begin{bmatrix} 0 & 2 \\ 0 & 5 \end{bmatrix}$ A^{-1} = ______________

does not exist

14. $A = \begin{bmatrix} 1 & 0 & 2 \\ 0 & 2 & 1 \\ -1 & 0 & 3 \end{bmatrix}$

Form [A|I].

$$\left[\begin{array}{ccc|ccc} 1 & 0 & 2 & \_\_ & \_\_ & \_\_ \\ 0 & 2 & 1 & \_\_ & \_\_ & \_\_ \\ -1 & 0 & 3 & \_\_ & \_\_ & \_\_ \end{array}\right]$$

1; 0; 0
0; 1; 0
0; 0; 1

We already have 0 in the second row, first column. To get ____ in the third row, first column, multiply each element in the first row by ____ and add the result to the third row.

0
1

$$\left[\begin{array}{ccc|ccc} 1 & 0 & 2 & 1 & 0 & 0 \\ 0 & 2 & 1 & 0 & 1 & 0 \\ \_\_ & \_\_ & \_\_ & \_\_ & \_\_ & \_\_ \end{array}\right]$$

0; 0; 5; 1; 0; 1

We have now completed the first column of I. Now multiply the second row by _____ to get 1 in the second row, second column.

1/2

$$\left[\begin{array}{ccc|ccc} 1 & 0 & 2 & 1 & 0 & 0 \\ \_\_\_ & \_\_\_ & \_\_\_ & \_\_\_ & \_\_\_ & \_\_\_ \\ 0 & 0 & 5 & 1 & 0 & 1 \end{array}\right]$$

0; 1; 1/2; 0; 1/2; 0

We have now completed the second column of I since we were fortunate to have _____ already in their proper place. Multiply each element in the _____ row by _____ to obtain a 1 in the third row, third element position.

zeros
third
1/5

$$\left[\begin{array}{ccc|ccc} 1 & 0 & 2 & 1 & 0 & 0 \\ 0 & 1 & 1/2 & 0 & 1/2 & 0 \\ \_\_\_ & \_\_\_ & \_\_\_ & \_\_\_ & \_\_\_ & \_\_\_ \end{array}\right]$$

0; 0; 1; 1/5; 0; 1/5

Finish off the row transformations; the inverse is

____

$$\begin{bmatrix} 3/5 & 0 & -2/5 \\ -1/10 & 1/2 & -1/10 \\ 1/5 & 0 & 1/5 \end{bmatrix}$$

15. $A = \begin{bmatrix} 2 & 0 & 1 \\ -1 & 1 & 0 \\ 0 & 0 & 1 \end{bmatrix}$ $A^{-1} =$ _______________

$$\begin{bmatrix} 1/2 & 0 & -1/2 \\ 1/2 & 1 & -1/2 \\ 0 & 0 & 1 \end{bmatrix}$$

16. $A = \begin{bmatrix} -1 & 1 & 0 \\ 0 & 1 & 2 \\ 1 & -1 & 1 \end{bmatrix}$ $A^{-1} =$ _______________

$$\begin{bmatrix} -3 & 1 & -2 \\ -2 & 1 & -2 \\ 1 & 0 & 1 \end{bmatrix}$$

17. $A = \begin{bmatrix} 3 & 1 & 2 \\ 1 & 0 & 3 \\ 6 & 2 & 4 \end{bmatrix}$ $A^{-1} =$ _______________

does not exist

18. To solve a system by using a matrix ________, we need A, the matrix of ________________, X, the matrix of ________________, and B, the matrix of ____________.

inverse
coefficients
variables
constants

19. Write the given system as the matrix equation

_______ .

AX = B

20. The solution of this system is

X = _______ .

$A^{-1}B$

Write the matrix of coefficients, A, the matrix of variables, X, and the matrix of constants, B, for each of the systems of Frames 21 and 22.

21. $5x + 3y = -7$
$2x + 2y = -3$

$A = \begin{bmatrix} \_\_ & \_\_ \\ 2 & 2 \end{bmatrix}$, $X = \begin{bmatrix} x \\ y \end{bmatrix}$, $B =$ _______

5; 3; $\begin{bmatrix} -7 \\ -3 \end{bmatrix}$

22. $x + 4y = 1$
$-2x - y = 5$

$A = \begin{bmatrix} \_\_ & \_\_ \\ \_\_ & \_\_ \end{bmatrix}$, $X = \begin{bmatrix} x \\ y \end{bmatrix}$, $B = \begin{bmatrix} \_\_ \\ \_\_ \end{bmatrix}$

1; 4; 1
-2; -1; 5

Solve each of the systems of two equations in two variables of Frames 23 and 24 using the inverse of the coefficient matrix.

23. $2x + y = 5$
$-x + 2y = -10$

First find A, X, and ____ .

B

$A = \begin{bmatrix} 2 & 1 \\ \_\_ & \_\_ \end{bmatrix}$, $X = \begin{bmatrix} x \\ y \end{bmatrix}$, $B = \begin{bmatrix} \_\_ \\ -10 \end{bmatrix}$

5
-1; 2

To find X, we must solve the equation _______ .

$X = A^{-1}B$

First find that

$A^{-1} = \begin{bmatrix} \_\_ & \_\_ \\ \_\_ & \_\_ \end{bmatrix}$.

2/5; -1/5
1/5; 2/5

Now find $A^{-1}B$.

$$x = A^{-1}B = \begin{bmatrix} 2/5 & -1/5 \\ 1/5 & 2/5 \end{bmatrix}\begin{bmatrix} 5 \\ -10 \end{bmatrix} = \begin{bmatrix} \_\_\_\_ \\ \_\_\_\_ \end{bmatrix}$$

4
-3

Then x = ____, y = ____. Check your answer by substituting in the original equations. The solution set is __________.

4; -3

{(4, -3)}

24. $x + 2y = 4$
$-x - y = -3$

$$A = \begin{bmatrix} 1 & 2 \\ -1 & -1 \end{bmatrix},\ X = \begin{bmatrix} x \\ y \end{bmatrix},\ \text{and } B = \begin{bmatrix} 4 \\ -3 \end{bmatrix}$$

Find A^{-1}:

$$A^{-1} = \begin{bmatrix} \_\_\_\_ & \_\_\_\_ \\ \_\_\_\_ & \_\_\_\_ \end{bmatrix}$$

-1; -2
1; 1

Finally,

$$X = A^{-1}B = \begin{bmatrix} -1 & -2 \\ 1 & 1 \end{bmatrix}\begin{bmatrix} 4 \\ -3 \end{bmatrix} = \begin{bmatrix} \_\_\_\_ \\ \_\_\_\_ \end{bmatrix}$$

2
1

and x = ____, y = ____. Check your solution by substituting in the original equations. The solution set is __________.

2; 1

{(2, 1)}

25. $2x + y + 3z = 6$
$-x + 2y - z = 0$
$x - y + 2z = 2$

First,

$$A = \begin{bmatrix} 2 & 1 & 3 \\ -1 & 2 & -1 \\ 1 & -1 & 2 \end{bmatrix},\ X = \begin{bmatrix} x \\ y \\ z \end{bmatrix},\ B = \begin{bmatrix} \_\_\_\_ \\ \_\_\_\_ \\ \_\_\_\_ \end{bmatrix}$$

6
0
2

Next, find A^{-1}. Start with the augmented matrix.	
$$\left[\begin{array}{ccc\|ccc} 2 & 1 & 3 & \_\_\_ & \_\_\_ & \_\_\_ \\ -1 & 2 & -1 & \_\_\_ & \_\_\_ & \_\_\_ \\ 1 & -1 & 2 & \_\_\_ & \_\_\_ & \_\_\_ \end{array}\right].$$	1; 0; 0 0; 1; 0 0; 0; 1
Interchanging the ______ and ______ rows gives	first; third
$$\left[\begin{array}{ccc\|ccc} 1 & -1 & 2 & 0 & 0 & 1 \\ -1 & 2 & -1 & 0 & 1 & 0 \\ 2 & 1 & 3 & 1 & 0 & 0 \end{array}\right].$$	
Adding the first row to the second row gives	
$$\left[\begin{array}{ccc\|ccc} 1 & -1 & 2 & 0 & 0 & 1 \\ \_\_\_ & \_\_\_ & \_\_\_ & \_\_\_ & \_\_\_ & \_\_\_ \\ 2 & 1 & 3 & 1 & 0 & 0 \end{array}\right].$$	0; 1; 1; 0; 1; 1
Multiplying the first row by ____ and adding to the ________ row gives	-2 third
$$\left[\begin{array}{ccc\|ccc} 1 & -1 & 2 & 0 & 0 & 1 \\ 0 & 1 & 1 & 0 & 1 & 1 \\ \_\_\_ & \_\_\_ & \_\_\_ & \_\_\_ & \_\_\_ & \_\_\_ \end{array}\right]$$	0; 3; -1; 1; 0; -2
Adding the second row to the first row gives	
$$\left[\begin{array}{ccc\|ccc} \_\_\_ & \_\_\_ & \_\_\_ & \_\_\_ & \_\_\_ & \_\_\_ \\ 0 & 1 & 1 & 0 & 1 & 1 \\ 0 & 3 & -1 & 1 & 0 & -2 \end{array}\right]$$	1; 0; 3; 0; 1; 2
Multiplying the second row by ____ and adding to the ________ row gives	-3 third
$$\left[\begin{array}{ccc\|ccc} 1 & 0 & 3 & 0 & 1 & 2 \\ 0 & 1 & 1 & 0 & 1 & 1 \\ \_\_\_ & \_\_\_ & \_\_\_ & \_\_\_ & \_\_\_ & \_\_\_ \end{array}\right]$$	0; 0; -4; 1; -3; -5
Finish the row transformations and check A^{-1} by multiplying AA^{-1}.	
A^{-1} = ________________________	$\begin{bmatrix} 3/4 & -5/4 & -7/4 \\ 1/4 & 1/4 & -1/4 \\ -1/4 & 3/4 & 5/4 \end{bmatrix}$

Then,

$$X = \begin{bmatrix} 3/4 & -5/4 & -7/4 \\ 1/4 & 1/4 & -1/4 \\ -1/4 & 3/4 & 5/4 \end{bmatrix} \begin{bmatrix} 6 \\ 0 \\ 2 \end{bmatrix} = \begin{bmatrix} 1 \\ \_\_\_\_ \\ \_\_\_\_ \end{bmatrix}.$$

Answer: 1; 1

The solution set is ______________.

Answer: {(1, 1, 1)}

8.4 Determinants

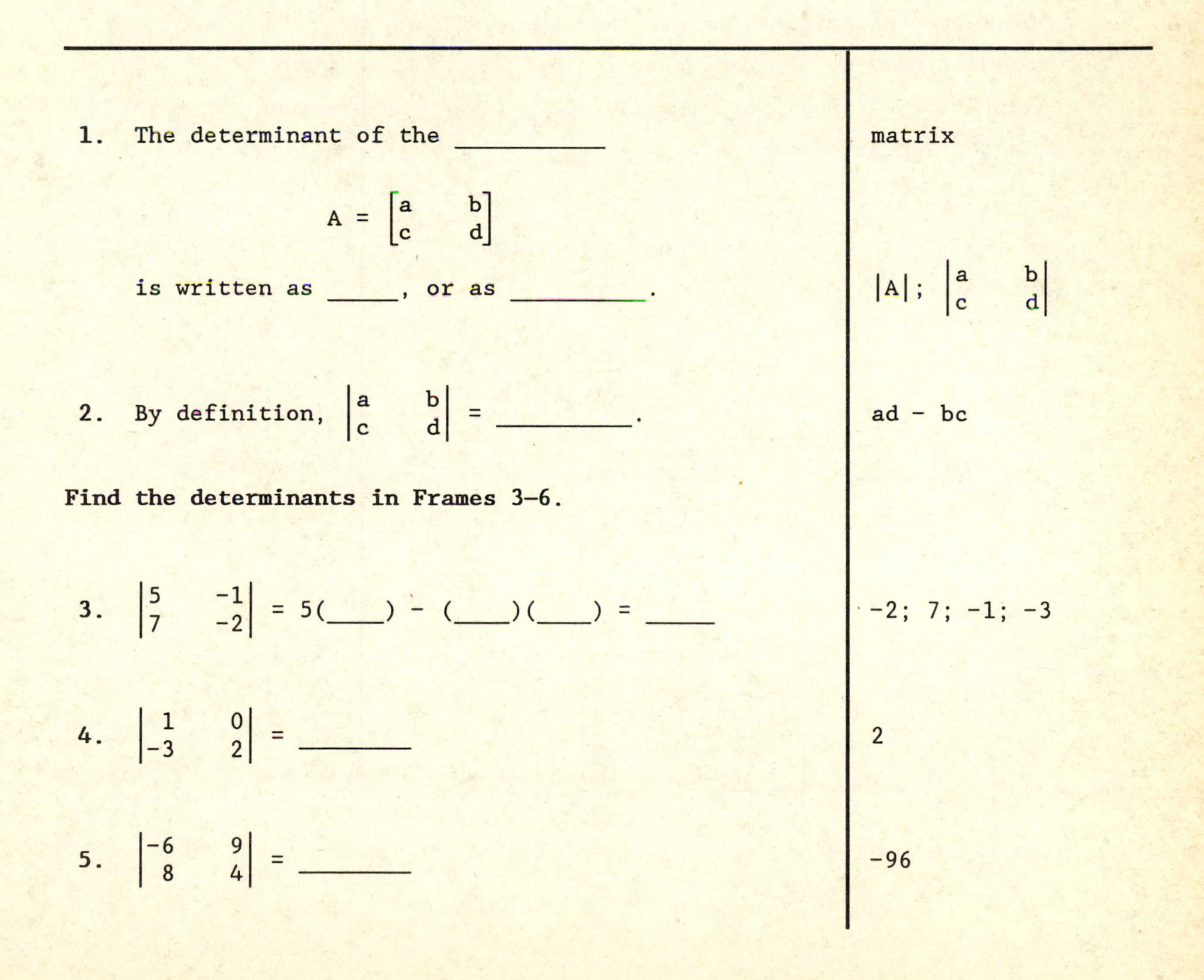

1. The determinant of the ______________

$$A = \begin{bmatrix} a & b \\ c & d \end{bmatrix}$$

is written as ______, or as ____________.

Answer: matrix; $|A|$; $\begin{vmatrix} a & b \\ c & d \end{vmatrix}$

2. By definition, $\begin{vmatrix} a & b \\ c & d \end{vmatrix}$ = ____________.

Answer: ad − bc

Find the determinants in Frames 3–6.

3. $\begin{vmatrix} 5 & -1 \\ 7 & -2 \end{vmatrix}$ = 5(____) − (____)(____) = _____

Answer: −2; 7; −1; −3

4. $\begin{vmatrix} 1 & 0 \\ -3 & 2 \end{vmatrix}$ = __________

Answer: 2

5. $\begin{vmatrix} -6 & 9 \\ 8 & 4 \end{vmatrix}$ = __________

Answer: −96

6. $\begin{vmatrix} 1 & b \\ -c & 3 \end{vmatrix} =$ ________ — 3 + bc

7. The determinant of a 3 × 3 matrix

$$A = \begin{bmatrix} a & b & c \\ d & e & f \\ g & h & i \end{bmatrix}$$

is defined as

$$\_\_\_\_ = \begin{bmatrix} a & b & c \\ d & e & f \\ g & h & i \end{bmatrix}$$ — |A|

$= (aei + \_\_\_ + cdh) - (\_\_\_ + hfa + \_\_\_)$. — bfg; gec; idb

8. To evaluate determinants larger than those of 2 × 2 matrices, we can use expansion by _______. A minor is the result when the row and __________ for a particular element are __________. — minors; column; deleted

9. Find the minor for each element in the first column of

$$\begin{vmatrix} 1 & -2 & 1 \\ -2 & 4 & 0 \\ 3 & 2 & 2 \end{vmatrix}.$$

Element	*Minor*	
1	$\begin{vmatrix} 4 & 0 \\ \_\_ & \_\_ \end{vmatrix}$	2; 2
-2	$\begin{vmatrix} \_\_ & \_\_ \\ 2 & 2 \end{vmatrix}$	-2; 1
3	$\begin{vmatrix} \_\_ & \_\_ \\ \_\_ & \_\_ \end{vmatrix}$	-2; 1 4; 0

The determinant is found by multiplying each ________ and its ________. Each product has a sign associated with it; the sign comes from the following ________ array.

____	____	____
____	____	____
____	____	____

In this example, the determinant is

$$\begin{vmatrix} 1 & -2 & 1 \\ -2 & 4 & 0 \\ 3 & 2 & 2 \end{vmatrix} = (\_\_\_\_)(1)\begin{vmatrix} 4 & 0 \\ 2 & 2 \end{vmatrix}$$

$$(\_\_\_\_)(-2)\begin{vmatrix} -2 & 1 \\ 2 & 2 \end{vmatrix}$$

$$(\_\_\_\_)(3)\begin{vmatrix} -2 & 1 \\ 4 & 0 \end{vmatrix}$$

$$= 1(\_\_\_) + 2(\_\_\_) + 3(\_\_\_)$$

$$= \_\_\_\_\_\_\_\_.$$

element; minor

sign

+; -; +

-; +; -

+; -; +

+

-

+

8; -6; -4

-16

10. The ________ of an element is the minor of the element with its associated sign. Find the following determinant using cofactors.

$$\begin{vmatrix} 1 & 3 & 1 \\ 4 & 2 & 0 \\ -1 & 3 & 1 \end{vmatrix} = \_\_\_\_\_$$

Element	*Minor*	*Cofactor*
1	$\begin{vmatrix} 2 & 0 \\ \_\_\_ & \_\_\_ \end{vmatrix} = \_\_\_\_$	(____) ____
4	$\begin{vmatrix} 3 & 1 \\ \_\_\_ & \_\_\_ \end{vmatrix} = \_\_\_\_$	(____) ____
-1	$\begin{vmatrix} 3 & 1 \\ \_\_\_ & \_\_\_ \end{vmatrix} = \_\_\_\_$	(____) ____

cofactor

4

3; 1; 2; +; 2

3; 1; 0; -; 0

2; 0; -2; +; -2

Multiply each element times the cofactor and add the products.

1(____) + 4(____) + -1(____) = ____

2; 0; -2; 4

Find each determinant in Frames 11–13.

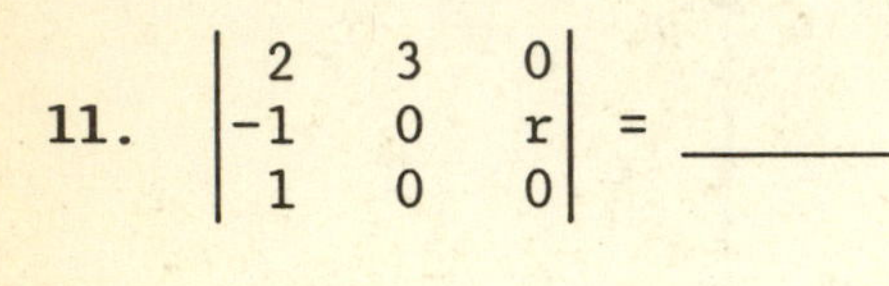

11. $\begin{vmatrix} 2 & 3 & 0 \\ -1 & 0 & r \\ 1 & 0 & 0 \end{vmatrix} =$ ________

3r

12. $\begin{vmatrix} 6 & 5 & -4 \\ 2 & 3 & 0 \\ -2 & 1 & 3 \end{vmatrix} =$ ________

-8

13. $\begin{vmatrix} -1 & 0 & 2 \\ 0 & 4 & 1 \\ -1 & 0 & -1 \end{vmatrix} =$ ________

12

Find the value of each of the 4 × 4 determinants in Frames 14 and 15.

14. $\begin{vmatrix} 2 & 0 & 0 & 3 \\ -5 & 1 & 0 & 0 \\ 0 & 2 & 1 & 1 \\ 0 & -3 & 0 & 1 \end{vmatrix}$

Expanding by minors about the first row we get

(____)(2) $\begin{vmatrix} 1 & 0 & 0 \\ 2 & 1 & 1 \\ -3 & 0 & 1 \end{vmatrix}$ (____)(0) $\begin{vmatrix} -5 & 0 & 0 \\ 0 & 1 & 1 \\ 0 & 0 & 1 \end{vmatrix}$

+; -

(____)(0) $\begin{vmatrix} -5 & 1 & 0 \\ 0 & 2 & 1 \\ 0 & -3 & 1 \end{vmatrix}$ (____)(3) $\begin{vmatrix} -5 & 1 & 0 \\ 0 & 2 & 1 \\ 0 & -3 & 0 \end{vmatrix}$.

+; -

Use the first row in the leftmost determinant and the third column in the rightmost determinant. The values of the other two parts will automatically be zero.

$$\left(2 \cdot 1\begin{vmatrix} 1 & 1 \\ 0 & 1 \end{vmatrix} - 0 + 0\right) + \left(0 + (-3)(-1)\begin{vmatrix} -5 & 1 \\ 0 & -3 \end{vmatrix} + 0\right)$$

Continuing the evaluation we get

$2 \cdot 1(1 - 0) + 3(15 - 0) =$ ______. | 47

15. $\begin{vmatrix} 1 & 0 & 1 & 0 \\ -1 & 1 & 0 & 0 \\ 2 & 0 & 0 & 1 \\ -2 & 1 & 0 & 0 \end{vmatrix} =$ ________ | -1

8.5 Properties of Determinants

---

1. Let us summarize the properties of ____________. | determinants

   (1) If every element in a _____ or column of a matrix is _____, then the determinant equals _____. | row, 0, 0

   (2) If corresponding rows and __________ of a matrix are __________________, then the ________________ is not changed. | columns, interchanged, determinant

   (3) If any two rows or columns of a matrix are ________________, the sign of the determinant is ___________. | interchanged, reversed

(4) If every element of a row (or column) of a matrix is multiplied by a ______ ________ k, then the determinant of the new matrix is k times the ______________ of the original matrix.

real; number

determinant

(5) If two rows (or columns) of a matrix are ______________, then the determinant is _____.

identical

0

(6) If a multiple of a row (or column) of a matrix is added to the corresponding elements of another row (or column), then the determinant of the matrix is ______________.

unchanged

Tell why each of the determinants of Frames 2–4 has a value of 0.

2. $\begin{vmatrix} 1 & 0 & 1 & 1 \\ 0 & 0 & 3 & 2 \\ -1 & 0 & 6 & 5 \\ 2 & 0 & 3 & 3 \end{vmatrix}$

Use Property ____. The value of this determinant is 0 because one (*row/column*) has elements that are all ____.

1

column

0

3. $\begin{vmatrix} -2 & 1 & 3 \\ 6 & 4 & 5 \\ 2 & -1 & -3 \end{vmatrix}$

Use Property 6 and add the third row to the first row, leaving the third row as is. We get

$$\begin{vmatrix} \_\_\_ & \_\_\_ & \_\_\_ \\ 6 & 4 & 5 \\ 2 & -1 & -3 \end{vmatrix}.$$

0; 0; 0

By Property ____ the value of this determinant is 0 because the elements of one ______ are all 0.	1 row
4. $\begin{vmatrix} 1 & -4 & 1 \\ 3 & 1 & 3 \\ 2 & 0 & 2 \end{vmatrix}$	
Use Property ____. The value of this determinant is 0 because two ____________ are identical.	5 columns
Use the appropriate properties from this section to tell why each of the statements of Frames 5–9 is true. Do not evaluate the determinants.	
5. $\begin{vmatrix} 3 & -1 \\ 5 & 7 \end{vmatrix} = \begin{vmatrix} 3 & 5 \\ -1 & 7 \end{vmatrix}$	
This is a direct result of Property ____ since the rows and columns have been _______________.	2 interchanged
6. $\begin{vmatrix} 1 & 4 \\ 1 & 4 \end{vmatrix} = 0$	
This determinant is 0 by Property ____.	5
7. $\begin{vmatrix} 6 & -3 & 1 \\ 4 & 0 & 2 \\ 7 & 1 & 2 \end{vmatrix} = -\begin{vmatrix} 1 & -3 & 6 \\ 2 & 0 & 4 \\ 2 & 1 & 7 \end{vmatrix}$	
This is a direct result of Property ______ since the first and third columns have been ___________________.	3 interchanged

8. $\begin{vmatrix} 2 & -6 & 4 \\ 1 & 0 & -1 \\ 7 & 2 & 3 \end{vmatrix} = 2\begin{vmatrix} 1 & -3 & 2 \\ 1 & 0 & -1 \\ 7 & 2 & 3 \end{vmatrix}$

This is a direct result of Property ____ since the elements of the first row of the left determinant is ____ times the elements of the first row of the right determinant.

4

2

9. $\begin{vmatrix} 4 & -1 \\ 3 & 1 \end{vmatrix} = \begin{vmatrix} 19 & 4 \\ 3 & 1 \end{vmatrix}$

This is a direct result of Property ____. Each element in the __________ row was multiplied by ____, with the result added to the elements of the __________ row.

6

second

5

first

Use Property 6 to find the value of each of the determinants of Frames 10–12.

10. $\begin{vmatrix} 7 & 28 \\ 3 & 12 \end{vmatrix}$

Using Property ____, multiply the elements of the second column by ______ and add the corresponding elements to the first column, leaving the second column intact. The determinant is ____.

6

-1/4

0

11. $\begin{vmatrix} -1 & 0 & 2 & 3 \\ 6 & 1 & 0 & 2 \\ -5 & 0 & 10 & 15 \\ 5 & 3 & 6 & 1 \end{vmatrix} =$ ____

0

12. $\begin{vmatrix} 5 & 3 & -1 & 0 \\ 2 & 1 & 6 & -1 \\ -3 & 2 & 5 & 0 \\ 2 & 1 & 4 & 1 \end{vmatrix}$

Our goal is to change some row or column to one in which every element but one is ____. The fourth column is the likely one to choose. Adding the corresponding elements of the second and fourth rows and leaving the fourth row intact, we get | 0

$$\begin{vmatrix} 5 & 3 & -1 & 0 \\ \_\_\_ & \_\_\_ & \_\_\_ & \_\_\_ \\ -3 & 2 & 5 & 0 \\ 2 & 1 & 4 & 1 \end{vmatrix}$$

4; 2; 10; 0

Expand by minors about the __________ column. | fourth

$$= 1\begin{vmatrix} 5 & 3 & -1 \\ 4 & 2 & 10 \\ -3 & 2 & 5 \end{vmatrix}.$$

Multiply the last row by _____ and add it to the __________ row. | -2; second

$$\begin{vmatrix} 5 & 3 & -1 \\ \_\_\_ & \_\_\_ & \_\_\_ \\ -3 & 2 & 5 \end{vmatrix}$$

10; -2; 0

Multiply the first row by ____ and add it to the _________ row. | 5; third

$$\begin{vmatrix} 5 & 3 & -1 \\ 10 & -2 & 0 \\ \_\_\_ & \_\_\_ & \_\_\_ \end{vmatrix}$$

22; 17; 0

$$= -1\begin{vmatrix} 10 & -2 \\ 22 & 17 \end{vmatrix} = -1(170 + 44) = \_\_\_\_$$

-214

8.6 Solution of Linear Systems by Cramer's Rule

---

1. Given the general system	
$a_1x + b_1y = c_1$	
$a_2x + b_2y = c_2$	
and given $D = \begin{vmatrix} a_1 & b_1 \\ a_2 & b_2 \end{vmatrix}$, $D_x = \begin{vmatrix} c_1 & b_1 \\ c_2 & b_2 \end{vmatrix}$,	
$D_y = \begin{vmatrix} \_\_\_ & \_\_\_ \\ a_2 & c_2 \end{vmatrix}$,	a_1; c_1
then x = _____ and y = _____. This theorem is	$\frac{D_x}{D}$; $\frac{D_y}{D}$
called __________ rule. Cramer's rule does not	Cramer's
apply if D = ____.	0

Use Cramer's Rule to solve the systems of Frames 2–6.

2. $4x + 3y = 2$ $2x + 4y = 6$	
$D = \begin{vmatrix} \_\_\_ & 3 \\ \_\_\_ & 4 \end{vmatrix} = 16 - 6 = 10$	4 2
$D_x = \begin{vmatrix} 2 & 3 \\ 6 & 4 \end{vmatrix} = \_\_\_ - \_\_\_ = \_\_\_$	8; 18; -10
D_y = _____.	20
Then $x = \frac{-10}{10}$ = ____ and y = ____. Check your	-1; 2
answer, as before, by substituting in both	
equations. The solution set is ____________.	{(-1, 2)}

3. $5x + 3y = 27$
$2x + 2y = 14$

$$D = \begin{vmatrix} 5 & 3 \\ 2 & 2 \end{vmatrix} = 4, \quad D_x = \underline{\qquad},$$

12

$$D_y = \begin{vmatrix} 5 & 27 \\ 2 & 14 \end{vmatrix} = 16, \quad x = \underline{\quad}, \quad y = \underline{\quad}.$$

3; 4

4. $2x + 3y = -6$
$x + 2y = -2$

$$x = \underline{\quad}, \quad y = \underline{\quad}$$

−6; 2

5. $2x + 3y + 2z = -2$
$3x + y + z = -5$
$x + 2y + 2z = 0$

Cramer's rule generalizes as follows:

$$D = \begin{vmatrix} 2 & 3 & 2 \\ 3 & 1 & 1 \\ 1 & 2 & 2 \end{vmatrix}, \quad D_x = \begin{vmatrix} -2 & 3 & 2 \\ -5 & 1 & 1 \\ 0 & 2 & 2 \end{vmatrix},$$

$$D_y = \begin{vmatrix} 2 & \underline{\quad} & 2 \\ 3 & \underline{\quad} & 1 \\ 1 & \underline{\quad} & 2 \end{vmatrix}, \quad D_z = \begin{vmatrix} 2 & 3 & \underline{\quad} \\ 3 & 1 & \underline{\quad} \\ 1 & 2 & \underline{\quad} \end{vmatrix}$$

−2; −2
−5; −5
0; 0

Now we need to find the values of the various determinants. To find D, we can multiply the last row by −2 and add to the first row, giving

$$\begin{vmatrix} \underline{\quad} & \underline{\quad} & \underline{\quad} \\ 3 & 1 & 1 \\ 1 & 2 & 2 \end{vmatrix}.$$

0; −1; −2

(Remember the shortcuts in evaluating determinants.) Multiply the last row by −3 and add to the second row; then complete the process of finding D.

$$\begin{vmatrix} 0 & -1 & -2 \\ 0 & -5 & -5 \\ 1 & 2 & 2 \end{vmatrix} = 1\begin{vmatrix} -1 & -2 \\ -5 & -5 \end{vmatrix} = \underline{\qquad}$$

−5

(Here we expanded about the first column).

Find D_x by expanding about the first column. | $-2(0) + 5(2)$

$$D_x = -2\begin{vmatrix}1 & 1\\2 & 2\end{vmatrix} - (-5)\begin{vmatrix}3 & 2\\2 & 2\end{vmatrix} + 0 = \underline{\qquad\qquad}$$
$$= \underline{\qquad\qquad}$$

10

Then x = _____ = _____.

$\frac{10}{-5}$; -2

Find D_y by expanding about the third row.

$$D_y = 1\begin{vmatrix}-2 & 2\\-5 & 1\end{vmatrix} - 0 + 2\begin{vmatrix}2 & -2\\3 & -5\end{vmatrix} = \underline{\qquad\qquad}$$
$$= \underline{\qquad\qquad}$$

$1(8) + 2(-4)$

0

Then y = _____ = _____. Find D_z and z. Check your answer by substituting into all three equations.

$\frac{0}{-5}$; 0

The solution set is ______________.

$\{(-2, 0, 1)\}$

6. $4x + y - z = 3$
   $x + 2y + z = 8$
   $2x - y + 2z = 6$

$$D = \begin{vmatrix}\_\_\_ & \_\_\_ & \_\_\_\\ \_\_\_ & \_\_\_ & \_\_\_\\ \_\_\_ & \_\_\_ & \_\_\_\end{vmatrix}$$

4; 1; -1

1; 2; 1

2; -1; 2

Expanding about the second row we get

$$-1\begin{vmatrix}1 & -1\\-1 & 2\end{vmatrix} + 2\begin{vmatrix}4 & -1\\2 & 2\end{vmatrix} - 1\begin{vmatrix}4 & 1\\2 & -1\end{vmatrix}$$
$$= -1(1) + 2(10) - 1(-6)$$
$$= \underline{\qquad}.$$

25

Then

$$D_x = \begin{vmatrix}\_\_\_ & 1 & -1\\ \_\_\_ & 2 & 1\\ \_\_\_ & -1 & 2\end{vmatrix}.$$

3

8

6

Adding the first row to the second we get

$$\begin{vmatrix}3 & 1 & -1\\ \_\_\_ & \_\_\_ & \_\_\_\\ 6 & -1 & 2\end{vmatrix}.$$

11; 3; 0

Multiply the first row by 2 and add to the third row.

$$\begin{vmatrix} 3 & 1 & -1 \\ 11 & 3 & 0 \\ \_\_\_\_ & \_\_\_\_ & \_\_\_\_ \end{vmatrix}$$

12; 1; 0

Expand about the third column.

$$-1\begin{vmatrix} 11 & 3 \\ 12 & 1 \end{vmatrix} + 0 + 0 = -1(-25) = \_\_\_\_.$$

25

Then, $x = 25/25 = 1$. Find y and z and check your answer.

Solution set: ________

$\{(1, 2, 3)\}$

Chapter 8 Test

The answers are at the back of this study guide.

1. Find the values of the variables in the following equation.

$$\begin{bmatrix} m - 4 & y + 1 \\ 3 & 7 \end{bmatrix} = \begin{bmatrix} 3 & 4 \\ r + 2 & a - 1 \end{bmatrix}$$

1. ______________

Perform each of the following operations, whenever possible.

2. $\begin{bmatrix} 5 & -1 & 6 \\ 2 & -3 & 2 \end{bmatrix} + \begin{bmatrix} -5 & 1 & -4 \\ 2 & -1 & 3 \end{bmatrix}$

2. ______________

3. $\begin{bmatrix} 4 & -1 & 2 \\ 2 & -2 & 0 \\ 0 & -5 & 1 \end{bmatrix} - \begin{bmatrix} -1 & 3 & 7 \\ -2 & -1 & 1 \\ 5 & 4 & 2 \end{bmatrix}$

3. ______________

4. Tom has 4 rock and roll records, 2 country western, and 5 classical. Joann has 3 rock and roll records, 7 country western, and 4 classical. Write this information first as a 3×2 matrix and then as 2×3 matrix.

4. ______________

Find each of the following matrix products, whenever possible.

5. $\begin{bmatrix} 2 & 5 \\ 7 & 4 \end{bmatrix} \begin{bmatrix} -1 \\ 3 \end{bmatrix}$

5. ______________

6. $\begin{bmatrix} 5 & 0 & 4 \\ -1 & 2 & -1 \end{bmatrix} \begin{bmatrix} 1 & 3 \\ -2 & 4 \\ 1 & 2 \end{bmatrix}$

6. ______________

7. Is there an identity matrix for multiplication of 4×4 matrices? If so, what is it?

7. ____________________

8. Decide whether or not the following pair of matrices are inverses.

$$\begin{bmatrix} 1 & 0 & 2 \\ 2 & 0 & 3 \\ 1 & 1 & 0 \end{bmatrix} \begin{bmatrix} -3 & 2 & 0 \\ 3 & -2 & 1 \\ 2 & -1 & 0 \end{bmatrix}$$

8. ____________________

Find the inverse, if it exists, for each of the following.

9. $\begin{bmatrix} 10 & 0 \\ 3 & 1 \end{bmatrix}$

9. ____________________

10. $\begin{bmatrix} 1 & 2 \\ 1 & 3 \end{bmatrix}$

10. ____________________

11. $\begin{bmatrix} 5 & 1 & 2 \\ 3 & 0 & 0 \\ 10 & 2 & 4 \end{bmatrix}$

11. ____________________

Use matrix inverses to solve each of the following.

12. $$\begin{aligned} x + 2y &= 5 \\ 2x - y &= 5 \end{aligned}$$

12. ____________________

13. $$\begin{aligned} x + y + z &= 1 \\ 2x - y + 2z &= -1 \\ -x - y + z &= -1 \end{aligned}$$

13. ____________________

Evaluate each of the following determinants.

14. $\begin{vmatrix} 2 & -8 \\ 7 & 2 \end{vmatrix}$

14. ____________________

15. $\begin{vmatrix} 5 & 0 & -2 \\ 1 & 1 & 2 \\ 3 & 0 & 5 \end{vmatrix}$

15. ____________________

16. Solve the following determinant equation for x.

$$\begin{vmatrix} x & -2 & 4 \\ 0 & 1 & 2 \\ -3x & 3 & -1 \end{vmatrix} = 51$$

16. ____________________

Give the property or properties that justify each of the following statements.

17. $\begin{vmatrix} 2 & 5 & 4 \\ 3 & 1 & 7 \\ 5 & 0 & 2 \end{vmatrix} = \begin{vmatrix} 2 & 3 & 5 \\ 5 & 1 & 0 \\ 4 & 7 & 2 \end{vmatrix}$

17. ____________________

18. $\begin{vmatrix} 5 & 7 & 4 \\ 9 & 8 & 3 \\ -1 & -4 & 7 \end{vmatrix} = -\begin{vmatrix} 5 & 7 & 4 \\ -1 & -4 & 7 \\ 9 & 8 & 3 \end{vmatrix}$

18. ____________________

Solve each of the following systems by Cramer's rule. Identify any systems with dependent equations or any inconsistent systems.

19. $$\begin{aligned} 2x + y &= 9 \\ x + 3y &= 2 \end{aligned}$$

19. ______________________

20. $$\begin{aligned} 3x + y + 2z &= -1 \\ 2x + 2y + z &= 2 \\ -2x - y + 2z &= 0 \end{aligned}$$

20. ______________________

CHAPTER 9 PRETEST

Pretest answers are at the back of this study guide.

Write the first five terms for each of the following sequences.

1. $a_n = (-1)^n(n - 1)$ 1. ______________________

2. $a_1 = -2$, for $n \geq 2$ $a_n = 2 \cdot a_{n-1}$ 2. ______________________

3. $a_n = 2n$ if n is odd and $a_n = 2n + 1$ if n is even 3. ______________________

4. Arithmetic, $a_1 = 2$, $d = 4$ 4. ______________________

5. Geometric, $a_1 = 2$, $r = 2$ 5. ______________________

6. A certain arithmetic sequence has $a_5 = 13$ and $a_7 = 7$. Find a_{11}. 6. ______________________

7. Find S_{10} for the arithmetic sequence with $a_1 = 5$ and $d = 2$. 7. ______________________

8. For a given geometric sequence, $a_1 = 4$ and $a_3 = 36$. Find a_5. 8. ______________________

9. Find a_7 for the geometric sequence with $a_2 = 2x^3$ and $a_4 = 8x^7$. 9. ______________________

10. Find S_4 for the geometric sequence with $a_1 = 16$ and $r = 1/2$. 10. ______________________

Evaluate each of the following sums that exist.

11. $\sum_{i=1}^{4} \frac{i+1}{i+2}$

11. ____________________

12. $\sum_{i=1}^{4} \left(-\frac{2}{3}\right)^{i}$

12. ____________________

13. $9 + 3 + 1 + \frac{1}{3} + \ldots$

13. ____________________

14. $\sum_{i=1}^{\infty} 30\left(\frac{1}{3}\right)^{i-1}$

14. ____________________

15. Carol Morley deposited \$40 in a new savings account on March 1. On the first of each month thereafter she deposited \$3 more than the previous month's deposit. Find the total amount of money she deposited in 12 months.

15. ____________________

16. The number of bacteria in a certain culture doubles every hour. If there are 100 bacteria present at 8:00 A.M. today, how many will be present at noon today?

16. ____________________

17. Prove that the following statement is true for every positive integer n. Use mathematical induction.

$$1^2 + 2^2 + 3^2 + \ldots + n^2 = \frac{n(n+1)(2n+1)}{6}$$

17. Write your proof on a separate piece of paper.

Evaluate the following.

18. P(8, 5) 18. ________

19. P(12, 3) 19. ________

20. $\binom{8}{3}$ 20. ________

21. $\binom{7}{4}$ 21. ________

22. A boat manufacturer offers 7 models in its 16-foot category with a choice of 5 colors and 4 motors. How many different varieties of the 16-foot boat are available?

22. ________

23. How many ways can 10 children hold 4 cats? Assume that each cat is held by one child.

23. ________

24. How many ways can a sample of 3 light bulbs be chosen from a carton that contains 60 bulbs?

24. ________

25. Discuss the difference between the probability that an event will occur and the odds that the event will occur.

25. ________

A single die is tossed. Find the probability that the face that is up shows each of the following.

26. A number more than 4

26. ______________________

27. 3 or an even number

27. ______________________

28. A number not divisible by 3

28. ______________________

29. In the preceding die-tossing experiment, what are the odds that the face that is up shows an odd number?

29. ______________________

30. A sample of 5 tapes are chosen. The probability of exactly 0, 1, 2, 3, 4, or 5 tapes being defective is given in the following table.

Number defective	0	1	2	3	4	5
Probability	.13	.38	.26	.12	.08	.03

Find the probability that less than three are defective.

30. ______________________

CHAPTER 9 SEQUENCES AND SERIES; PROBABILITY

9.1 Sequences; Arithmetic Sequences

---

1. If the domain of a function is the set of ________ __________, then the ________ elements can be ordered, as f(1), f(2), f(3), and so on. This ordered list of numbers is called a __________. It is customary to use the letter ____ to represent the variable and _____ to represent the functional value instead of f(n).	positive integers; range sequence n a_n
2. The elements of the sequence are called the _______ of the sequence.	terms
3. For example, $a_n = 5n + 3$ represents a __________. Here a_1 = ____, a_2 = ____, a_3 = ____, a_4 = ____, and so on. The range of such a function when written in the order _____, _____, _____, _____, ..., is called a __________. The range of this sequence is the infinite sequence _____, _____, _____, _____, ...	sequence 8; 13; 18; 23 a_1; a_2; a_3; a_4 sequence 8; 13; 18; 23

For the sequences of Frames 4–12, find the first four terms, a_1, a_2, a_3, and a_4.

4. $a_n = 7n - 5$	
$a_1 = 7($____$) - 5 =$ ____	1; 2
$a_2 = 7($____$) - 5 =$ ____	2; 9
$a_3 = 7($____$) - 5 =$ ____	3; 16
$a_4 = 7($____$) - 5 =$ ____	4; 23

5. $a_n = -3n + 4$	
$a_1 = -3(\_\_\_\_) + 4 = \_\_\_\_$	1; 1
$a_2 = -3(\_\_\_\_) + 4 = \_\_\_\_$	2; -2
$a_3 = -3(\_\_\_\_) + 4 = \_\_\_\_$	3; -5
$a_4 = -3(\_\_\_\_) + 4 = \_\_\_\_$	4; -8
6. $a_n = \dfrac{n}{2n + 1}$	
$a_1 = \dfrac{\_\_\_\_}{2(\_\_\_\_) + 1} = \_\_\_\_$	$\frac{1}{1}$; 1/3
$a_2 = \_\_\_\_$, $a_3 = \_\_\_\_$, $a_4 = \_\_\_\_$	2/5; 3/7; 4/9
7. $a_n = n^2 + 3n$	
$a_1 = (\_\_\_\_)^2 + 3(\_\_\_\_) = \_\_\_\_$	1; 1; 4
$a_2 = \_\_\_\_$, $a_3 = \_\_\_\_$, $a_4 = \_\_\_\_$	10; 18; 28
8. $a_n = \dfrac{n^2 - 1}{n^2 + 1}$	
$a_1 = \_\_\_\_$, $a_2 = \_\_\_\_$, $a_3 = \_\_\_\_$, $a_4 = \_\_\_\_$	0; 3/5; 4/5; 15/17
9. $a_n = 4^n$	
$a_1 = \_\_\_\_$, $a_2 = \_\_\_\_$, $a_3 = \_\_\_\_$, $a_4 = \_\_\_\_$	4; 16; 64; 256
10. $a_n = 2^{-n}$	
$a_1 = \_\_\_\_$, $a_2 = \_\_\_\_$, $a_3 = \_\_\_\_$, $a_4 = \_\_\_\_$	1/2; 1/4; 1/8; 1/16
11. $a_n = (n - 1)(n + 1)$	
$a_1 = \_\_\_\_$, $a_2 = \_\_\_\_$, $a_3 = \_\_\_\_$, $a_4 = \_\_\_\_$	0; 3; 8; 15
12. $a_n = (-1)^n(2n - 1)$	
$a_1 = \_\_\_\_$, $a_2 = \_\_\_\_$, $a_3 = \_\_\_\_$, $a_4 = \_\_\_\_$	-1; 3; -5; 7
13. A __________ formula defines the _____ term of a sequence in terms of the previous term.	recursion; nth

For the sequences of Frames 14–17, find the first four terms, a_1, a_2, a_3, and a_4.

14. $a_1 = 4;\ a_n = 3a_{n-1} - 2$

$a_1 = 4$

$a_2 = 3a_1 - 2 = 3(\_\_\_\_) - 2 = \_\_\_\_$ — 4; 10

$a_3 = 3a_2 - 2 = 3(\_\_\_\_) - 2 = \_\_\_\_$ — 10; 28

$a_4 = 3a_3 - 2 = 3(\_\_\_\_) - 2 = \_\_\_\_$ — 28; 82

15. $a_1 = 3;\ a_n = a_{n-1} + n^2$

$a_1 = 3$

$a_2 = (\_\_\_\_) + (\_\_\_\_)^2 = \_\_\_\_$ — 3; 2; 7

$a_3 = (\_\_\_\_) + (\_\_\_\_)^2 = \_\_\_\_$ — 7; 3; 16

$a_4 = (\_\_\_\_) + (\_\_\_\_)^2 = \_\_\_\_$ — 16; 4; 32

16. $a_1 = \frac{1}{2};\ a_n = \frac{1}{a_{n-1} + 1}$

$a_1 = \frac{1}{2}$

$a_2 = \frac{1}{\_\_\_\_ + 1} = \frac{1}{\_\_\_\_} = \_\_\_\_$ — 1/2; 3/2; 2/3

$a_3 = \frac{1}{\_\_\_\_ + 1} = \frac{1}{\_\_\_\_} = \_\_\_\_$ — 2/3; 5/3; 3/5

$a_4 = \frac{1}{\_\_\_\_ + 1} = \frac{1}{\_\_\_\_} = \_\_\_\_$ — 3/5; 8/5; 5/8

17. $a_1 = 5;\ a_n = 2n - a_{n-1}$

$a_1 = 5$

$a_2 = 2(\_\_\_\_) - \_\_\_\_ = \_\_\_\_$ — 2; 5; -1

$a_3 = 2(\_\_\_\_) - \_\_\_\_ = \_\_\_\_$ — 3; -1; 7

$a_4 = 2(\_\_\_\_) - \_\_\_\_ = \_\_\_\_$ — 4; 7; 1

18. A sequence in which each term after the first is obtained by adding a fixed number to the preceding term is called an ____________ sequence (or arithmetic ____________). The fixed number is called the ____________ ____________.

arithmetic
progression
common; difference

19. 3, 5, 7, 9, 11, 13, 15 is an ______________ sequence having first term a_1 = ____ and common difference d = ____. (To find the common difference when given an arithmetic sequence such as this, select any term except the _______ and ____________ the preceding term.)	arithmetic 3 2 first subtract

Find the first term and common difference for any of the sequences of Frames 20–24 that are arithmetic.

20. 9, 17, 25, 33, 41, 49 a_1 = ____, d = 33 - ____ = ____	9; 25; 8
21. 2, 4, 8, 16, 32 Since 16 - ____ = ____, while 32 - ____ = ____, each term cannot be obtained by adding the same number to the preceding term. Hence this sequence (*is/is not*) an arithmetic sequence.	8; 8; 16; 16 is not
22. 12, 9, 6, 3, 0, -3 a_1 = ____, d = ____	12; -3
23. 3, $3\frac{1}{2}$, 4, $4\frac{1}{2}$, 5 a_1 = ____, d = ____	3; 1/2
24. 1/2, 1/3, 1/4, 1/5, 1/6 a_1 = ____, d = ____	Not arithmetic
25. The nth term, or general term, of an arithmetic sequence is given by the formula a_n = ________________.	$a_1 + (n - 1)d$

26. The tenth term of the arithmetic sequence having first term 8 and common difference 3 is given by

$a_{\_\_\_\_}$ = ____ + (____ − 1)(____)
= ____ + ____
= ____.

Answers: 10; 8; 10; 3
8; 27
35

27. To find a_{15} for the arithmetic sequence 9, 11, 13, 15, 17, ..., first find that a_1 = ____ and d = ____. Then

a_{15} = ____ + (____ − 1)(____)
= ____.

Answers: 9
2
9; 15; 2
37

In Frames 28–32, find the indicated term for the given arithmetic sequence.

28. −8, −5, −2, 1, ...; find a_{18}.

a_1 = ____, d = ____
a_{18} = ____ + (____)(____) = ____

Answers: −8; 3
−8; 17; 3; 43

29. $a_1 = -5$, $d = -3$; find a_{21}.

a_{21} = ____ + (____)(____) = ____

Answers: −5; 20; −3; −65

30. $a_1 = 10$, $d = 1/2$; find a_{15}.

a_{15} = ____ + (____)(____) = ____

Answers: 10; 14; 1/2; 17

31. $a_8 = 6$, $a_9 = 10$; find a_{31}.

First find ____ and ____. Here

d = a_9 − ____ = ____ − ____ = ____.

Use the formula for the nth term to find a_1. We know a_8 = ____, d = ____, and, for a_8, n = ____.

Hence

____ = a_1 + (____)(____),
a_1 = ____.

Now find a_{31}.

a_{31} = ____ + (____)(____) = ____

Answers: a_1; d
a_8; 10; 6; 4
6; 4; 8
6; 7; 4
−22
−22; 30; 4; 98

32. $a_{11} = 12$, $a_{12} = 9$; find a_{26}.

$d =$ ____, $a_1 =$ ____, $a_{26} =$ ____ | -3; 42; -33

9.2 Geometric Sequences

1. A sequence in which each term after the first is some constant multiple of the preceding term is called a __________ sequence. The constant multiplier is called the ________ ratio. | geometric; common

2. The sequence 1, 2, 4, 8, 16 is a __________ sequence with first term ____ and common ratio ____. | geometric; 1; 2

Find the first term and common ratio of any of the sequences of Frames 3–7 that are geometric.

3. 10, 5, 5/2, 5/4

$a_1 =$ ____ To find the constant ratio, r, take any term except the first and ________ it by the preceding term. If we choose the term 5 here, we can find r by writing

$r = (\_\_\_) \div (\_\_\_) = \_\_\_$. | 10; divide; 5; 10; 1/2

4. 8, -8, 8, -8, 8

$a_1 =$ ____, $r =$ ____ | 8; -1

5. 6, 8, 12, 18

$a_1 =$ ____, $r =$ ____ | Not geometric

6. 1, -2, 4, -8, 16

$a_1 =$ ____, $r =$ ____ | 1; -2

7. 1/2, 1/4, 1/8, 1/16

a_1 = ____, r = ____ | 1/2; 1/2

8. The nth term of the geometric sequence having first term a_1 and common ratio r is given by

a_n = ____ • r——. | a_1; n - 1

9. We can use this formula to find the nth term of the sequence of Frame 7. This sequence has a_1 = ____ and r = ____, so that the nth term is | 1/2; 1/2

a_n = ____ • (____)$^{n-1}$. | 1/2; 1/2

Simplify to get

a_n = ______. | $(1/2)^n$

10. Find the nth term of the geometric sequence with first term 2/3 and common ratio 9. The nth term is

a_n = ____ • (____)$^{n-1}$. | 2/3; 9

11. The formula for the nth term can be used to find any specific term. For example, the fifth term of the geometric sequence having first term 3 and common ratio 2 is given by

a_5 = ____(____)—— | 3; 2; 5 - 1 (or 4)

= ____. | 48

Find the indicated term for the geometric sequences of Frames 12–19.

12. 3, 6, 12, 24; find a_8.

Here a_1 = ____ and r = ____. Thus a_8 is given by | 3; 2

a_8 = ____(____)—— | 3; 2; 7

= _____. | 384

13. 1/5, 1/15, 1/45, ...; find a_6.	
a_1 = _____, r = _____	1/5; 1/3
a_6 = ____(____)——	1/5; 1/3; 5
= __________	1/1215
14. $a_1 = -2$, $r = -2$; find a_5.	
a_5 = ____(____)——	-2; -2; 4
= _____	-32
15. $a_1 = 5$, $r = 3/4$; find a_6. (Set up only; do not calculate.)	
a_6 = ____(____)——	5; 3/4; 5
16. $a_1 = -5$, $r = -1/3$; find a_8. (Set up only.)	
a_8 = ____(____)——	-5; -1/3; 7
17. $a_4 = 12$, $a_5 = 9$; find a_{12}. (Set up only.) To find r we write	
r = (____) ÷ (____) = _____.	9; 12; 3/4
To find a_1 use the formula $a_n = a_1 \cdot r^{n-1}$. Here we can use a_4 as a_n.	
____ = a_1(____)——	12; 3/4; 3
a_1 = ______	256/9
Now set up a_{12}.	
a_{12} = ______(____)——	256/9; 3/4; 11
18. $a_6 = 25$, $a_7 = 50$; find a_9.	
a_9 = _____	200
19. $a_3 = 9$, $a_7 = 16/9$; find a_{11}.	
a_{11} = ________	256/729

9.3 Series; Applications of Sequences and Series

1. The sum of the terms of a sequence is called a ________. The Greek letter ________, denoted ______, is used to mean "_______." The letter _____ is commonly used as the ________ of summation.	series; sigma Σ; sum i; index

Evaluate the sums in Frames 2–5.

2. $\sum_{i=1}^{4} 2 + i^2$	
$= (2 + \_\_\_^2) + (2 + \_\_\_^2) + (2 + \_\_\_^2) + (2 + \_\_\_^2)$	1; 2; 3; 4
$= (2 + \_\_\_) + (2 + \_\_\_) + (2 + \_\_\_) + (2 + \_\_\_)$	1; 4; 9; 16
$= \_\_\_\_ + \_\_\_\_ + \_\_\_\_ + \_\_\_\_$	3; 6; 11; 18
$= \_\_\_\_$	38
3. $\sum_{i=4}^{6} 5i - 2$	
$= (5 \cdot \_\_\_ - 2) + (5 \cdot \_\_\_ - 2) + (5 \cdot \_\_\_ - 2)$	4; 5; 6
$= (\_\_\_\_ - 2) + (\_\_\_\_ - 2) + (\_\_\_\_ - 2)$	20; 25; 30
$= \_\_\_\_ + \_\_\_\_ + \_\_\_\_$	18; 23; 28
$= \_\_\_\_$	69
4. $\sum_{i=1}^{4} \frac{1}{i} = \_\_\_\_$	$\frac{25}{12}$
5. $\sum_{i=2}^{5} \frac{i + 1}{12 - 2i} = \_\_\_\_$	$\frac{127}{24}$

6. If an arithmetic sequence has first term a_1 and common difference d, then the sum of the first _____ terms is given by the formulas

S_n = ______________

or S_n = ______________.

The first of these formulas is used if we know the first and ______ terms. The second formula is used if we know the first term and the _________ ______________.

n

$\frac{n}{2}(a_1 + a_n)$

$\frac{n}{2}[2a_1 + (n - 1)d]$

last

common; difference

7. To find the sum of the first 10 terms of the arithmetic sequence having $a_1 = 8$ and $a_{10} = 30$, use the (*first/second*) formula from above.

$$S_{10} = \frac{\_\_\_}{2}(\_\_\_ + \_\_\_)$$

$$= \_\_\_$$

first

10; 8; 30

190

8. To find the sum of the first 10 terms of the arithmetic sequence having $a_1 = 12$ and $d = -3$, use the (*first/second*) formula from above.

$$S_{10} = \frac{\_\_\_}{2}[2(\_\_\_) + (\_\_\_)(\_\_\_)]$$

$$= \_\_\_(\_\_\_ + \_\_\_)$$

$$= \_\_\_$$

second

10; 12; 9; -3

5; 24; -27

-15

9. The sum $\sum_{i=1}^{6} (2i - 3)$ represents the series

____ + ____ + ____ + ____ + ____ + ____

This is an arithmetic sequence having a_1 = ____ and a_6 = ____. The sum is thus given by

$$S_6 = \frac{\_\_\_}{2}(\_\_\_ + \_\_\_) = \_\_\_.$$

-1; 1; 3; 5; 7; 9

-1

9

6; -1; 9; 24

10. Any sum of the form

$$\sum_{i=1}^{n} (mi + p),$$

where m and p are real numbers, represents the

sum of the terms of an arithmetic sequence having ______ ______ d = ____. Then,	common; difference; m
$a_1 = m(\_\_\_\_) + \_\_\_\_$	1; p
$= \_\_\_\_ + \_\_\_\_.$	m; p
To find, for example,	
$\sum_{i=1}^{8} (3i + 7),$	
m = ____, p = ____,	3; 7
$a_1 = \_\_\_\_ + \_\_\_\_ = \_\_\_\_$	3; 7; 10
and d = ____. Hence,	3
$\sum_{i=1}^{8} (3i + 7) = \frac{\_\_\_\_}{2}[2(\_\_\_\_) + \_\_\_\_(\_\_\_\_)].$	8; 10; 7; 3
$= \_\_\_\_(\_\_\_\_ + \_\_\_\_)$	4; 20; 21
$= \_\_\_\_$	164

Find the sums indicated in Frames 11–14.

11. $\sum_{i=1}^{12} (3i - 11)$	
$a_1 = \_\_\_\_$, $d = \_\_\_\_$	-8; 3
The sum is ____.	102
12. $\sum_{i=1}^{10} (5 - 3i)$	
$a_1 = \_\_\_\_$, $d = \_\_\_\_$	2; -3
The sum is ____.	-115
13. $\sum_{i=1}^{15} (2i + 5) = \_\_\_\_$	315
14. $\sum_{i=1}^{10} \left(\frac{2}{3}i + 2\right) = \_\_\_\_$	$\frac{170}{3}$
15. To find the sum of the first 200 positive integers, write them as a series.	
____ + ____ + ____ + . . . + ____	1; 2; 3; 200

Here we have an arithmetic sequence with a_1 = ____ and a_{200} = _____. Thus the sum of the first 200 terms is

$$\frac{\_\_\_}{2}(\_\_\_\_\_ + \_\_\_\_\_) = \_\_\_\_\_\_\_\_\_\_$$

1; 200

200; 1; 200; 20,100

16. Find the sum of the first 500 positive integers.

____

125,250

17. The sum of the first n terms of a geometric sequence having first term a_1 and common ratio r ($r \neq 1$) is given by the formula

$$S_n = \frac{\_\_\_\_(1 - \_\_\_\_)}{\_\_\_\_}.$$

a_1; r^n
$1 - r$

18. To find the sum of the first 7 terms of the geometric sequence 3, 6, 12, 24, ..., we note that a_1 = ____ and r = ____. Thus

$$S_7 = \frac{\_\_\_\_(1 - \_\_\_\_)}{1 - \_\_\_}$$

$$= 3(\_\_\_\_)$$

$$= \_\_\_\_.$$

3; 2

3; 2^7
2

127

381

Find the sum of the first six terms for each of the geometric sequences of Frames 19 and 20. Set up only; do not evaluate.

19. $a_1 = -4$, $r = 3/2$

$$S_6 = \frac{\_\_\_\_(1 - \_\_\_\_)}{1 - \_\_\_}$$

-4; $(3/2)^6$
$3/2$

20. $a_2 = 10$, $a_3 = 5$

First find r: r = (____) ÷ (____) = _____.

Then find a_1 by the formula $a_n = a_1 \cdot r^{n-1}$. If we use a_2 as a_n, we find that a_1 = _____. Thus

$$S_6 = \frac{\_\_\_\_(1 - \_\_\_\_)}{1 - \_\_\_}$$

5; 10; 1/2

20

20; $(1/2)^6$
$1/2$

21. A sum of the form $\sum_{i=1}^{n} m \cdot p^i$ represents the sum of the first n terms of a geometric sequence having first term a_1 = _____ and common ratio r = _____.

mp

p

22. To find the sum $\sum_{i=1}^{8} 3 \cdot 2^i$, we first find a_1:

$a_1 = 3(2)^{\_\_\_}$ = _____. Also, r = ____. Hence

$$S_8 = \frac{\_\_\_\_(1 - \_\_\_\_)}{\_\_\_\_}$$

= ______.

1; 6; 2

6; 2^8

-1

1530

Find the sums indicated in Frames 23–25.

23. $\sum_{i=1}^{4} 2 \cdot 3^i$

a_1 = ____, r = ____

S_4 = _____

6; 3

240

24. $\sum_{i=1}^{6} 3\left(\frac{2}{3}\right)^i$ (Set up only.)

a_1 = ____, r = ____

S_6 = ______________

2; 2/3

$$\frac{2[1 - (2/3)^6]}{1 - 2/3}$$

25. $\sum_{i=1}^{6} (-1)^i \cdot 2$

a_1 = ____, r = ____

Sum = ____

-2; -1

0

Solve the applied problems in Frames 26–28.

26. A gardener wants to plant a triangular flower bed. He uses 17 plants for the bottom row. Each row has 4 fewer plants than the previous row. Suppose that there are 5 rows in the flower bed.

(a) How many plants are planted in the top row?

This is a(n) (*arithmetic/geometric*) sequence	arithmetic
with a_1 = _____ and (d/r) = _____. Use the	17; d; -4
formula a_n = _____________.	$a_1 + (n - 1)d$
a_5 = _____	1
There (*is/are*) _____ plant(s) planted in the top row.	is; 1

(b) **What is the total number of plants in the flower bed?**

Find the sum of the terms of the sequence.	
Use the formula S_n = _______________.	$\frac{n}{2}(a_1 + a_n)$
S_5 = _____	45
There are _____ plants altogether.	45

27. **Health care workers find that each year .75 as many cases of a certain disease are recorded than the previous year. In 1992, 1600 cases were recorded.**

(a) **How many cases will be recorded in 1994?**

This is a(n) (*arithmetic/geometric*) sequence	geometric
with a_1 = ______ and (d/r) = _____. Use the	1600; r; .75 or 3/4
formula a_n = ________.	a_1r^{n-1}
Here, find $a_{\_\_}$.	3
a_3 = ______(_____)$^{\_\_\_}$	1600; 3/4; 2
= _______	900
There will be ______ cases recorded in 1994.	900

(b) **How many cases will be recorded altogether in the years 1992, 1993, and 1994?**

Find the sum of the first three terms.	
S_n = __________	$\frac{a_1(1 - r^n)}{1 - r}$
a_1 = ______, r = ______, n = ______	1600; 3/4; 3
S_3 = ________	3700
______ cases will be recorded during the three years.	3700

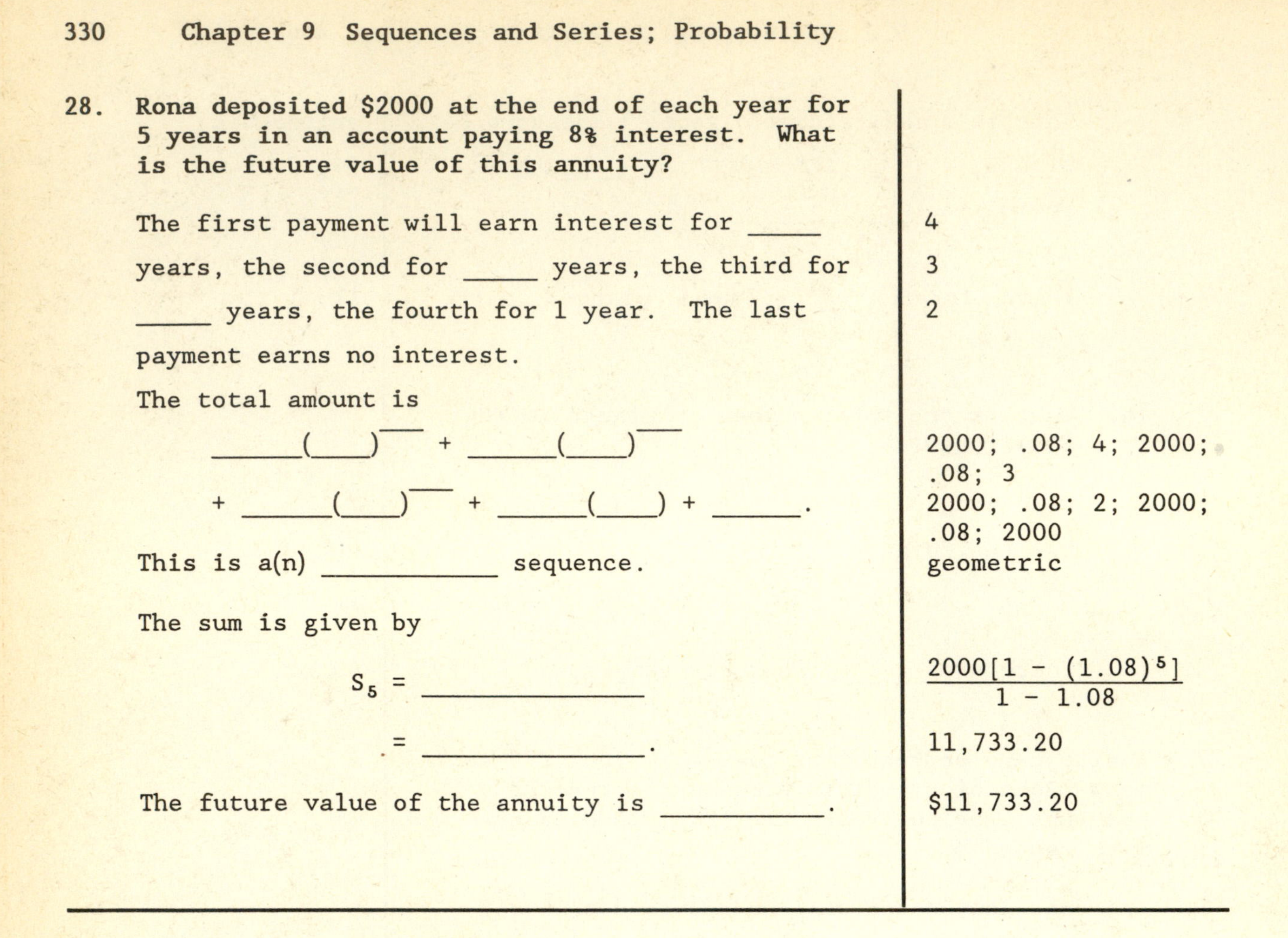

28. Rona deposited \$2000 at the end of each year for 5 years in an account paying 8% interest. What is the future value of this annuity?	
The first payment will earn interest for _____	4
years, the second for _____ years, the third for	3
_____ years, the fourth for 1 year. The last	2
payment earns no interest.	
The total amount is	
______(____)$^{\_\_\_}$ + ______(____)$^{\_\_\_}$	2000; .08; 4; 2000; .08; 3
+ ______(____)$^{\_\_\_}$ + ______(____) + ______.	2000; .08; 2; 2000; .08; 2000
This is a(n) ____________ sequence.	geometric
The sum is given by	
S_5 = ______________	$\frac{2000[1 - (1.08)^5]}{1 - 1.08}$
= ______________.	11,733.20
The future value of the annuity is ___________.	\$11,733.20

9.4 Sums of Infinite Geometric Sequences

1. 3, 2, 4/3, 8/9, 16/27, ... is an example of an	
___________ geometric sequence. If the constant	infinite
ratio r satisfies the condition _____________,	$\|r\| < 1$
the sum of an infinite geometric sequence can be found by the formula	
__________________.	$S_\infty = \frac{a_1}{1 - r}$
The sequence at the beginning of this frame has	
a_1 = _____ and r = _____. Since 2/3 < 1, the	3; 2/3
sum is given by	
$\frac{\_\_\_\_}{1 - \_\_\_\_}$ = _____.	3; $\frac{2}{3}$; 9

Find each of the sums indicated in Frames 2-6.

2. $\frac{2}{3}, \frac{1}{3}, \frac{1}{6}, \frac{1}{12}, \ldots$

a_1 = ______, r = ______

S_∞ = ______

Answer: $\frac{2}{3}$; $\frac{1}{2}$; $\frac{4}{3}$

3. $12, -3, \frac{3}{4}, -\frac{3}{16}, \ldots$

a_1 = ______, r = ______

S_∞ = ______

Answer: 12; $-\frac{1}{4}$; $\frac{48}{5}$

4. $\sum_{i=1}^{\infty} \left(\frac{3}{8}\right)^i$ = ______

Answer: $\frac{3}{5}$

5. $\sum_{i=1}^{\infty} 4\left(\frac{4}{3}\right)^i$ = ______

Answer: $|r| = 4/3 \nless 1$, so the sum does not exist

6. $\sum_{i=1}^{\infty} \frac{3}{4}\left(\frac{2}{3}\right)^i$ = ______

Answer: $\frac{3}{2}$

7. To write .343434... as p/q where p and q are integers, write .343434... as

.34 + .0034 + ________ + ...

This is the sum of the terms of an infinite geometric sequence having a_1 = _____ and r = _____.

$$S_\infty = \frac{\_\_\_}{1 - \_\_\_} = \frac{\_\_\_}{\_\_\_} = \frac{\_\_\_}{\_\_\_}$$

Answers: .000034 ; .34; .01 ; $\frac{.34}{.01}$; $\frac{.34}{.99}$; $\frac{34}{99}$

Write each repeating decimal in Frames 8-10 as a fraction of the form p/q where p and q are integers.

8. .2525... = ______

Answer: $\frac{25}{99}$

9. .104104104... = ______

Answer: $\frac{104}{999}$

10. .636363... = ______

Answer: $\frac{7}{11}$

9.5 Mathematical Induction

1. Let S_n be a statement concerning the positive integer ___. Suppose that	n
(a) S__ is true.	1
(b) For any positive integer k, $k \le n$, S__ implies S____.	k; k+1
Then _____ is true for every positive integer n.	S_n
2. In order to write a proof by mathematical induction, ______ steps are required.	two
Step 1 Prove that the statement is true for n = _____.	1
Step 2 Show that, for any positive integer k, $k \le n$, _____ implies _______.	S_k; S_{k+1}

Prove the theorems of Frames 3 and 4 by mathematical induction.

3. $1 + 3 + 5 + \ldots + (2n - 1) = n^2$	
Let ______ represent this statement.	S_n
Write S_1: 1 = _____	1^2
Is this true? (*yes/no*)	yes
Assume the statement is true for some natural number k so that	
$1 + 3 + 5 + \ldots + (2k - 1) =$ _____.	k^2
The next number in the sequence after $2k - 1$ is _______. Add this number to both sides.	$2k + 1$
$1 + 3 + 5 + \ldots + (2k - 1) +$ (________)	$2k + 1$
$= k^2 +$ (________)	$2k + 1$

On the right we have just ________, so that $1 + 3 + 5 + \ldots + (2k + 1) =$ ________. Thus, the truth of S_k implies the truth of _______, and the theorem has been proven by mathematical ____________.	$(k + 1)^2$ $(k + 1)^2$ S_{k+1} induction
4. $\frac{1}{1 \cdot 2} + \frac{1}{2 \cdot 3} + \frac{1}{3 \cdot 4} + \ldots + \frac{1}{n(n + 1)} = \frac{n}{n + 1}$	
Let $n = 1$. Then $\frac{1}{1(1 + 1)} = \frac{1}{1 \cdot 2} = \frac{1}{1 + 1}$.	
Therefore, this statement (*is/is not*) true for $n = 1$.	is
Suppose the statement is true for _____.	k
Then $\frac{1}{1 \cdot 2} + \frac{1}{2 \cdot 3} + \frac{1}{3 \cdot 4} + \ldots + \frac{1}{k(k + 1)} =$ ______.	$\frac{k}{k + 1}$
The next term on the left is ______________. Add this to both sides of the equation.	$\frac{1}{(k + 1)(k + 2)}$
$\frac{1}{1 \cdot 2} + \frac{1}{2 \cdot 3} + \frac{1}{3 \cdot 4} + \ldots +$ ______________	$\frac{1}{(k + 1)(k + 2)}$
$= \frac{1}{k + 1} +$ ______________	$\frac{1}{(k + 1)(k + 2)}$
Simplify on the right.	
$\frac{k}{k + 1} + \frac{1}{(k + 1)(k + 2)} = \frac{\_\_\_\_\_\_ + 1}{(k + 1)(k + 2)}$	$k(k + 2)$
$= \frac{\_\_\_\_\_\_}{(k + 1)(k + 2)}$	$k^2 + 2k + 1$
$=$ ______	$\frac{k + 1}{k + 2}$
The truth of S_k implies the truth of _________, proving the theorem.	S_{k+1}

9.6 Permutations

1. The multiplication principle of counting states that if one event can occur in m ways, and a second event can occur in n ways, then both events can occur in ______ ways, provided the outcome of the first does not influence the outcome of the second. Such events are called ______________ events.	mn independent
2. The multiplication principle of counting can be extended to ______ ___________ of events.	any; number
Solve the problems of Frames 3–4.	
3. **How many different mobile homes are available if the manufacturer offers a choice of 6 basic plans, 4 different exteriors, and 3 different woods for the interior paneling?** Use the ____________________ ________________ of ________________. We can break this down into events. The first event is the selection of a basic plan, the second event is the selection of type of exterior, and the third event is the selection of interior paneling. None of these events depends on another. Therefore the number of different mobile homes is $6 \cdot 4 \cdot$ _____ = ______.	multiplication; principle counting 3; 72

4. The Sharp family has decided to get a new dog, with all family members adding their preferences to the list. They have narrowed their choices down to 3 breeds and are still undecided as to the sex of the dog. They also have a list of 5 names. How many possibilities are there for the breed, sex, and name of the dog? ________

30

5. For any positive integer n, the product

$$n(n - 1)(n - 2) \ldots (2)(1)$$

is symbolized by _____, which is read "________________."

n!; n factorial

6. 0! is defined to be equal to ____.

1

7. The number of ________________ of n things taken ____ at a time is written __________.

permutations

r; P(n, r)

8. To find P(n, r), use the formula

$$P(n, r) = \underline{\hspace{3cm}}.$$

$$\frac{n!}{(n - r)!}$$

Find the permutations in Frames 9 and 10.

9. P(6, 4)

Use the formula of Frame 8.

$$P(6, 4) = \frac{\_\_\_!}{(\_\_\_\_\_)!}$$

$$= \frac{6 \cdot 5 \cdot 4 \cdot 3 \cdot 2 \cdot 1}{\_\_\_\_\_}$$

$$= \_\_\_\_\_\_$$

6

6 - 4

2 • 1

360

10. P(7, 2) = _______

42

11. There is a special formula for P(n, n):	
$P(n, n) =$ ______.	n!
For example,	
$P(5, 5) =$ ______	5!
$= 5 \cdot 4 \cdot 3 \cdot 2 \cdot 1 =$ ______.	120
12. Find $P(4, 4) =$ ______.	4! or 24
13. In how many ways can 7 children swing in 4 swings? This represents P(7, 4), which is ______.	840
14. In how many ways can 5 people ride 3 horses? ______	60
15. **How many 4-letter radio-station call letters can be made if the first 3 letters are KEQ and no letter can be repeated? If repetitions are allowed?**	
If we use KEQ as the first three letters then there are ______ letters left to use that are	23
not KEQ. Therefore, there are ______ ways if	23
no letter can be repeated. If repetitions are allowed, there are ______ ways.	26
16. **The Child Care Co-op has a "dress up" box of clothes. There are 7 hats, 4 long dresses, and 6 pairs of shoes. How many different outfits, which include 1 hat, 1 long dress, and 1 pair of shoes, are possible?** ______	$7 \cdot 4 \cdot 6 = 168$

17. How many ways can a three-digit area code be formed if the middle digit must be 0 or 1 and the first and third digits cannot be 0 or 1? (Repetitions can occur in the first and third digit.)	
There are ____ choices for the first digit, since 0 or 1 cannot be used.	8
There are ____ choices for the middle digit.	2
There are ____ choices for the third digit.	8
Thus, there are ____ • ____ • ____ = _______ ways to form the area code with the given restrictions.	8; 2; 8; 128
18. How many ways can the seven days of the week be arranged if Saturday and Sunday must be in the last two positions and no repetitions are allowed?	
___ choices for the first day	5
___ choices for the second day	4
___ choices for the third day	3
___ choices for the fourth day	2
1 choice for the fifth day	
___ choices for the sixth day	2
1 choice for the seventh day	
There are ___ • ___ • ___ • ___ • 1 • ___ • 1 = _____ ways to arrange the days of the week given the restrictions.	5; 4; 3; 2; 2; 240

9.7 Combinations

1. Permutations are used to find the number of ways a group of items can be arranged in ________. Combinations are used if order (*is/is not*) important.	order is not
2. The number of combinations of n things, taken r at a time is written ______.	$\binom{n}{r}$
This number is found with the formula $\binom{n}{r}$ = ____________.	$\dfrac{n!}{r!(n - r)!}$

Find the number of combinations in Frames 3–5.

3. $\binom{7}{3} = \dfrac{7!}{\_\_\_!(\_\_\_\_)!} =$ _____	3; 7 - 3; 35
4. $\binom{9}{5} =$ _____	126
5. $\binom{20}{3} =$ _____	1140

Solve the problems of Frames 6–12.

6. **How many samples of 10 fish can be netted from an aquarium holding 20 fish of different varieties?**

   Since the order of the fish in each sample does not matter, use ________________.

   combinations

   We want

   $$\binom{20}{10} = \frac{20!}{(20-10)!10!} = \frac{20!}{10!10!} = \_\_\_\_\_\_\_\_.$$

   184,756

7. **There are 25 students in Ms. Hugh's ninth-grade class and 3 are to be selected to be on the student council. How many ways can they be selected?**

   Here again _________ does not matter, so use combinations.

   order

   ____

   $\binom{25}{3} = 2300$

8. **Sundance Natural Food Restaurant makes sandwiches with cheese, avocado, tomato, onion, alfalfa sprouts, and nuts. How many different sandwiches can they make with any 4 of the ingredients?**

   The number of combinations is

   $$\binom{6}{4} = \frac{6!}{(6-4)!4!} = \_\_\_\_\_\_.$$

   15

9. **If a seashell collection has 12 olives, 7 sand dollars, and 4 nautilus shells, how many samples of 3 can be drawn in which all 3 are olive shells?**

   $$\binom{12}{3} = \_\_\_\_\_\_\_$$

   220

10. The English department faculty has decided to form a committee to mediate between administration and the students. There are 40 faculty members. The committee will consist of 7 people.

(a) In how many ways can the committee be chosen?

$\binom{40}{7}$ = ________ | 18,643,560

(b) If there are 15 women and 25 men on the faculty, how many committees could be chosen containing 4 women and 3 men?

First, find the combinations possible that have 4 women out of 15 women.

$\binom{\_\_}{4}$ = ________ | 15; 1365

Now, once the possible combinations of women are found, consider the possible combinations of men.

$\binom{\_\_}{3}$ = ________ | 25; 2300

Given any one of the ________ combinations of women, any one of the possible combinations of men could be picked, so (1365)(2300) = ________ possible combinations of 4 women and 3 men that may be chosen from 15 women and 25 men. | 1365 | 3,139,500

11. Permutations are required to solve problems when different orderings or ________ of objects are to be found. Clue words for permutations are order, ________. and ________. | arrangements | arrangement; schedule

12. ________ are required when the order of objects does not matter. Clue words for ________ are group, ________, and ________. | Combinations | combinations; committee sample

Solve the problems of Frames 11 and 12 by using combinations or permutations as required.

13. How many ways can the letters of the word ZEBRA be rearranged?
    A clue word is "____________________". A ____________________ is required.
    There are _________ ways the letters can be arranged.

rearranged
permutation
120

14. How many different samples of 3 out of 5 books may be selected?
    A clue word is "______________". A _____________ is required. There are _____ different samples.

samples; combination
10

15. How many committees of 12 students can be formed from a class of 12 students?

$$\binom{12}{12} = \frac{\underline{\hspace{2em}}!}{\underline{\hspace{2em}}!\,\underline{\hspace{2em}}!} = \frac{\underline{\hspace{2em}}!}{\underline{\hspace{2em}}!} = \underline{\hspace{3em}}$$

In general, $\binom{n}{n}$ = ____, since there is only ____ way to select a group of n from a set of n objects when order is disregarded.

12; 12
12; 0; 12; 1
1; 1

16. How many speakers can be chosen from an oratory group of 8 members?

$$\binom{8}{1} = \frac{\underline{\hspace{2em}}!}{\underline{\hspace{2em}}!\,\underline{\hspace{2em}}!} = \underline{\hspace{3em}}$$

In general, $\binom{n}{1}$ = ____, since there are ____ ways of choosing 1 object from a set of n objects.

8
1; 7; 8
n; n

9.8 An Introduction to Probability

1. To define probability, start with an experiment having one or more __________, each of which is __________ likely to occur. — outcomes; equally

2. The set of all possible __________ for an experiment makes up the __________ space for the experiment. — outcomes; sample

Write the sample space for each experiment in Frames 3 and 4.

3. Tossing a coin __________ — $\{H, T\}$

4. Rolling a die __________ — $\{1, 2, 3, 4, 5, 6\}$

5. The probability of an event E, written _______, is the ratio of the number of __________ in sample space S that belong to event E, n(E), to the total number of outcomes in sample space S, ______. In symbols, the ratio is written — P(E); outcomes; n(S)

$$P(E) = \_\_\_\_\_\_.$$

— $\frac{n(E)}{n(S)}$

Find the probability of each event.

6. Heads on a single toss of a coin.
The sample space is ________, with ____ elements. — $\{H, T\}$; 2
The event "toss of heads" is _______, with _____ — $\{H\}$; 1
element. The probability of heads is __________. — 1; 1/2

7. A die is rolled and the result is at least 5.	
Sample space: ______	$\{1, 2, 3, 4, 5, 6\}$
Event: ______	$\{5, 6\}$
Probability: ______	2/6 = 1/3
8. A diamond is drawn from a deck of 52 cards. (There are 13 diamonds in the deck.) ______	13/52 = 1/4
9. The complement of event E, written _____, is	E′
made up of all outcomes in the sample space that (*do/do not*) belong to E.	do not
10. Since E ∪ E′ = S, then P(E) + P(E′) = _____.	1
11. Write this result in an alternate way: P(E) = ______.	1 − P(E′)
12. If P(E) = 8/11, find P(E′). ______	3/11
13. The odds in favor of an event E is the ratio of P(E) and ______.	P(E′)

Find the odds in favor of each event in Frames 14–16.

14. P(E) = $\frac{2}{5}$	
If P(E) = 2/5, then P(E′) = ______, and the odds in favor of E are	3/5
$\dfrac{\frac{2}{5}}{\rule{1em}{0.4pt}} =$ _____.	$\frac{3}{5}$; $\frac{2}{3}$
Odds of 2/3 are often written ______.	2 to 3

15. $P(E) = \frac{1}{7}$

Odds in favor of E: ___________ | 1 to 6

16. $P(E') = \frac{9}{19}$

Odds in favor of E: ___________ | 10 to 9

17. Suppose the odds in favor of E are 8 to 5. Find P(E).
Odds of 8 to 5 means 8 __________ outcomes out of _____ total outcomes, so | favorable; 13

$P(E) =$ ________. | $\frac{8}{13}$

18. Find P(E) if the odds against E are 7 to 9.

________ | $\frac{9}{16}$

19. Two events that cannot occur at the same time are __________ exclusive events. | mutually

20. For example, rolling a 5 and a 2 on a die at the same time are mutually __________ events. | exclusive

21. Suppose a card is chosen from a standard deck of cards. It (*is/is not*) possible to choose a queen and a diamond at the same time. "Choosing a queen" and "choosing a diamond" (*are/are not*) mutually exclusive events. | is; are not
The event "choosing a queen and a diamond" is the ____________________ of these events. | intersection
The event "choosing a queen or a diamond" is the __________ of the events. | union

22. To find the probability of the union of two events, use the formula

$P(E \text{ or } F) = P(\_\_\_\_\_)$ | $E \cup F$

$= P(E) + \_\_\_\_\_ - \_\_\_\_\_\_\_.$ | $P(F)$; $P(E \cap F)$

23. In Frame 21, let E be the event "choosing a queen" and F the event "choosing a diamond." Then P(E) = 4/52 = _____ and P(F) = ____ = ____.

1/13; 13/52/ 1/4

Also, the probability of choosing the queen of diamonds, or P(_____), is ________. Use the formula to find the probability of choosing a queen or a diamond.

$E \cap F$; 1/52

$P(E \text{ or } F) = P(\_\_\_\_) = \_\_\_\_ + \_\_\_\_ - P(E \cap F)$

$E \cup F$; P(E); P(F)

$= \frac{1}{13} + \_\_\_\_ - \_\_\_\_$

$\frac{1}{4}$; $\frac{1}{52}$

$= \_\_\_\_\_\_\_\_$

$\frac{16}{52} = \frac{4}{13}$

The probability of choosing a queen or a diamond is ________.

4/13

24. Find P(E or F) if P(E) = 1/2, P(F) = 1/3, and $P(E \cap F) = 1/12$.
The probability is

$P(E \text{ or } F) = \frac{1}{2} + \_\_\_\_\_ - \_\_\_\_\_$

$\frac{1}{3}$; $\frac{1}{12}$

$= \_\_\_\_.$

$\frac{3}{4}$

25. Find P(M or N) if P(M) = 3/10, P(N) = 2/5, and $P(M \cap N) = 1/20$.

____

$\frac{13}{20}$

26. Complete the following properties of probability. For any events E and F in a sample space S:

$0 \leq \_\_\_\_\_\_\_ \leq 1$

P(S) = _____

P(Ø) = _____

P(E) = ____________

P(E or F) = P(E) + P(F) - __________

P(E)

1

0

1 - P(E′)

$P(E \cap F)$

CHAPTER 9 TEST

The answers for these questions are at the back of this study guide.

Write the first five terms for each of the following sequences.

1. $a_n = 3(n - 1)$ 1. ______________

2. $a_1 = 4$, for $n \geq 2$ $a_n = 3a_{n-1}$ 2. ______________

3. $a_n = n + 3$ if n is odd, and $a_n = 2a_{n-1}$ if n is even. 3. ______________

4. Arithmetic, $a_1 = 9$, $d = -2$ 4. ______________

5. Geometric, $a_4 = 12$, $r = 1/3$ 5. ______________

6. A certain arithmetic sequence has $a_6 = 23$ and $a_8 = 31$. Find a_{20}. 6. ______________

7. Find S_5 for the arithmetic sequence with $a_3 = 2$ and $d = -6$. 7. ______________

8. For a given geometric sequence, $a_7 = 16$ and $a_8 = 32$. Find a_5. 8. ______________

9. Find a_6 for the geometric sequence with $a_1 = 3x^2$ and $a_3 = 12x^8$. 9. ______________

10. Find S_4 for the geometric sequence with $a_1 = 8$ and $r = 1/4$. 10. ______________

Evaluate each of the following sums that exist.

11. $\sum_{i=1}^{5} (4i - 3)$ 11. ______________

12. $\sum_{i=1}^{4} \left(\frac{1}{3}\right)(9^i)$ — 12. ______

13. $\sum_{i=1}^{\infty} -3\left(\frac{1}{2}\right)^i$ — 13. ______

14. $128 + 48 + 18 + \frac{27}{4} + \ldots$ — 14. ______

15. Monique harvested 16 tomatoes from her greenhouse one week. Each week thereafter she harvested 10 more tomatoes than the previous week's harvest. Find the total number of tomatoes she harvested after 8 weeks. — 15. ______

16. Suppose that a bacteria colony doubles in size each day. If there are 100 bacteria in the colony today, how many will there be exactly one week from today? — 16. ______

17. Use mathematical induction to prove that the following statement is true for every positive integer n.

$$2 + 4 + 8 + \ldots + 2^n = 2^{n+1} - 2$$

17. Write your proof on a separate piece of paper.

Evaluate the following.

18. P(9, 4) — 18. ______

19. P(8. 1) — 19. ______

20. $\binom{7}{3}$ — 20. ______

21. $\binom{9}{9}$ — 21. ______

22. A bakery sells pies in 4 different sizes. There are 6 fillings and 3 types of crust available. How many different pies can be made? 22. ____________

23. In how many ways can 6 apprentices be assigned to 4 master workers? 23. ____________

24. How many different 5 card "hands" can be dealt from a deck of 12 cards? 24. ____________

A single die is tossed. Find the probability that the face that is up shows each of the following.

25. A number less than 5 25. ____________

26. 6 or an odd number 26. ____________

27. A number not divisible by 2 27. ____________

28. In the preceding die-tossing experiment, what are the odds that the face that is up shows a 5 or a 6? 28. ____________

29. A sample of 5 light bulbs is chosen. The probability of exactly 0, 1, 2, 3, 4, or 5 bulbs having broken filaments is given in the following table.

Number broken	0	1	2	3	4	5
Probability	.15	.21	.33	.19	.08	.04

Find the probability that at most 3 filaments are broken. 29. ____________

30. Write the formula for finding the probability of the union of two mutually exclusive events E and F. 30. ____________

APPENDIX A SETS

1. A set is a collection of objects called ________ or ________.	elements members.
2. The set containing the elements a, b, or c is written as ____________. The set $\{b, a, c\}$ (*is/is not*) the same set as $\{a, b, c\}$.	$\{a, b, c\}$ is
3. To show that 3 is an element of the set $\{1, 3, 5, 7\}$, we write 3 _____ $\{1, 3, 5, 7\}$. To show that 4 is not an element of $\{1, 3, 5, 7\}$, we write 4 _____ $\{1, 3, 5, 7\}$	$\in$ $\notin$
4. __________ letters name sets.	Capital
5. The set $\{9, 11, 12, 13, 10\}$ has a limited number of elements so it is called a(n) ________ set.	finite
6. Three dots are used following a list of elements to show that the list continues in the same ________. For example, the set of even numbers may be written $\{2, 4, 6, 8, \ldots\}$. Such a set has a(n) __________ list of elements. It is called a(n) ____________ set.	pattern unending infinite

7. Letters are sometimes used to represent numbers. A letter used to represent an element from a set of numbers is called a __________. For example, the set of all __________ between 0 and 1 is represented ________________.

variable

numbers

{x|x is a number between 0 and 1}

List the elements belonging to the sets in Frames 8 and 9.

8. {x|x is a counting number less than 7}

____

{1, 2, 3, 4, 5, 6}

9. {x|x is a counting number greater than or equal to 4 and less than or equal to 6}

____

{4, 5, 6}

10. We can usually identify a set that contains all the elements in any set used in a given problem. This set is called the ______________ set.

universal

11. A set containing no elements is called the _____ set or ______ set, written as ______ or ______.

null

empty; Ø; { }

12. The set of all numbers in the alphabet is ______.

Ø (or { })

13. Set A is a __________ of set B, or A _____ B, if every element of A is an element of B.

subset; ⊆

Answer *true* or *false* for Frames 14–17.

14. $\{3, 5, 9\} \subseteq \{1, 3, 5, \ldots\}$

true

15. $\{1/2, 3/2\} \not\subseteq \{x|$ x is a number between 0 and 2$\}$

false

16. Every set is a subset of itself.	true
17. The empty set is a subset of every set.	true
18. Two sets A and B are _______ if $A \subseteq B$ and $B \subseteq A$.	equal
19. For a given set A and a universal set U, the set of all elements of U that do not belong to A is called the _____________ of set A, written ______.	complement; A′
Let $U = \{a, b, c, d, e, f\}$, $A = \{a, c, e\}$, $B = \{a, c, d, f\}$. Find the sets in Frames 20–23.	
20. A′ _______________	$\{b, d, f\}$
21. B′ _______________	$\{b, e\}$
22. ∅′ _______________	U or $\{a, b, c, d, e, f\}$
23. U′ _______________	∅ or $\{\ \}$
24. The only elements belonging to both $A = \{a, e, c\}$ and $B = \{a, c, d, f\}$ are ____ and ____. The set $\{a, c\}$ is called the _______________ of sets A and B, written A ____ B.	a; c intersection ∩
25. The sets $C = \{a, f\}$ and $D = \{b, c, d, e\}$ have ____ element(s) in common. These sets are called ___________ sets. $C \cap D$ = __________.	no disjoint; ∅
26. The set of all elements belonging to either set $A = \{a, e, c\}$ or set $D = \{b, c, d, e\}$ is called the ________ of sets A and D, written A ____ D. The elements of the set $A \cup D$ are _________________.	union ∪ $\{a, b, c, d, e\}$

Let $M = \{1, 2, 3, 7, 9, 10\}$, $N = \{2, 4, 6, 10\}$, and $U = \{1, 2, 3, 4, 5, 6, 7, 8, 9, 10\}$. List all the elements of the sets in Frames 27–32.

Frame	Answer
27. M' __________	$\{4, 5, 6, 8\}$
28. $M \cap N$ __________	$\{2, 10\}$
29. $M \cup N$ __________	$\{1, 2, 3, 4, 6, 7, 9, 10\}$
30. $(M \cap N)'$ __________	$\{1, 3, 4, 5, 6, 7, 8, 9\}$
31. $(M \cup N')'$ __________	$\{4, 6\}$
32. $M' \cap N$ __________	$\{4, 6\}$

ANSWERS
TO
PRETESTS AND TESTS

CHAPTER 1 PRETEST

1. 0, 1
2. −1/4, 0, 1, 25/53
3. $-\sqrt{5}$, $\sqrt{3}$
4. All
5. 20
6. 1
7. 46
8. 10/3
9. Inverse
10. Commutative
11. Associative
12. Distributive
13. Commutative
14. Inverse
15. $-18m + 45x$
16. $-3y + 8$
17. $-5/2$, $-\sqrt{3}$, -1, $-1/2$
18. $-|-5 + 1|$, -3, $-|-2|$, $|4|$
19. 2
20. $\sqrt{42} - 5$
21. $2y - 8$
22. Transitive
23. Multiplication property
24. $r^3 - 11r^2 + 11r + 7$
25. $2x^3 + x^2 + 2x + 4$
26. $5m^2 + 31m - 27$
27. $27y^2 - 3yx - 14x^2$
28. $8r^4 - 34r^3 + 37r^2 - 16r + 15$
29. $3r^{2y} - 10r^y - 8$
30. $64w^6 + 576w^5z + 2160w^4z^2 + 4320w^3z^3 + 4860w^2z^4 + 2916wz^5 + 729z^6$
31. $d^5 - 10d^4f + 40d^3f^2 - 80d^2f^3 + 80df^4 - 32f^5$
32. $589{,}824x^7y^2$
33. $-6048c^2d^5$
34. $(x + 6)(x - 3)$
35. $(2m + 3r)(m - r)$
36. $(3x + 4m^2)(3x - 4m^2)$
37. $(3 + 2x)(9 - 6x + 4x^2)$
38. $(5y^2 - 4z^4)(25y^4 + 20y^2z^4 + 16z^8)$
39. $(2y - 7)(8r + s)$
40. $(x + y + 3)(x + y - 3)$
41. $\dfrac{x + 2}{x - 2}$
42. $-2n$
43. $1 - 4/x^2$
44. $\dfrac{5m^2 - 13}{(2m - 3)(m + 2)}$
45. $\dfrac{-r^2 - 8r}{(r + 3)^2(r - 2)}$
46. $\dfrac{2m - 2}{2m^2 + 1}$
47. $\dfrac{x^2}{y + x}$
48. 25/81
49. $24m^2/n$

50. $4y^7r^2/27$

51. $p^3q^9/20$

52. $z^{2/3}/(x^4y)$

53. $t^{1/2}/(r^3s^{3/2})$

54. $1/m^{7/2}$

55. $6y - 4$

56. $10\sqrt{10}$

57. $2\sqrt[5]{2}$

58. $2rs\sqrt{7rt}$

59. $\dfrac{x(\sqrt{x} - 3)}{x - 9}$

60. 14

CHAPTER 1 TEST

1. 0, 5, 39/3 (or 13)

2. $-\sqrt{11}$, $\sqrt{28}$

3. -4

4. Commutative

5. Inverse

6. $5r^2 - 4r - 5$

7. $6k^2 - 17kz - 14z^2$

8. $6r^4 - 17r^3 + 8r^2 - 4r + 1$

9. $243a^5 - 810a^4b + 1080a^3b^2 - 720a^2b^3 + 240ab^4 - 32b^5$

10. $-560k^4m^3$

11. $(5r - 3x)(2r + 3x)$

12. $(6a^3 - 5r)(36a^6 + 30a^3r + 25r^2)$

13. $(3x - 1 + y)(3x - 1 - y)$

14. The first step is wrong. Numerator and denominator should be factored before writing fraction in simplest form.

15. $3(z + 5y)$

16. $\dfrac{2y^2 - 4y - 4}{(y + 2)(y - 2)}$

17. $\dfrac{m^2 - 4m + 2}{(m - 3)(m - 1)(m - 2)}$

18. $\dfrac{12r - 8}{4r^2 + 5}$

19. m^7n^5

20. $18z^7$

21. $10\sqrt{5}$

22. $2\sqrt[4]{3}$

23. $2x^2t\sqrt{3st}$

24. $\dfrac{x + 2\sqrt{x}}{x - 4}$

25. $2r - 3s - 5\sqrt{rs}$

CHAPTER 2 PRETEST

1. {-5/7}
2. {-1/5}
3. $x = \frac{Mka}{k - M}$
4. 24/7 days
5. -7 + i
6. 41 - i
7. 116
8. -3i
9. {2, -3/2}
10. $\left\{\frac{5 \pm \sqrt{13}}{6}\right\}$
11. {1/2, -4}
12. -151; two imaginary solutions
13. 60 mph
14. $\{\pm 2, \pm\sqrt{2}\}$
15. {3/2, 1}
16. {6}
17. {1, -2/3}
18. $(-\infty, -2/3]$
19. [-7/2, 5/2]
20. $(-\infty, -3] \cup [5, \infty)$
21. [-3, -2)
22. {8, -2}
23. {7/4, 3/4}
24. $(-\infty, -7] \cup [7, \infty)$
25. (1, 7)

CHAPTER 2 TEST

1. {1/3}
2. ∅
3. $r = \frac{2yz}{8 - y}$
4. 2 oz
5. $133\frac{1}{3}$ mi
6. -2 - 13i
7. 34 + 13i
8. $\frac{29}{17} + \frac{20}{17}i$
9. -i
10. {1/4, -2/3}
11. {2/3, -3/2}
12. 0; exactly one rational solution
13. 15 hr
14. $\{\pm\sqrt{5}, \pm i\sqrt{2}\}$
15. {-3, 4/3}
16. ∅
17. {5}
18. $(-\infty, -7)$
19. [-3/5, 4/5]
20. $(-\infty, -2] \cup [3, \infty)$
21. (-1, 4]
22. {4, 8}
23. Because 1/2 makes the denominator zero, and division by zero is not defined.
24. $(-\infty, -2] \cup [2, \infty)$
25. (-6, 3)

CHAPTER 3 PRETEST

1. (a) $3\sqrt{2}$
   (b) $(-1/2, 13/2)$

2. 3

3. 4/3

4. -1

5. $2x - y = 7$

6. $y = -4$

7. $2x - y = 0$

8.

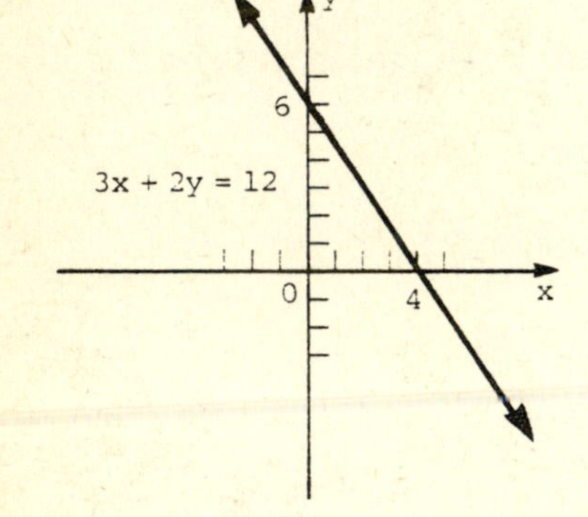

9.

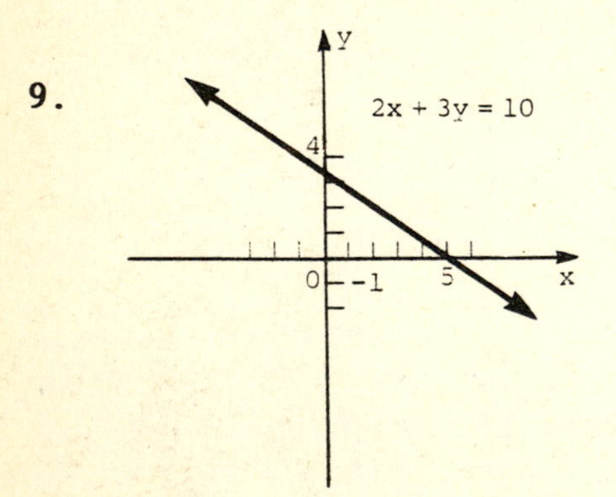

10.

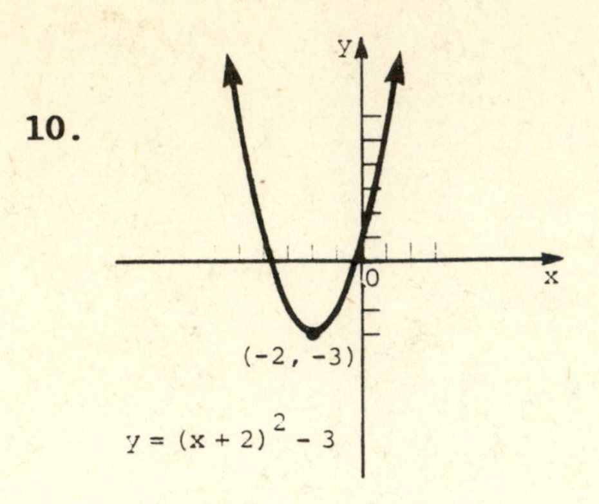

Axis: $x = -2$;
vertex: $(-2, -3)$;
domain: $(-\infty, \infty)$;
range: $[-3, \infty)$;

11.

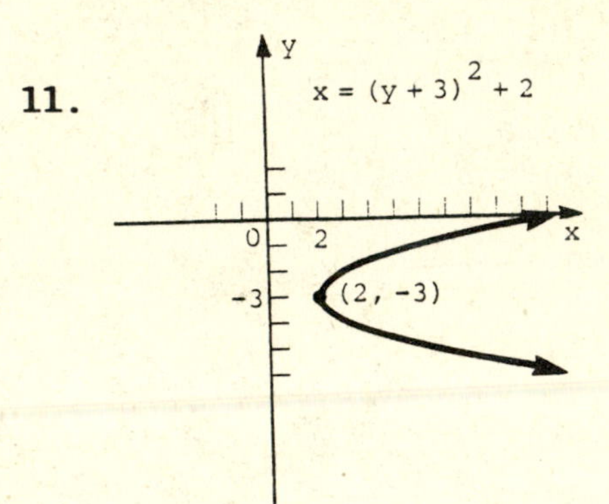

Axis: $y = -3$;
vertex: $(2, -3)$;
domain: $[2, \infty)$;
range: $(-\infty, \infty)$;

12. Vertex: $(3, 2)$; axis: $x = 3$

13. 40 meals; \$1700

14. $(x + 1)^2 + (y - 5)^2 = 3$

15. Center: $(2, -3)$; radius: $\sqrt{31}$

16. Origin

17.

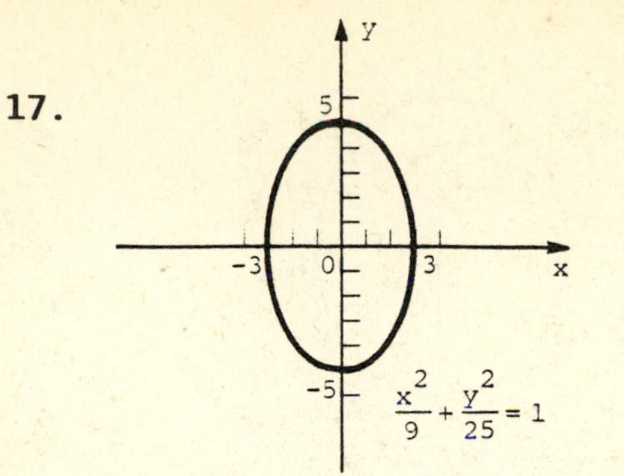

Domain: [-3, 3];
range: [-5, 5]

19.

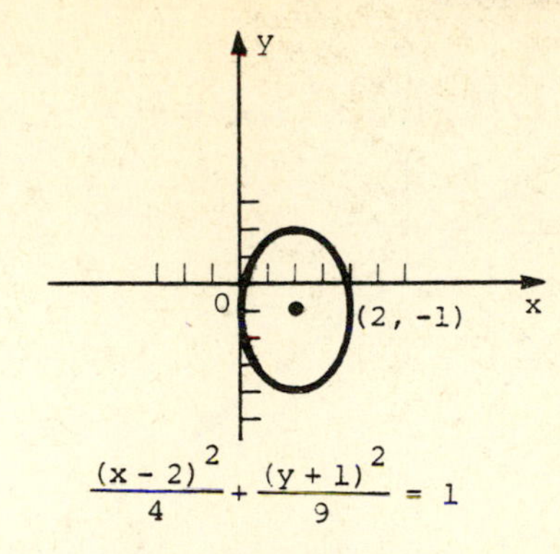

Domain: [0, 4];
range: [-4, 2]

18.

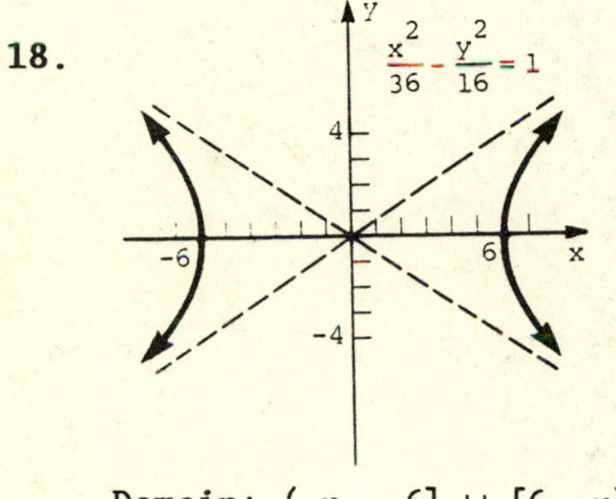

Domain: $(-\infty, -6] \cup [6, \infty)$;
range: $(-\infty, \infty)$

20.

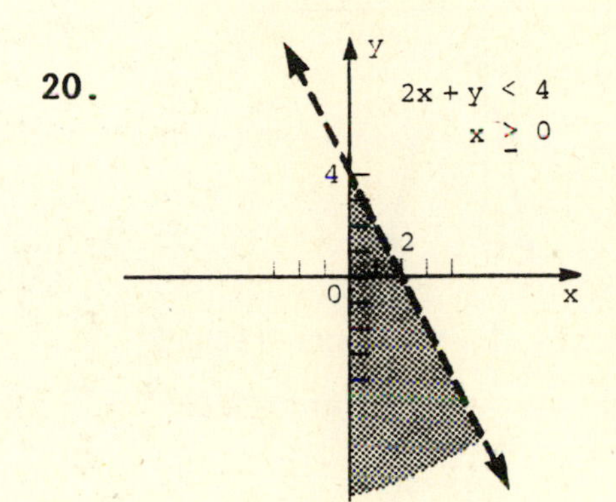

CHAPTER 3 TEST

1. (a) $2\sqrt{34}$ (b) (1, -2)

2. -2

3. 2/7

4. A line with zero slope is horizontal; a line with undefined slope is vertical.

5. $11x + 9y = -25$

6. $x = 3$

7. $7x - 2y = -35$

8.

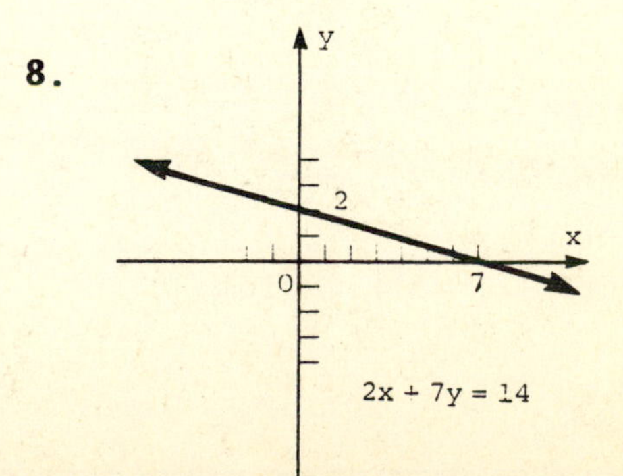

9.

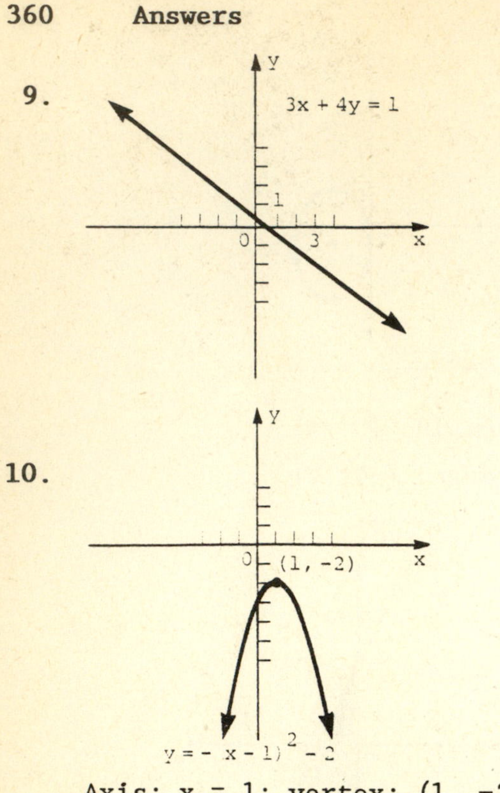

10.

Axis: $x = 1$; vertex: $(1, -2)$; domain: $(-\infty, \infty)$; range: $(-\infty, -2]$

11.

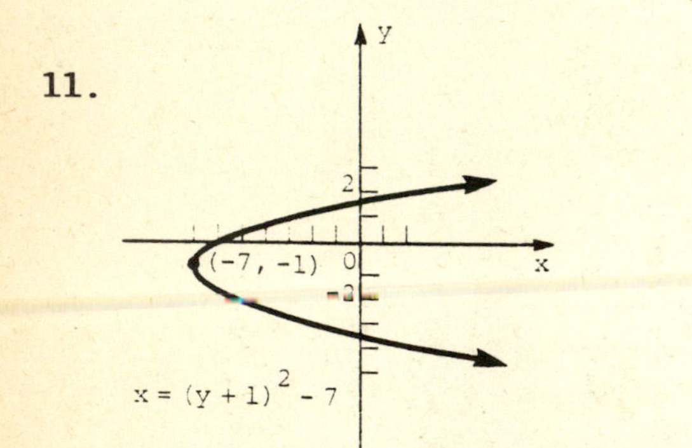

Axis: $y = -1$; vertex: $(-7, -1)$; domain: $[-7, \infty)$; range: $(-\infty, \infty)$

12. Vertex: $(2, -2)$; axis: $x = 2$

13. 2; \$8

14. $(x + 3)^2 + (y - 2)^2 = 25$; domain: $[-8, 2]$; range: $[-3, 7]$

15. Center: $(2, -5)$; radius: 3

16. y–axis

17.

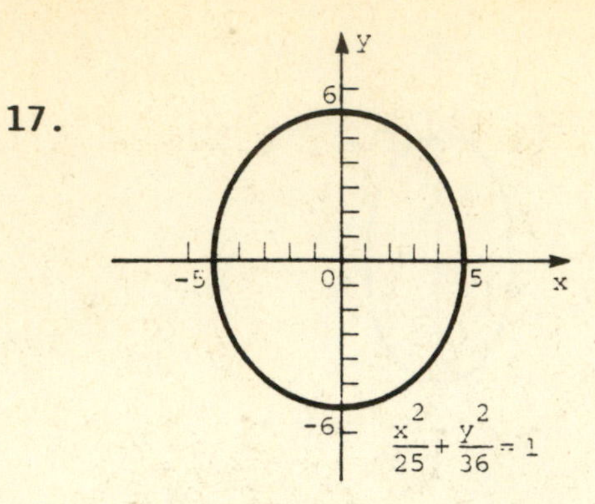

Domain: $[-5, 5]$; range: $[-6, 6]$

18.

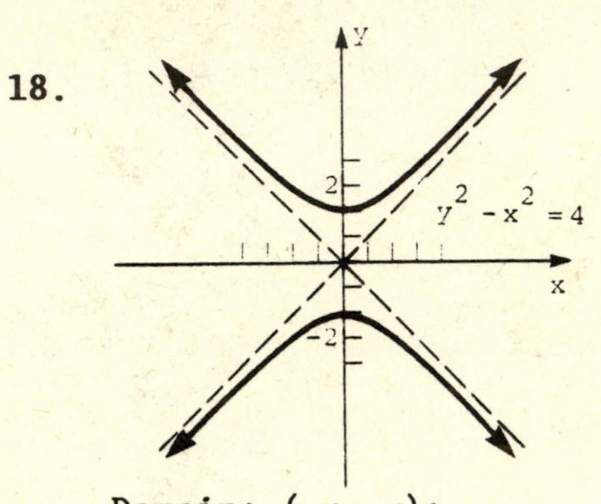

Domain: $(-\infty, \infty)$; range: $(-\infty, -2] \cup [2, \infty)$

19.

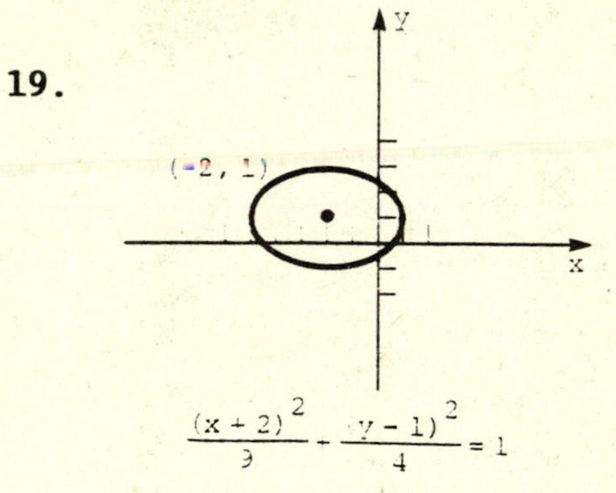

Domain: $[-5, 1]$; range: $[-1, 3]$

20.

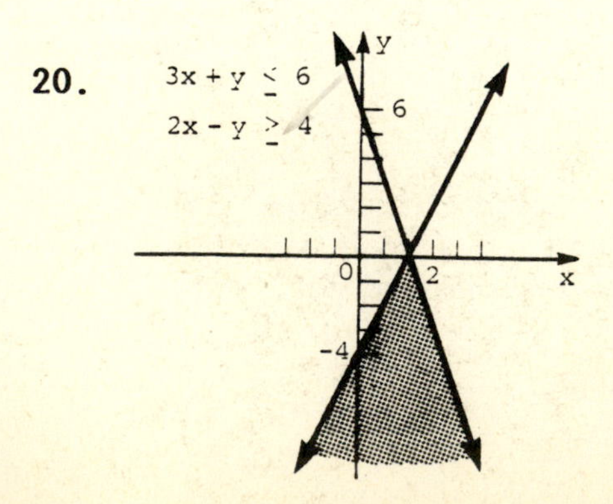

CHAPTER 4 PRETEST

1. Domain: $(-\infty, \infty)$
   Range: $[1, \infty)$

2. Domain: $[5, \infty)$
   Range: $[0, \infty)$

3. Domain: $(-\infty, -3] \cup [3, \infty)$
   Range: $[0, \infty)$

4. $4x^2 + 3x - 8$

5. -24

6. 15

7. -8

8. $12x^2 - 14$

9. $36x^2 - 120x + 97$

10. 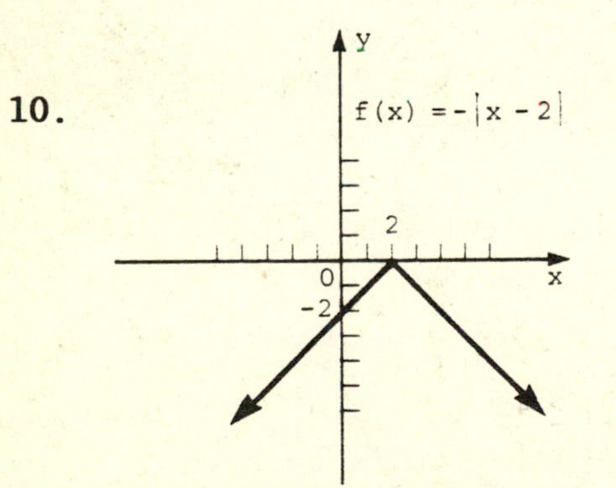

11. 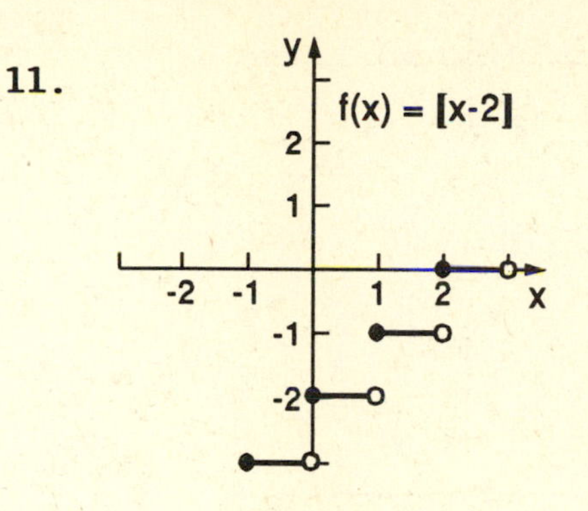

12. 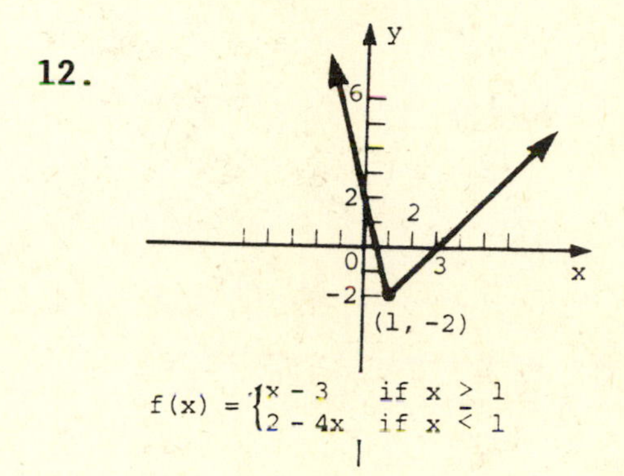

13. One-to-one

14. One-to-one

15. Not one-to-one

16. $f^{-1}(x) = -\sqrt{5 - x}$, domain $(-\infty, 5]$

17. $f^{-1}(x) = \dfrac{5x}{2x + 4}$

18. $f^{-1}(x) = 2 - x^2$, domain $(-\infty, 0]$

19. Yes; explanations may vary.

20. 35/2 candela

CHAPTER 4 TEST

1. Domain: $(-\infty, \infty)$
   Range: $(-\infty, \infty)$

2. Domain: $(-\infty, \infty)$
   Range: $(-\infty, 0]$

3. Domain: $(-\infty, \infty)$
   Range: $[-7, \infty)$

4. $2x^2 + 5x + 4$

5. 20

6. -77

7. -16/5

8. $46 - 10x^2$

9. $50x^2 - 220x + 235$

10.

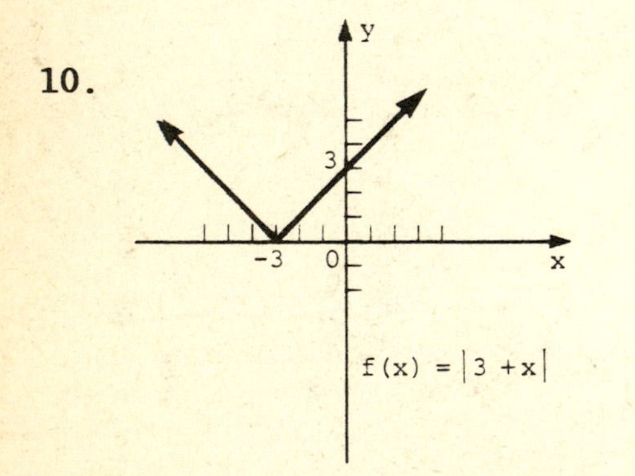

11.

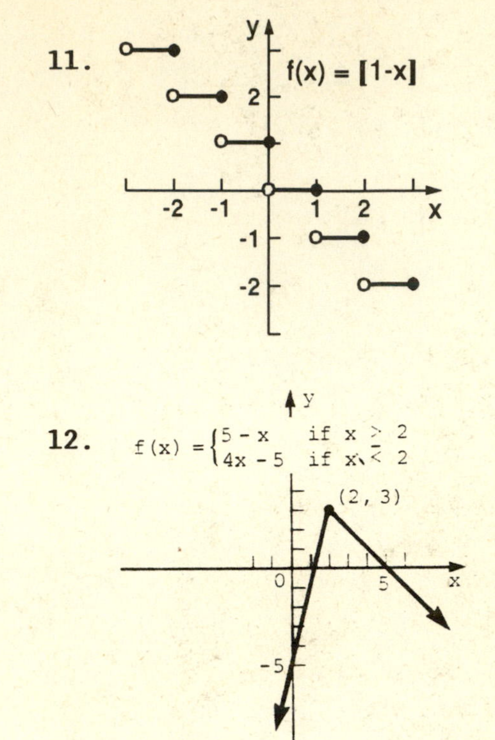

12.

13. One-to-one

14. One-to-one

15. Not one-to-one

16. $f^{-1}(x) = \sqrt{x + 3}$, domain $[-3, \infty)$

17. $f^{-1}(x) = \dfrac{2x + 1}{x + 3}$

18. $f^{-1}(x) = x^2 - 4$, domain $(-\infty, 0]$

19. f(x) has no inverse because it is not one-to-one.

20. 12,500/3 lb

CHAPTER 5 PRETEST

1. {-2}

2. {3}

3. $\log_4 \frac{1}{64} = -3$

4. $\ln 59.1 = m$

5. $8^{5/6} = \sqrt{32}$

6. $e^v = 17$

7. 1

8. $f(x) = 2^{-x}$

9.

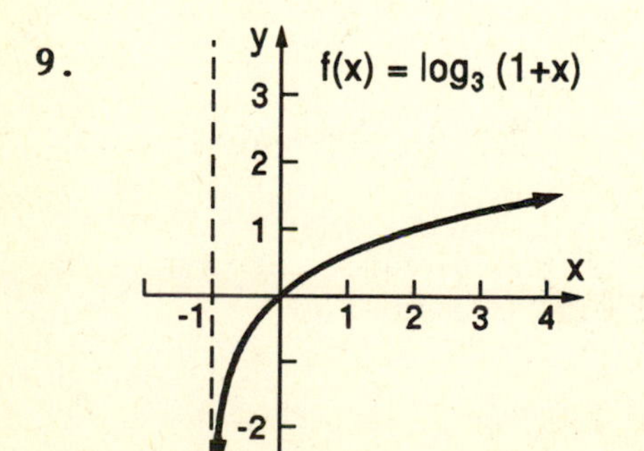

10. $3 \log_5 p + \frac{5}{2} \log_5 q - \log_5 3 - 2 \log_5 r$

11. 8.667

12. 1.760

13. $(-\infty, -7/3]$

14. {-2.596}

15. {-63}

16. {5}

17. (a) 250 (b) 412

18. $3392.61

19. 2.8 mo

20. 8.6 yr

CHAPTER 5 TEST

1. $\{5/7\}$
2. $\{3\sqrt[3]{5}\}$
3. $\log_2 w = 6$
4. $\ln 3.19 = t$
5. $7^{4/3} = \sqrt[3]{2401}$
6. $e^q = 24$
7. 1
8.

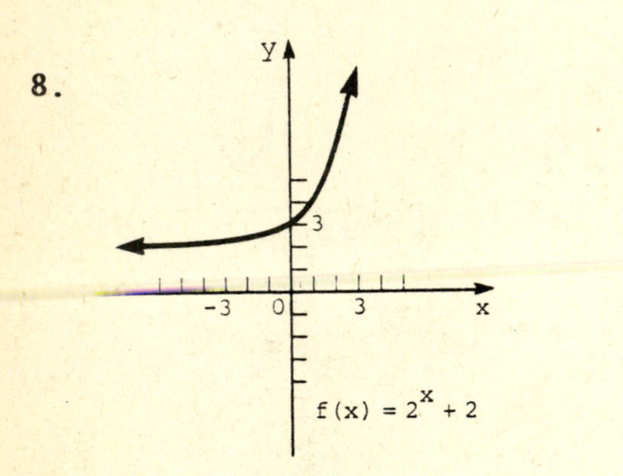

9.

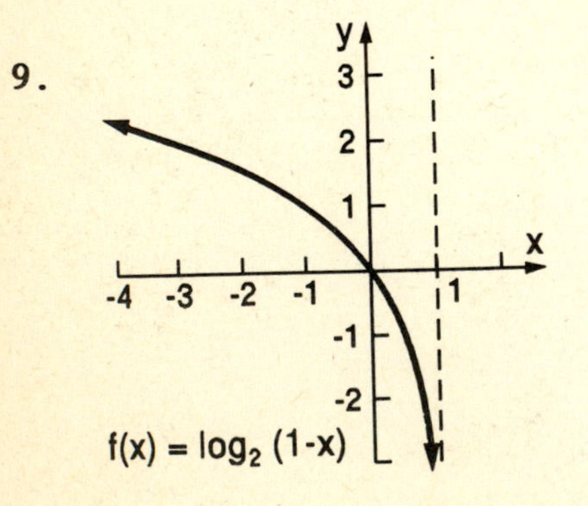

10. $2 \log_2 m + \frac{3}{2} \log_2 n - 3 \log_2 r - \log_2 y$
11. 8.966
12. 1.577
13. $[4/3, \infty)$
14. $\{1.764\}$
15. $\{8\}$
16. $\{8\}$
17. (a) 73.6 g (b) 3.5 hr
18. $5770.52
19. 7.8 yr
20. 17.5 yr

CHAPTER 6 PRETEST

1.

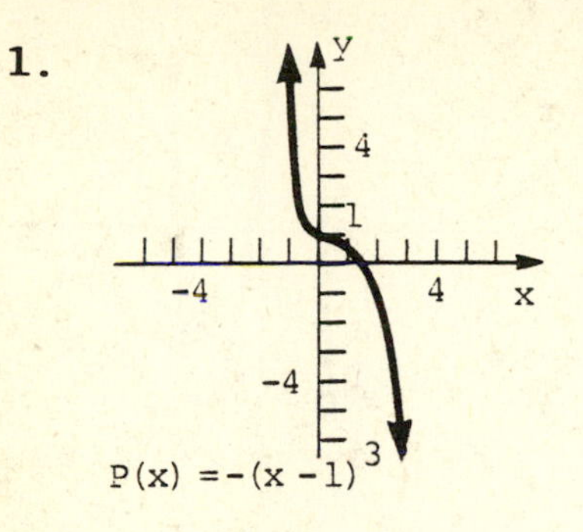

2.

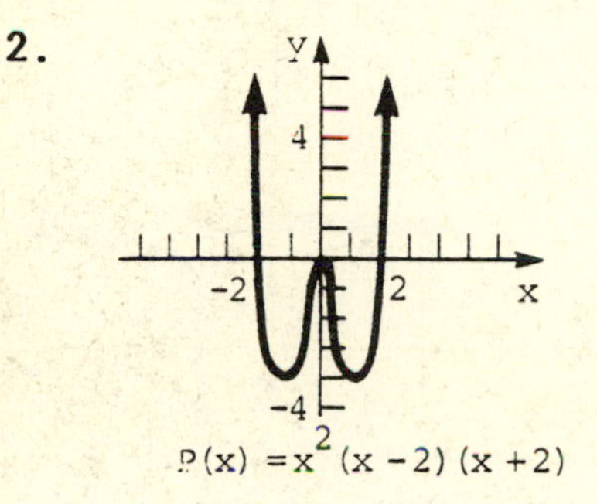

3. $5x^2 - 4x + 5 - \dfrac{8}{x + 2}$

4. $x^3 + 3x^2 - x - 3 - \dfrac{17}{x - 3}$

5. 89

6. $P(x) = x^4 + 2x^3 - 9x^2 - 18x$

7. Yes, because when $P(x)$ is divided by $x + 2$, the remainder is 0.

8. No, because $P(3) = 33$.

9. $P(x) = -\frac{2}{3}(x^3 - 4x^2 - 2x + 8)$

10. $P(x) = (x + 2)(x - 1)(x + 1)$

11. (a) $\pm1, \pm3, \pm5, \pm15$ (b) $-1, -3, 5$

12. The polynomial has no variations in sign.

13. If $P(x)$ is divided synthetically by $x - 3$, all numbers in the bottom row are nonnegative, and by $x + 4$, the numbers in the bottom row alternate in sign.

14. 2 or 0 positive; 1 negative

15.

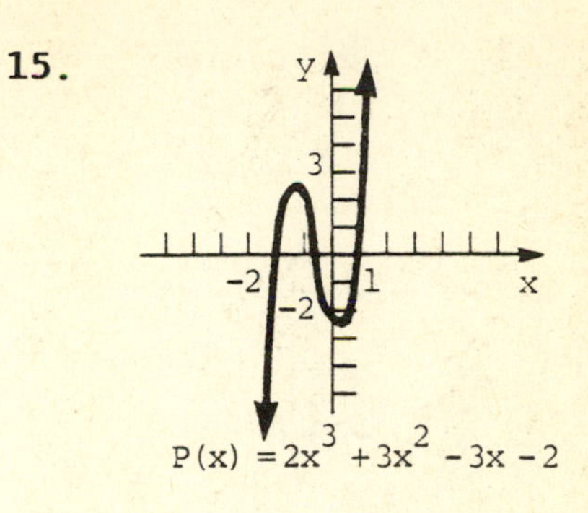

16.

y
2
-1
x
x = 3
$f(x) = \dfrac{1}{x - 3}$

17.

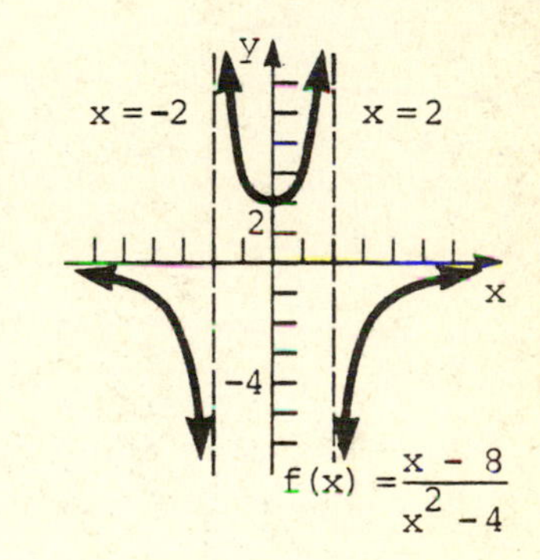

18.

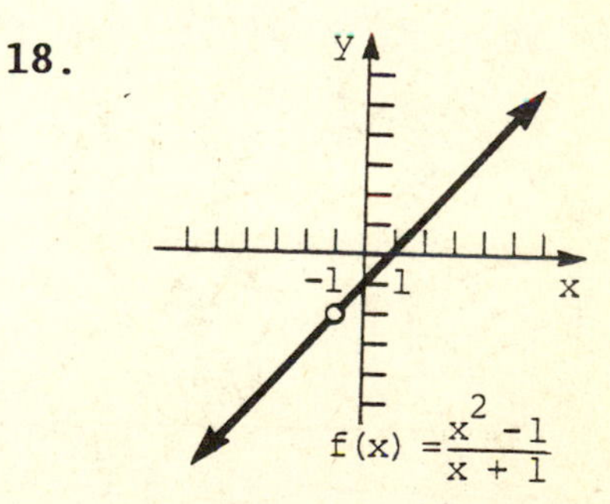

19. $y = 3x + 5$

20. (b)

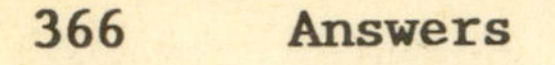

CHAPTER 6 TEST

1.

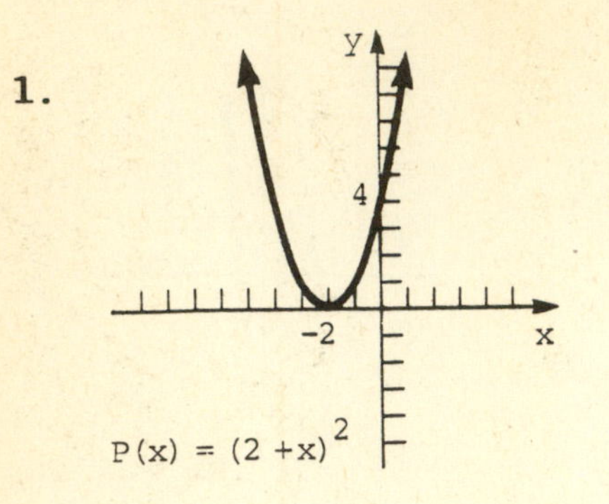

2.

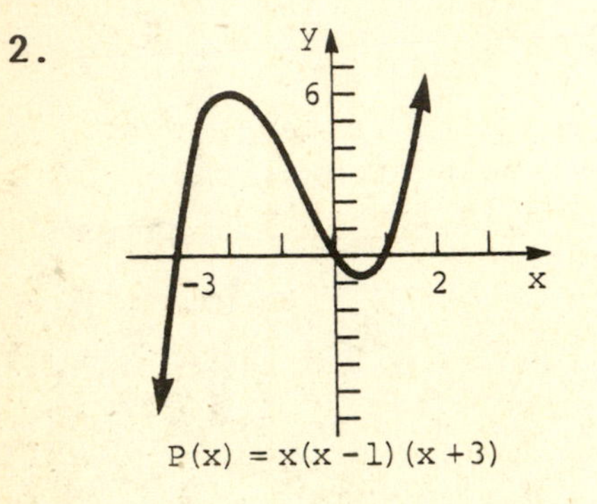

3. $4x^2 - 11x + 28 - \dfrac{57}{x + 2}$

4. $x^4 + 3x^3 + 9x^2 + 7x + 21 + \dfrac{1}{x - 3}$

5. 6

6. $P(x) = x^5 + 4x^4 - x^3 - 16x^2 - 12x$

7. No, because when $P(x)$ is divided by $x - 4$, the remainder is 152.

8. Yes, because $P(-3) = 0$.

9. $P(x) = 2x^4 - 6x^3 - 6x^2 - 6x - 4$

10. $P(x) = (x - 2)(x - 3)(x + 1)$

11. (a) $\pm 2, \pm 1, \pm 1/2$ (b) $1, -2, -1/2$

12. $P(-x)$ has only one variation in sign.

13. If $P(x)$ is divided synthetically by $x - 4$, all numbers in the bottom row are nonnegative, and by $x + 2$, the numbers in the bottom row alternate in sign.

14. 3 or 1 positive; 1 negative

15.

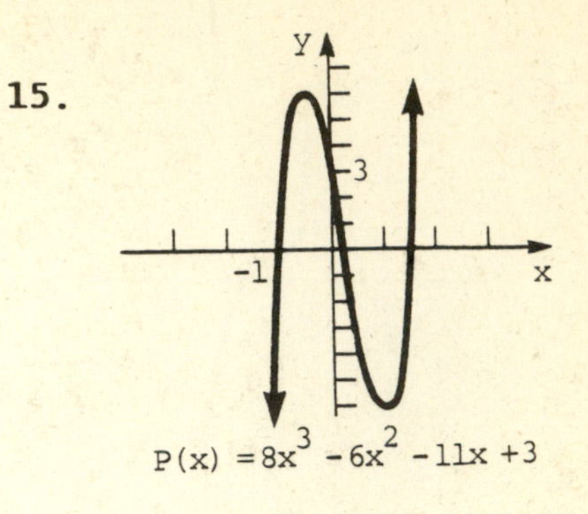

16.

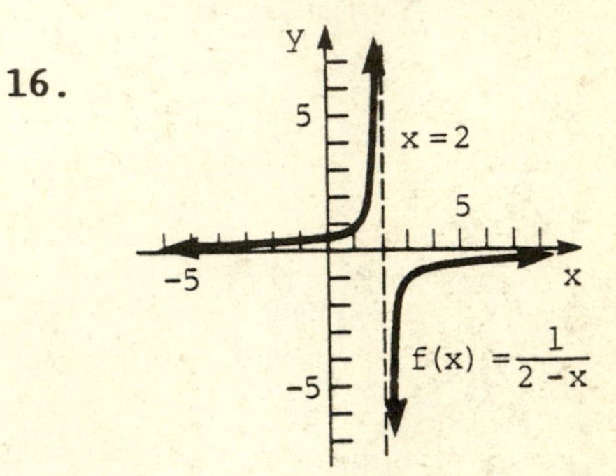

17.

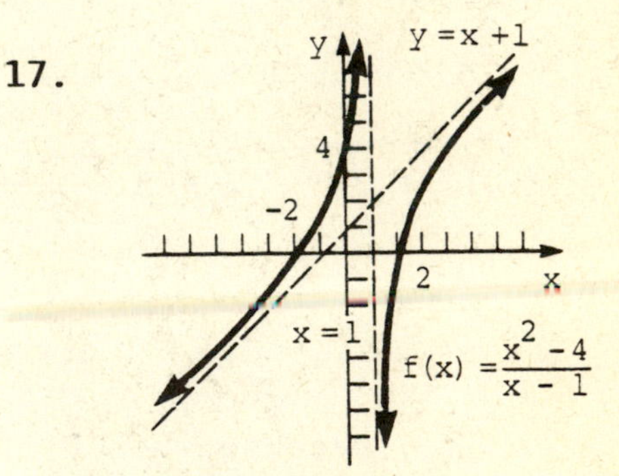

18.

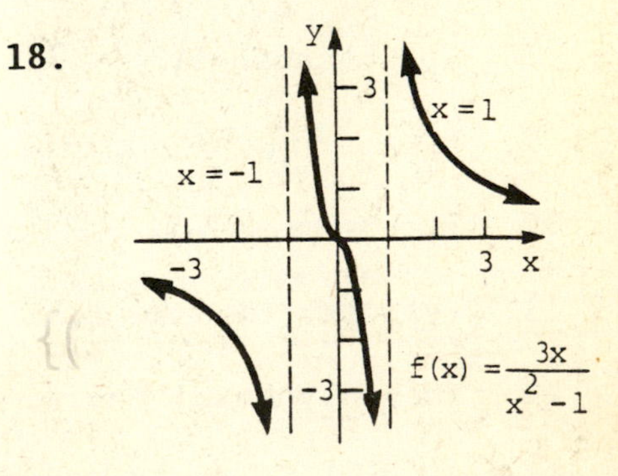

19. $y = 3x + 1$

20. (c)

CHAPTER 7 PRETEST

1. $\{(-2, 3)\}$

2. $\{(-4, -1)\}$

3. Ø; inconsistent

4. $\{(-4 - 3y, y)\}$; dependent

5. $\{(0, -1)\}$

6. $\{(1, 2)\}$

7. No, because the equations are dependent and have infinitely many solutions, including (1, 1).

8. Plant: \$3.50; basket: \$6.00

9. $\{(1, -2, 1)\}$

10. $\{(0, 3, -2)\}$

11. $\left\{\left(\frac{4z + 5}{5}, \frac{z + 5}{5}, z\right)\right\}$

12. 1, -2, 3

13. $\{(6, -2)\}$

14. $\{(-2, 1, 0)\}$

15. $\{(1, -1), (-2, -4)\}$

16. $\{(4, 0)\}$

17. Yes, the line is tangent to the parabola at a point or the line is the axis of symmetry.

18. 11, -11

19.

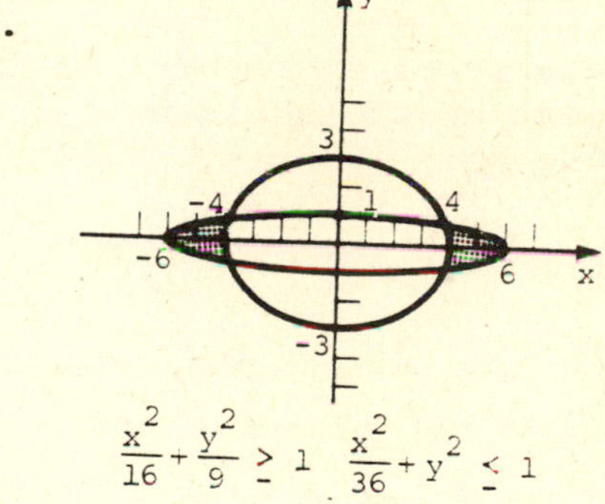

$\frac{x^2}{16} + \frac{y^2}{9} \geq 1$ $\frac{x^2}{36} + y^2 \leq 1$

20. Model A: 3; model B: 4; \$84

CHAPTER 7 TEST

1. $\{(-1, 4)\}$

2. $\{(6, -4)\}$

3. $\left\{\left(\frac{4 + y}{3}, y\right)\right\}$; dependent

4. $\{(2, -2)\}$

5. $\emptyset$; inconsistent

6. $\{(0, 4)\}$

7. No, because every solution must be of the form $\left(\frac{1 + 4y}{3}, y\right)$.

8. Small: $.49; large: $.83

9. $\{(1, 3, 2)\}$

10. $\{(2, 0, -1)\}$

11. $\left\{\left(\frac{2z + 2}{5}, \frac{13z + 8}{10}, z\right)\right\}$

12. 0, 0, 2

13. $\{(7, -3)\}$

14. $\{(-4, 1, 0)\}$

15. $\{(2, 2), (-1, -1)\}$

16. $\{(0, 3), (3, 0)\}$

17. Yes, the circles must be tangent to each other.

18. 5 and −3 or −6 and 8

19.

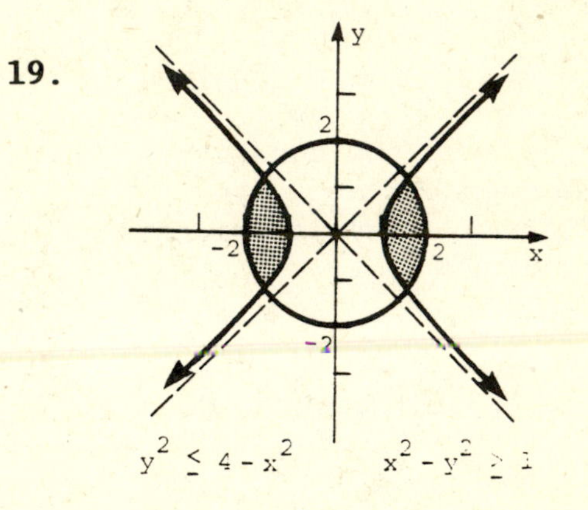

20. Model X: 3; Model Y: 4; $204

CHAPTER 8 PRETEST

1. $x = -6$; $y = 3$; $q = 5$; $z = 1$

2. $\begin{bmatrix} 3 & -3 & 6 \\ 0 & -14 & 1 \end{bmatrix}$

3. $\begin{bmatrix} 1 & 1 & 1 \\ 5 & 5 & 5 \\ -6 & -6 & -6 \end{bmatrix}$

4. $\begin{bmatrix} 2 & 3 \\ 4 & 2 \\ 1 & 2 \end{bmatrix}$; $\begin{bmatrix} 2 & 4 & 1 \\ 3 & 2 & 2 \end{bmatrix}$

5. $\begin{bmatrix} 23 \\ 30 \end{bmatrix}$

6. $\begin{bmatrix} -2 & 6 \\ 0 & -8 \end{bmatrix}$

7. Yes

8. Yes

9. $\begin{bmatrix} 2/21 & -1/7 \\ 1/7 & 2/7 \end{bmatrix}$

10. No inverse

11. $\begin{bmatrix} 0 & 1 & 0 \\ -1 & 3 & 1 \\ 2 & -9 & -1 \end{bmatrix}$

12. $\{(1, -1)\}$

13. $\{(-1, -2, -3)\}$

14. 23

15. 5

16. $x = -2$

17. Property 5 (The first and third rows are identical so the determinant is 0.)

18. Property 4 (Two times the determinant of the matrix on the right equals the determinant of the matrix whose third row elements equal 2 times the third row elements of the original matrix.)

19. $\{(-1, 2)\}$

20. $\{(2, 0, -1)\}$

CHAPTER 8 TEST

1. m = 7; y = 3; r = 1; a = 8

2. $\begin{bmatrix} 0 & 0 & 2 \\ 4 & -4 & 5 \end{bmatrix}$

3. $\begin{bmatrix} 5 & -4 & -5 \\ 4 & -1 & -1 \\ -5 & -9 & -1 \end{bmatrix}$

4. $\begin{bmatrix} 4 & 3 \\ 2 & 7 \\ 5 & 4 \end{bmatrix}$; $\begin{bmatrix} 4 & 2 & 5 \\ 3 & 7 & 4 \end{bmatrix}$

5. $\begin{bmatrix} 13 \\ 5 \end{bmatrix}$

6. $\begin{bmatrix} 9 & 23 \\ -6 & 3 \end{bmatrix}$

7. Yes; $\begin{bmatrix} 1 & 0 & 0 & 0 \\ 0 & 1 & 0 & 0 \\ 0 & 0 & 1 & 0 \\ 0 & 0 & 0 & 1 \end{bmatrix}$

8. Yes

9. $\begin{bmatrix} 1/10 & 0 \\ -3/10 & 1 \end{bmatrix}$

10. $\begin{bmatrix} 3 & -2 \\ -1 & 1 \end{bmatrix}$

11. No inverse

12. {(3, 1)}

13. {(0, 1, 0)}

14. 60

15. 31

16. x = 3

17. Property 2 (Corresponding rows and columns of the matrix are interchanged to give the same determinant.)

18. Property 3 (Two rows of the matrix are interchanged to reverse the sign of the determinant.)

19. {(5, -1)}

20. {(-1, 2, 0)}

CHAPTER 9 PRETEST

1. 0, 1, -2, 3, -4
2. -2, -4, -8, -16, -32
3. 2, 5, 6, 9, 10
4. 2, 6, 10, 14, 18
5. 2, 4, 8, 16, 32
6. -5
7. 140
8. 324
9. $64x^{13}$
10. 30
11. 61/20
12. -26/81
13. 27/2
14. 45
15. $678
16. 3100
17. (i) Does $1^2 = \frac{1(1 + 1)(2 \cdot 1 + 1)}{6}$?

$$1^2 = 1$$

$$\frac{1(1 + 1)(2 \cdot 1 + 1)}{6} = \frac{1(2)(3)}{6} = 1$$

Therefore S_1 is true.

(ii) Assume true for $n = k$.

$$1^2 + 2^2 + 3^2 + \ldots + k^2 = \frac{k(k + 1)(2k + 1)}{6}$$

$$1^2 + 2^2 + 3^2 + \ldots + k^2 + (k + 1)^2$$

$$= \frac{k(k + 1)(2k + 1)}{6} + (k + 1)^2 = \frac{2k^3 + 3k^2 + k}{6} + k^2 + 2k + 1$$

$$= \frac{2k^3 + 3k^2 + k + 6k^2 + 12k + 6}{6} = \frac{2k^3 + 9k^2 + 13k + 6}{6}$$

$$= \frac{(k + 1)(k + 2)(2k + 3)}{6} = \frac{(k + 1)[(k + 1) + 1][2(k + 1) + 1]}{6}$$

Therefore, if S_k is true, then S_{k+1} is true. By mathematical induction, S_n is true for all positive integers.

18. 6720
19. 1320
20. 56
21. 35
22. 140
23. 5040
24. 34,220
25. In symbols, $P(E) = \frac{n(E)}{n(E) + n(E')}$

and $\text{odds} = \frac{n(E)}{n(E')}$.

26. 1/3
27. 2/3
28. 2/3
29. 1 to 1
30. .77

CHAPTER 9 TEST

1. 0, 3, 6, 9, 12
2. 4, 12, 36, 108, 324
3. 4, 8, 6, 12, 8
4. 9, 7, 5, 3, 1
5. 324, 108, 36, 12, 4
6. 79
7. 10
8. 4
9. $96x^{17}$
10. 85/8
11. 45
12. 2460
13. -3
14. 1024/5
15. 408
16. 12,800
17. (i) Does $2^1 = 2^{1+1} - 2$?

$$2^1 = 2$$

$$2^{1+1} - 2 = 4 - 2 = 2$$

Therefore, S_1 is true.

(ii) Assume true for $n = k$:

$$2 + 4 + \ldots + 2^k = 2^{k+1} - 2$$

$$2 + 4 + \ldots + 2^k + 2^{k+1} = 2^{k+1} - 2 + 2^{k+1} = 2 \cdot 2^{k+1} - 2$$

$$= 2^{k+2} - 2 = 2^{(k+1)+1} - 2$$

Therefore, if S_k is true, then S_{k+1} is true. By mathematical induction, S_n is true for all positive integers.

18. 3024
19. 1
20. 35
21. 1
22. 72
23. 360
24. 792
25. 2/3
26. 2/3
27. 1/2
28. 1 to 2
29. .88
30. $P(E \cup F) = P(E) + P(F)$

NOTES

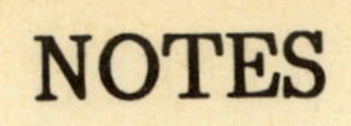
NOTES